Theory and Practice
of Direct Methods
in Crystallography

Theory and Practice of Direct Methods in Crystallography

Edited by

M. F. C. Ladd
University of Surrey
Guildford, Surrey, England

and

R. A. Palmer
Birkbeck College
University of London
London, England

Plenum Press · New York and London

Library of Congress Cataloging in Publication Data

Main entry under title:

Theory and practice of direct methods in crystallography.

Includes index.
1. Crystallography – Methodology. I. Ladd, Marcus Frederick Charles. II. Palmer,
Rex Alfred, 1936-
QD907.T48   549'.1   79-10566
ISBN-13: 978-1-4613-2981-7   e-ISBN-13: 978-1-4613-2979-4
DOI: 10.1007/ 978-1-4613-2979-4

© 1980 Plenum Press, New York
Softcover reprint of the hardcover 1st edition 1980
A Division of Plenum Publishing Corporation
227 West 17th Street, New York, N.Y. 10011

Contributors

G. Allegra, Politecnico di Milano, Istituto di Chimica, Piazza Leonardo da Vinci, Milan, Italy

Paul T. Beurskens, Crystallography Laboratory, Toernooiveld, Nijmegen, The Netherlands

Patrick Argos, Department of Biological Sciences, Purdue University, West Lafayette, Indiana 47907

George DeTitta, Medical Foundation of Buffalo, Inc., Buffalo, New York 14203

William Duax, Medical Foundation of Buffalo, Inc., Buffalo, New York 14203

Edward Green, Medical Foundation of Buffalo, Inc., Buffalo, New York 14203

Herbert Hauptman, Medical Foundation of Buffalo, Inc., Buffalo, New York 14203

Gert Kruger, National Physical Research Laboratory, Pretoria, South Africa

M. F. C. Ladd, Department of Chemical Physics, University of Surrey, Guildford, Surrey GU2 5XH, United Kingdom

David Langs, Medical Foundation of Buffalo, Inc., Buffalo, New York 14203

R. A. Palmer, Department of Crystallography, Birkbeck College, London WC1E 7HX, United Kingdom

D. Rogers, Imperial College of Science and Technology, London SW7 2AY, United Kingdom

Michael G. Rossmann, Department of Biological Sciences, Purdue University, West Lafayette, Indiana 47907

David Sayre, Department of Mathematical Sciences, IBM Thomas J. Watson Research Center, Yorktown Heights, New York 10598

G. Tsoucaris, Laboratoire de Physique, Université de Paris—Sud, Centre Pharmaceutique, 92290 Chatenay-Malabry, France

Th. E. M. van den Hark, Crystallography Laboratory, Toernooiveld, Nijmegen, The Netherlands

Charles Weeks, Medical Foundation of Buffalo, Inc., Buffalo, New York 14203

Preface

Direct methods of crystal structure determination are usually associated with techniques in which phases for a set of structure factors are determined from the corresponding experimental amplitudes by probabilistic calculations. It is thus implied that such *ab initio* phase calculations do not require a knowledge of atomic positions, and this basis distinguishes direct methods from other techniques for structure determination. An acceptably wider interpretation of the term direct methods leads to other important applications involving, *inter alia,* the use of heavy atoms, resolution-limited phase data for large molecules, rotation functions, and Fourier series. These topics are discussed in the later chapters of this book.

Although some earlier theoretical investigations were made by Harker and Kaspar, direct methods may be considered to have begun around the year 1950. Important landmarks in the development of the subject include the book by Hauptmann and Karle, *The Centrosymmetric Crystal* (1953), the definitive paper by Karle and Karle in *Acta Crystallographica* (1966), and the recent (1978) sophisticated program package MULTAN 78 produced mainly by Germain, Main, and Woolfson. Woolfson's book, *Direct Methods in Crystallography,* was published in 1961, but because of the rapid progress in direct methods, much of it soon became outmoded.

It is interesting to note that direct methods nearly came into being many years earlier. Certainly the Σ_2 relationship was used implicitly by Lonsdale in 1928 in determining the crystal structure of hexamethylbenzene. Had the power of this relationship been realized at that time, the history of this subject would almost certainly have been significantly different.

Direct methods still form a challenging field of study and application for structure analysts. Some of the areas of the subject have been brought together in this book, which we believe will form a valuable addition to the literature and prove interesting to research workers in this field.

The difficulties attending the production of a contributed volume have not passed by the present work, and some contributors have felt discomfiture over the time that has elapsed between the submission of manuscripts and the publication of this book. We regret the delay, and hope that the contributors and other readers alike will find the result pleasing.

We are grateful to Plenum Publishing Corporation for their assistance in preparing the book and bringing it to a state of completion.

M. F. C. Ladd
University of Surrey

R. A. Palmer
Birkbeck College,
University of London

Contents

3. Symbolic Addition and Multisolution Methods
M. F. C. Ladd and R. A. Palmer

4. Probabilistic Theory of the Structure Seminvariants
Herbert Hauptman

5. Application of Calculated Cosine Invariants in Phase Determination

William Duax, Charles Weeks, Herbert Hauptman, George DeTitta, David Langs, Edward Green, and Gert Kruger

6. Phase Correlation with Calculated Cosine Invariants for Routine Structure Analysis

Paul T. Beurskens and Th. E. M. van den Hark

7. Application of Direct Methods to Difference Structure Factors

Paul T. Beurskens and Th. E. M. van den Hark

8. Phase Extension and Refinement Using Convolutional and Related Equation Systems

David Sayre

9. Maximum Determinant Method

G. Tsoucaris

10. Molecular Replacement Method

Patrick Argos and Michael G. Rossmann

Principles of Direct Methods of Phase Determination in Crystal Structure Analysis

G. ALLEGRA

1.1. Introduction

It is well known that a three-dimensional image of any crystal structure at the atomic level could be readily obtained if, in addition to the experimentally measured intensities, the phases of the diffracted X-ray waves were known. Herein lies the origin of the *phase problem* in X-ray crystallography, since the phases cannot be obtained directly from physical measurements. For many crystallographers, the very existence of this problem makes the structure solution of a crystalline substance a fascinating adventure, to be tackled with chemical and physical intuition, imagination, and mathematical expertise. A substantial change has taken place during the last 30 years, namely, the extensive development of analytical methods to derive, directly, phases from the observed amplitudes. Although these techniques may not necessarily be successful in every case, they have extended considerably the limit above which a structure becomes sufficiently complex to make its solution a really difficult task. In spite of the relatively automatic nature of recent methods, the very possibility of obtaining the three-dimensional structure of a reasonably complex crystal through completely objective procedures is, in itself, a great and beautiful scientific achievement.

G. ALLEGRA • Politecnico di Milano, Istituto di Chimica, Piazza Leonardo da Vinci, Milan, Italy.

Direct methods represent a substantial proportion of the techniques used for phase determination at present. We may define direct methods as those mathematical techniques that attempt to solve the phase problem from the observed amplitudes through purely mathematical techniques, with no recourse to structural chemical information. In particular, methods that utilize the Patterson function in order to locate one or more heavy atoms in the crystal, from which approximate phases may be obtained, will not be taken into consideration.

The general expression for the electron density is

$$\varrho(x, y, z) = \frac{1}{V} \sum_h \sum_k \sum_l^{+\infty} F(hkl) \exp[-2\pi i(hx + ky + lz)] \qquad (1.1)$$

where x, y, and z are the fractional coordinates of any point in the unit cell and $F(hkl)$ is the structure factor associated with the reflection having indices hkl. It will be convenient to represent any point in direct space by the vector $\mathbf{r} = x\mathbf{a} + y\mathbf{b} + z\mathbf{c}$ with respect to the origin of a unit cell with vector edges $\mathbf{a}$, $\mathbf{b}$, and $\mathbf{c}$, and of volume V. Similarly any point in reciprocal space is represented by the vector $\mathbf{h} = h\mathbf{a}^* + k\mathbf{b}^* + l\mathbf{c}^*$, where $\mathbf{a}^*$, $\mathbf{b}^*$, and $\mathbf{c}^*$ are the reciprocal vectors associated with $\mathbf{a}$, $\mathbf{b}$, and $\mathbf{c}$, respectively. Since the structure factor $F(hkl)$, or $F(\mathbf{h})$, is, in general, a complex number with phase angle $\phi(\mathbf{h})$, we may write $F(\mathbf{h})$ as $|F(\mathbf{h})| \exp[i\phi(\mathbf{h})]$, and so

$$\varrho(\mathbf{r}) = \frac{1}{V} \sum_h |F(\mathbf{h})| \exp[i(\phi(\mathbf{h}) - 2\pi\mathbf{h} \cdot \mathbf{r})] \qquad (1.2)$$

It is reasonable to inquire whether the phases are in any way related to the experimentally derived $|F(\mathbf{h})|$ values. To give a preliminary answer to this question, consider the effect of assigning completely arbitrary values to the phases, and then calculating $\varrho(\mathbf{r})$ from (1.2). It is extremely unlikely that the three-dimensional function thus obtained would be physically meaningful. For instance, $\varrho(\mathbf{r})$ would probably have negative values in some regions of space, contrary to physical expectation, which requires $\varrho(\mathbf{r}) \geq 0$ for every $\mathbf{r}$. Therefore, we may stipulate a first requirement for the $\phi(\mathbf{h})$ values.

First criterion: Given the set of observed amplitudes $|F(\mathbf{h})|$, the corresponding phases must be such as to produce (or be consistent with) non-negative electron density values everywhere.

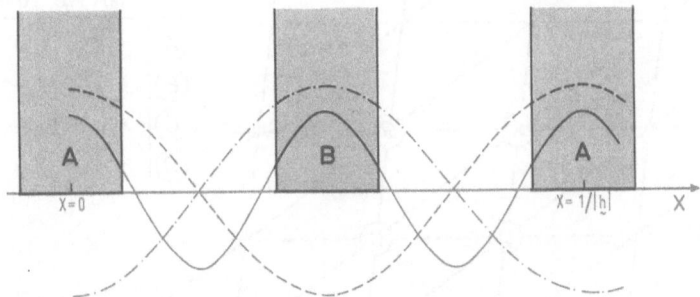

FIGURE 1.1. Graph showing that, whether the electron density is concentrated around region A [$F(\mathbf{h})$ positive, dashed line] or around region B [$F(\mathbf{h})$ negative, dash–dot line], $F(2\mathbf{h})$ is always positive (solid line).

The following two examples may give a qualitative illustration of the above criterion. First, suppose that in a centrosymmetric crystal both $|F(\mathbf{h})|$ and $|F(2\mathbf{h})|$ are large. It follows, as we shall see, that $F(2\mathbf{h})$ is more probably positive than negative. In fact (Fig. 1.1), if $F(\mathbf{h})$ is large, the electron density must cluster preferentially around either region A or region B (the x coordinate is taken to be parallel to the reciprocal vector $\mathbf{h}$), so that $F(\mathbf{h})$ is positive in the former case and negative in the latter. It is interesting to note that in both cases $F(2\mathbf{h})$ is positive. Of course, this conclusion must be regarded on a probability basis, inasmuch as it will be more probably correct, the larger the values of $|F(\mathbf{h})|$ and $|F(2\mathbf{h})|$.

Considering still the centrosymmetric case, Fig. 1.2 shows the alternate positive and negative planes of three reflections with reciprocal-lattice vectors $\mathbf{h}$, $\mathbf{k}$, and $\mathbf{h} + \mathbf{k}$; the figure is a projection in the plane containing the three vectors. As long as the three amplitudes $|F(\mathbf{h})|$, $|F(\mathbf{k})|$, and $|F(\mathbf{h} + \mathbf{k})|$ are all large, the electron density must cluster preferentially around any of the four areas centered around the points A, B, C, D. It is easy to recognize (see legend to Fig. 1.2) that at each point the product of the three structure factors is positive, and we may therefore write

$$s(\mathbf{h})s(\mathbf{k})s(\mathbf{h} + \mathbf{k}) \approx +1 \qquad (1.3)$$

where $s(\mathbf{h})$ represents the sign of $F(\mathbf{h})$ and $\approx +1$ means probably positive, the probability increasing with an increase in the amplitudes. The two examples depend on the same philosophy; in fact, is it easy to see that the second example reduces to the first if we take $\mathbf{k} = +\mathbf{h}$.

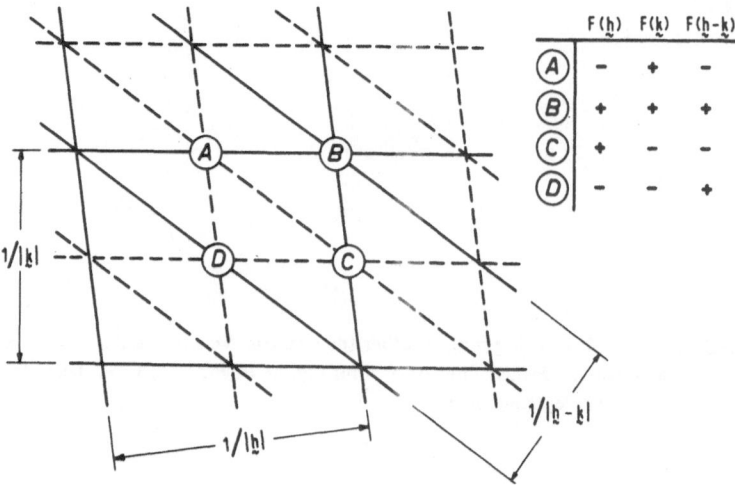

FIGURE 1.2. Large amplitudes of $F(\mathbf{h})$, $F(\mathbf{k})$, $F(\mathbf{h} - \mathbf{k})$ may be produced if the electron density is concentrated mainly around regions A, B, C, and D. In each case the product of the three structure factors is positive (see top right). Solid and dashed lines correspond, respectively, to positive and negative maxima for the three reflections.

1.2. Spherical Symmetry of Atoms: Sayre's Equation

So far, we have considered some implications of the inherent positivity of the electron density. However, it is evident that the correct phases should lead to images of real atoms displaying spherically symmetrical electron density (at least approximately). A beautiful application of the criterion of the spherical symmetry of the atoms which has fundamental consequences is the famous Sayre equation (1952); it is strictly valid only if all the atoms are identical.

Let N atoms in a unit cell be considered at rest (the same result would be obtained if they have identical isotropic temperature factors), and consider a hypothetical structure where all the electron densities are squared (squared structure). Obviously, if the atoms are sufficiently far apart that their electron densities do not overlap to an appreciable extent, the squared structure also consists of equal, spherically symmetrical atoms whose scattering factor may be calculated exactly as a function of $\mathbf{h}$. The general structure factor $G(\mathbf{h})$ of the squared structure is given by

$$\gamma \sum_{j=1}^{N} \exp(-2\pi i \mathbf{h} \cdot \mathbf{r}_j)$$

where γ is the scattering factor of the squared atom and depends only on the magnitude of the reciprocal vector $\mathbf{h}$. Since the structure factor $F(\mathbf{h})$ is of the same form except for the replacement of f by γ, we have

$$G(\mathbf{h}) = (\gamma/f)F(\mathbf{h}) \qquad (1.4)$$

On the other hand, introducing the general expression for $G(\mathbf{h})$, and using (1.2), we have

$$
\begin{aligned}
G(\mathbf{h}) &= \int_V \varrho^2(\mathbf{r}) \exp(2\pi i \mathbf{h} \cdot \mathbf{r})\, dV \\
&= \frac{1}{V^2} \int_V \sum_{\mathbf{k}} \sum_{\mathbf{l}} F(\mathbf{k})F(\mathbf{l}) \exp[-2\pi i(\mathbf{k}+\mathbf{l}) \cdot \mathbf{r}] \exp(2\pi i \mathbf{h} \cdot \mathbf{r})\, dV \\
&= \frac{1}{V} \sum_{\mathbf{k}} F(\mathbf{k})F(\mathbf{h}-\mathbf{k}) \qquad (1.5)
\end{aligned}
$$

where the last equality derives from the orthogonality of the exponential functions. Sayre's equation is now obtained by comparison of (1.5) with (1.4):

$$F(\mathbf{h}) = \frac{1}{V}\, \frac{f}{\gamma} \sum_{\mathbf{k}} F(\mathbf{k})F(\mathbf{h}-\mathbf{k}) \qquad (1.6)$$

It is important to bear in mind that, since the amplitudes $|F(\mathbf{h})|$ are the experimental data and the factors f and γ are known quantities depending on $|\mathbf{h}|$, the above equation provides an exact linkage among the phases: other equations may be obtained using similar ideas. It is remarkable that the above result is not restricted, in principle, by the assumption of positive electron density. Provided that the atoms are represented by identical, spherically symmetrical electron densities which do not overlap, (1.6) is valid. The particular type of electron density distribution will be reflected in the f and γ factors. Therefore, we may state a second criterion on which a correlation between phases and amplitudes is based.

Second criterion: If the atoms in a structure are identical and spherically symmetrical, and do not overlap in space, diffracted amplitudes and phases are connected by exact equations.

The above criterion may be extended in a probabilistic sense to the case of unequal atoms, although (1.6) is then no longer strictly valid. Incidentally, it is now possible to prove the probability relationship (1.3) in a more convincing way. By multiplying both sides of (1.6) by $F(-\mathbf{h})$, or $F^*(\mathbf{h})$, we obtain

$$|F(\mathbf{h})|^2 = \frac{1}{V}\frac{f}{\gamma}\sum_{\mathbf{k}} F(-\mathbf{h})F(\mathbf{k})F(\mathbf{h}-\mathbf{k}) \tag{1.7}$$

If three particular amplitudes $|F(\mathbf{h})|$, $|F(\mathbf{k})|$, and $|F(\mathbf{h}-\mathbf{k})|$ all have large values, the corresponding product will probably be a leading term in the above sum. Since this must be large and positive, the product will be also, the probability being larger, the greater the absolute value of the product. Therefore, we obtain the probability relation

$$\phi(-\mathbf{h}) + \phi(\mathbf{k}) + \phi(\mathbf{h}-\mathbf{k}) \approx 0 \tag{1.8}$$

which is known as the "sum of angles" formula, and which reduces to (1.3) for a centrosymmetric crystal.

1.3. Unitary and Normalized Structure Factors

The probability relations between amplitudes and phases may be strengthened and simplified through the introduction of either the unitary, $U(\mathbf{h})$, or the normalized, $E(\mathbf{h})$, structure factors. The reader is referred to Chapter 2 for the details concerning their actual evaluation and use: both of these modified structure factors relate to hypothetical diffracted waves which would be obtained if the atoms were pointlike entities at rest. Obviously, they bear a close resemblance to the actual structure factors with which they have the phase angle in common; they are expressed as

$$U(\mathbf{h}) = \varepsilon_{\mathbf{h}}^{-1/2} \sum_{j=1}^{N} (Z_j/\sigma_1) \exp(2\pi i \mathbf{h} \cdot \mathbf{r}_j)$$

$$E(\mathbf{h}) = \varepsilon_{\mathbf{h}}^{-1/2} \sum_{j=1}^{N} (Z_j/\sigma_2^{1/2}) \exp(2\pi i \mathbf{h} \cdot \mathbf{r}_j) \tag{1.9}$$

where

$$\sigma_n = \sum_{j=1}^{N} Z_j^n$$

Z_j is the atomic number of the jth atom, and the assumption is made that its scattering factor varies in reciprocal space as $Z_j\bar{f}$, $\bar{f}$ being a universal function which may be defined as the normalized atomic scattering factor, averaged over all the atoms in the unit cell. Obviously, the largest possible value of $|U(\mathbf{h})|$ is unity, which is always true for $\mathbf{h} = 0$; its actual amplitude represents the weighted fraction of the atomic centers that scatter in phase.

Although the structure factor $E(\mathbf{h})$, for a given $\mathbf{h}$, is proportional to $U(\mathbf{h})$, it is defined in such a way that the average value of $|E^2(\mathbf{h})|$ over all the reciprocal vectors is unity. The term $\varepsilon_{\mathbf{h}}$ is the average intensity multiple of the hth reflection (see Chapter 2).

The use of either U or E instead of the usual structure factor avoids explicit reference to the atomic scattering factors, so that the resulting equations are simpler and more powerful. It should be pointed out that these modified structure factors do not show any decrease, statistically speaking, with increasing $s = 2(\sin\theta)/\lambda$. In terms of E, Sayre's equation (1.6), for example, may be written in the equivalent form, first obtained by Hughes (1953),

$$E(\mathbf{h}) = N^{1/2}\langle E(\mathbf{k})E(\mathbf{h} - \mathbf{k})\rangle_{\mathbf{k}} \tag{1.10}$$

where $\langle \ \rangle_{\mathbf{k}}$ means the average over all values of $\mathbf{k}$.

1.4. Karle–Hauptman Determinants

The positivity requirement of the electron density (first criterion) may be expressed by a very elegant and general mathematical formulation. Karle and Hauptman (1950) were able to prove that under the above condition the determinant of any matrix constructed as in (1.11)

$$D = \begin{bmatrix} U(0) & U(\mathbf{h}_1) & U(\mathbf{h}_2) & \cdots & U(\mathbf{h}_n) \\ U(-\mathbf{h}_1) & U(0) & U(\mathbf{h}_2 - \mathbf{h}_1) & \cdots & U(\mathbf{h}_n - \mathbf{h}_1) \\ U(-\mathbf{h}_2) & U(\mathbf{h}_1 - \mathbf{h}_2) & U(0) & \cdots & U(\mathbf{h}_n - \mathbf{h}_2) \\ \vdots & \vdots & \vdots & \cdots & \vdots \\ U(-\mathbf{h}_n) & U(\mathbf{h}_1 - \mathbf{h}_n) & U(\mathbf{h}_2 - \mathbf{h}_n) & \cdots & U(0) \end{bmatrix} \tag{1.11}$$

is necessarily nonnegative, no matter how large the value of n or how the basic reciprocal vectors $\mathbf{h}_1, \ldots, \mathbf{h}_n$ are selected. A large variety of different inequalities among unitary structure factors may be constructed, two of which will be reported here. They correspond to the following determinants of the third order ($n = 2$):

$$D' = \begin{bmatrix} 1 & U(\mathbf{h}) & U(2\mathbf{h}) \\ U(-\mathbf{h}) & 1 & U(\mathbf{h}) \\ U(-2\mathbf{h}) & U(-\mathbf{h}) & 1 \end{bmatrix}, \quad D'' = \begin{bmatrix} 1 & U(\mathbf{h}) & U(\mathbf{k}) \\ U(-\mathbf{h}) & 1 & U(-\mathbf{h}+\mathbf{k}) \\ U(-\mathbf{k}) & U(\mathbf{h}-\mathbf{k}) & 1 \end{bmatrix} \tag{1.12}$$

Assuming a centrosymmetric structure, [i.e., $U(\mathbf{h}) = U(-\mathbf{h})$], and imposing the conditions $D' \geq 0$ and $D'' \geq 0$, we obtain the following inequalities, since $U(0) = 1$:

$$U(2\mathbf{h}) \geq 2\,|\,U(\mathbf{h})\,|^2 - 1$$

$$U(-\mathbf{h})U(\mathbf{k})U(\mathbf{h} - \mathbf{k}) \geq \tfrac{1}{2}(|\,U(\mathbf{h})\,|^2 + |\,U(\mathbf{k})\,|^2 + |\,U(\mathbf{h} - \mathbf{k})\,|^2 - 1) \qquad (1.13)$$

The first of the above inequalities was obtained by Harker and Kasper in 1948 by making use of Cauchy's inequality. Qualitatively speaking, we have arrived at the same conclusions already obtained at the beginning and summarized in (1.3); namely that if both $|\,U(\mathbf{h})\,|$ and $|\,U(2\mathbf{h})\,|$ are large amplitudes, $U(2\mathbf{h})$ is probably positive, and if $|\,U(\mathbf{h})\,|$, $|\,U(\mathbf{k})\,|$, and $|\,U(\mathbf{h} - \mathbf{k})\,|$ are all large, the product of the three structure factors is also probably positive. However, in favorable circumstances we may be *certain* about the sign of a structure factor or of a triple product, since the inequalities *must* hold by necessity. As an example, if $|\,U(\mathbf{h})\,| = 0.5$ and $|\,U(2\mathbf{h})\,| = 0.6$, $U(2\mathbf{h})$ must be positive since the inequality $-0.6 > -0.5$ is wrong. In the more general case of a noncentrosymmetric structure, expansion of the determinant D'', remembering that

$$U(\mathbf{h}) = U^*(-\mathbf{h}) = |\,U(\mathbf{h})\,|\,\exp[i\phi(\mathbf{h})]$$

gives

$$1 - |\,U(\mathbf{h})\,|^2 - |\,U(\mathbf{k})\,|^2 - |\,U(\mathbf{h} - \mathbf{k})\,|^2 + 2\,|\,U(-\mathbf{h})U(\mathbf{k})U(\mathbf{h} - \mathbf{k})\,|$$
$$\times \cos[\phi(-\mathbf{h}) + \phi(\mathbf{k}) + \phi(\mathbf{h} - \mathbf{k})] \geq 0 \qquad (1.14)$$

and Fig. 1.3 shows how the sum of the three phases is restricted in order to produce a positive value for the determinant. It is unfortunate, however, that we need very large unitary amplitudes in order that the inequalities may come into play and provide useful information on the phases. As an example, if none of the $|\,U\,|$ values involved in inequality (1.14) exceeds 0.5, the forbidden region vanishes: in other words, the inequality is always verified, and, therefore, is of no use. Since the root-mean-square value of $|\,U\,|$ is $N^{-1/2}$, the above condition is rarely fulfilled in most real structures, where the number of atoms in the unit cell is seldom less than 10.

So far, combining the two criteria of positive electron density and of spherical symmetry of the atoms, we have obtained three different types of information about the phases from the observed amplitudes: (a) probability relations, although in the qualitative form expressed by (1.3) and (1.8), (b) exact equations, such as (1.6) or the equivalent (1.10), (c) inequalities,

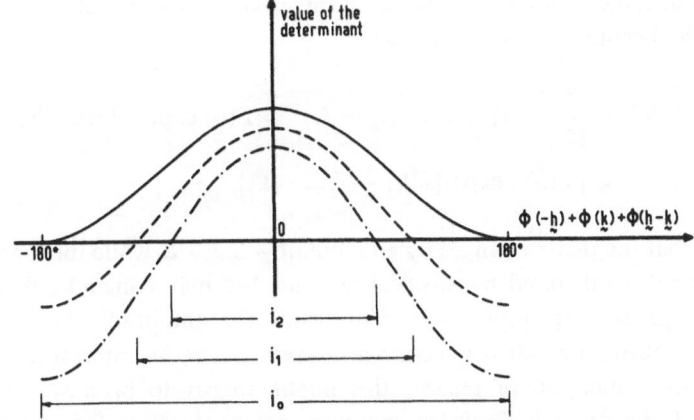

FIGURE 1.3. Value of the determinant D'' plotted against $[\phi(-\mathbf{h}) + \phi(\mathbf{k}) + \phi(\mathbf{h}-\mathbf{k})]$ for $|U(\mathbf{h})| = |U(\mathbf{k})| = |U(\mathbf{h}-\mathbf{k})| = 0.5$; 0.55; 0.6 (solid, dashed, and dash–dot lines, respectively). The phase sum is restricted to lie within the angular intervals i_0, i_1, and i_2 respectively, in the three cases.

exemplified by (1.13) and (1.14). Since the exact equations are very difficult to apply, at least in the early stages of a structure determination, because too many unknowns are involved, and the inequalities may seldom be utilized for reasons given above, the probability relations turn out to be the most powerful tool for direct phase determination. We shall give a concise preliminary account of different mathematical techniques, all of which may be referred to essentially as probability methods; they will be discussed extensively in the remainder of this book. However, a brief introduction to the concept of phase invariants and seminvariants is first desirable.

1.5. Structure Invariants and Seminvariants

It is easy to see that the phase of a single structure factor is not only dependent on the crystal structure. In fact, a source of indeterminacy is the choice of the origin of the unit cell. Let a structure factor $F(\mathbf{h})$ be referred to a given origin

$$F(\mathbf{h}) = |F(\mathbf{h})| \exp[i\phi(\mathbf{h})] = \sum_{j=1}^{N} f_j \exp(2\pi i \mathbf{h} \cdot \mathbf{r}_j) \qquad (1.15)$$

where f_j is the temperature-corrected scattering factor of the jth atom.

If we shift the origin by a vector Δ, all the atomic vectors change to $\mathbf{r}_j - \Delta$ and $F(\mathbf{h})$ becomes $F'(\mathbf{h})$, given by

$$F'(\mathbf{h}) = \sum_{j=1}^{N} f_j \exp[2\pi i \mathbf{h} \cdot (\mathbf{r}_j - \Delta)] = F(\mathbf{h}) \exp(-2\pi i \mathbf{h} \cdot \Delta)$$

$$= |F(\mathbf{h})| \exp\{i[\phi(\mathbf{h}) - 2\pi \mathbf{h} \cdot \Delta]\} \qquad (1.16)$$

We see that the phase changes by an amount $-2\pi \mathbf{h} \cdot \Delta$ while the amplitude is invariant, as dictated by physical reasons. We may conclude, therefore, that the phases are origin dependent, while the amplitudes are not; the latter quantities are called *structure invariants*. Since our interest is focused on the determination of phases, this might appear to be a very serious drawback. Fortunately, however, many important structure-factor products are, generally speaking, complex numbers with *origin-independent* phase angles; therefore, they are also structure invariants. It is easy to show that products of the type $F(\mathbf{h}_1)F(\mathbf{h}_2) \cdots F(\mathbf{h}_n)$ are structure invariant if and only if the condition

$$\mathbf{h}_1 + \mathbf{h}_2 + \cdots + \mathbf{h}_n = 0 \qquad (1.17)$$

is fulfilled. This follows from (1.16) since the combined phase change of the product upon a shift Δ of the origin is given by $-2\pi\Delta \cdot \sum_i \mathbf{h}_i$, or zero if (1.17) holds. Consequently the phase sum $\sum_i \phi(\mathbf{h}_i)$ is also a structure invariant, although the individual phases are not, and we may deal with it as with a quantity related solely with the structure. The simplest invariant structure-factor product is $|F(\mathbf{h})F(-\mathbf{h})| = |F(\mathbf{h})|^2$. The next simplest product is of the type $F(-\mathbf{h})F(\mathbf{k})F(\mathbf{h} - \mathbf{k})$, which is the general term of the sum in (1.7): it is not merely by chance that the structure invariant on the left-hand side of (1.7) is also given by a sum of phases that are structure invariants. The probability relationship (1.8) applies to a sum of phases which is a phase invariant. Expansion of the general Karle–Hauptman determinant (1.11) shows that each term is a product of $U(\mathbf{h})$ values where $\sum_i \mathbf{h}_i = 0$; consequently, all the related inequalities, such as (1.13) and (1.14), involve only structure invariants. Any derivation of phases from the amplitudes must lead *per se* to phase invariants, and not to single phases. This conclusion could have been predicted earlier on the basis of the following simple argument: any piece of information about phases that be may be derived from the amplitudes, which are invariant with respect to our subjective choice of origin, must lead to something that is also origin independent, or structure invariant.

An important point to be noted here concerns those noncentrosymmetric structures that have enantiomorphic forms. Since the amplitudes of each enantiomorphic structure are indistinguishable, except possibly for small anomalous scattering effects, we cannot derive information on the choice of enantiomorph from the amplitudes alone. In particular, the two enantiomorphs differ only in the *signs* of their phases $\phi(\mathbf{h})$ once their origin is fixed. Hence, the amplitudes may provide information about the *absolute value* of the phase invariants, $\sum_i \phi(\mathbf{h}_i)$, but not about their signs. The enantiomorph will be chosen arbitrarily by fixing the sign of one phase invariant, since structure-factor algebra connects the signs of all others.

Now the question arises, how shall we derive the individual phases once the phase invariants are known? As a general criterion, the problem may be solved by assigning arbitrary values to the phases of three suitably chosen structure factors $F(\mathbf{h}_1)$, $F(\mathbf{h}_2)$, and $F(\mathbf{h}_3)$, provided that $\mathbf{h}_1, \mathbf{h}_2, \mathbf{h}_3$ are not coplanar. In fact, it is possible to show (see Chapter 2) that this arbitrary choice can locate the origin in an arbitrary way, all other phases becoming defined in relation to that origin.

The above conclusion may need modification if we deem it useful to place the origin on some particular symmetry elements of the crystal. As an example, in centrosymmetric space groups it is convenient to take the origin on an inversion center. Then, the values of $\phi(\mathbf{h})$ are limited to 0 or π: the structure factor $F(\mathbf{h})$ is a real number with either a positive or a negative sign. As a consequence, three signs will be needed, generally, for origin specification. Space groups having a particular symmetry, for example, a nonprimitive unit cell, need a number of phases, or signs, to select the origin, which may be less than three (see Chapter 2). If the origin is located on a symmetry element, the phase of some particular reflections may be unambiguously specified. This case arises for a reflection having all three indices even; in a centrosymmetric space group, we shall denote its reciprocal vector as $2\mathbf{h}$. Confining our attention to primitive unit cells for simplicity, the component along $\mathbf{a}$, for example, of the vector Δ connecting any two inversion centers is restricted to the two possible values 0 and $a/2$. Consequently, the scalar product $2\mathbf{h} \cdot \Delta$ is always an integer, and there is no effective phase change in shifting the origin. We say that the structure factor $F(2\mathbf{h})$ or its sign is a *structure seminvariant*, since it has a fixed value provided we take the origin on an inversion center. We can obtain structure seminvariants in several ways; thus, for example, if the origin is on a center of symmetry, any product of the type $F(\mathbf{h}_1)F(\mathbf{h}_2) \cdots F(\mathbf{h}_n)$ is a structure seminvariant provided $\mathbf{H} = \sum_{i=1}^{n} \mathbf{h}_i$ is an even reciprocal vector, that is, it has even values for all of its indices.

In conclusion, the problem of direct phase determination is related strictly with the actual symmetry of the unit cell; obviously, the more symmetrical the space group, the greater is the amount of phase information obtainable from the same number of independent observed amplitudes. A full knowledge of the problem of origin definition and of the role of structure invariants and seminvariants is extremely important to this purpose; a complete theory covering this subject is given in a series of papers by Hauptman and Karle (1953, 1956, 1959) and Karle and Hauptman (1961). The subject forms the basis of Chapter 2 of this book.

1.6. Probability Theory

The mathematical problem of finding a probability distribution for the phases $\phi(\mathbf{h})$, or the signs $s(\mathbf{h})$ in the case of a centrosymmetric structure, once the amplitudes are given, may be formulated in one or other of the following two ways.

(a) Let a set of n-normalized amplitudes $|E(\mathbf{h})|$ be given, corresponding to a set of n reciprocal vectors $\mathbf{h}_1, \mathbf{h}_2, \ldots, \mathbf{h}_n$. Considering a set of random configurations of the atoms within the unit cell, let us select the subset of the configurations producing those amplitudes. Within this subset, let us evaluate the fraction of the configurations for which the phase of $E(\mathbf{h}_1)$ lies between $\phi(\mathbf{h}_1)$ and $\phi(\mathbf{h}_1) + d\phi(\mathbf{h}_1)$, the phase of $E(\mathbf{h}_2)$ between $\phi(\mathbf{h}_2) + d\phi(\mathbf{h}_2)$, and so on. The joint conditional probability that the several phases lie within the intervals given above is defined by

$$P(|E_1|, |E_2|, \ldots, |E_n| / \phi_1, \phi_2, \ldots, \phi_n) \, d\phi_1 \, d\phi_2 \ldots d\phi_n$$

where $E_i = E(\mathbf{h}_i)$, $\phi_i = \phi(\mathbf{h}_i)$.

(b) Let the same set of amplitudes be given, but the reciprocal vectors $\mathbf{h}_i$ be unspecified. The structure is taken as fixed, and the subset of the reciprocal vectors producing the observed amplitudes takes the role of the subset of atomic configurations considered in (a): the joint phase conditional probability is defined as above.

The two approaches differ because of the choice of the primitive random variables; in the first case they are the atomic coordinates, in the second the reciprocal-lattice vectors. In spite of their different philosophies, the two approaches lead to equivalent results in most cases. In the following treatment we shall refer to results obtained mainly by adoption of the first, and more usual approach.

There is no doubt that the triple-phase relationship (1.8) has proved to be a most important phase-determining formula. The following quantitative expressions for the related probability have been developed by Cochran and Woolfson (1955) for centrosymmetric and by Cochran (1955) for noncentrosymmetric structures.

Centrosymmetric case:

$$P_+(\mathbf{hk}) = \tfrac{1}{2} + \tfrac{1}{2}\tanh[\tfrac{1}{2}K(\mathbf{hk})]$$

Noncentrosymmetric case:

$$P[\phi(\mathbf{hk})] = \frac{\exp\{K(\mathbf{hk})\cos[\phi(\mathbf{hk})]\}}{2\pi I_0 K(\mathbf{hk})} \qquad (1.18)$$

where

$$\phi(\mathbf{hk}) = \phi(-\mathbf{h}) + \phi(\mathbf{k}) + \phi(\mathbf{h} - \mathbf{k})$$

and

$$K(\mathbf{hk}) = 2\sigma_3\sigma_2^{-3/2} \,|\, E(-\mathbf{h})E(\mathbf{k})E(\mathbf{h} - \mathbf{k}) \,|;$$

σ_n is defined under (1.9) and I_0 is the modified Bessel function of the second type. $P_+(\mathbf{hk})$ is the probability of the triple product $E(\mathbf{h})E(\mathbf{k})E(\mathbf{h} - \mathbf{k})$ being positive, while $P[\phi(\mathbf{hk})]\,d\phi(\mathbf{hk})$ is the probability that the value of $\phi(\mathbf{hk})$ lies between $\phi(\mathbf{hk})$ and $\phi(\mathbf{hk}) + d\phi(\mathbf{hk})$. Figures 1.4 and 1.5 are illustrative graphs, while Fig. 1.6 gives the variance of $P[\phi(\mathbf{hk})]$ around its zero value, as obtained by Karle and Karle (1966). Figures 1.4 and 1.6 show convincingly that the probability expectation of the triple-phase relationships given by (1.3) and (1.8) increases both with an increasing product of the three amplitudes and with increasing $\sigma_3\sigma_2^{-3/2}$. This result is true also of the probability of $E(2\mathbf{h})$ being positive in the centrosymmetric

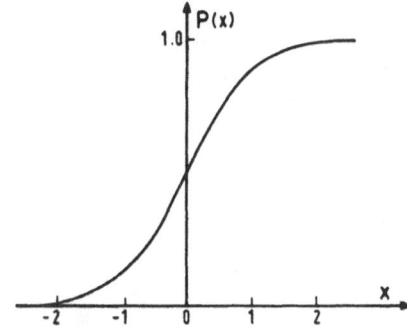

FIGURE 1.4. The probability P_+ plotted against $X\,[= \tfrac{1}{2}K(\mathbf{hk})]$.

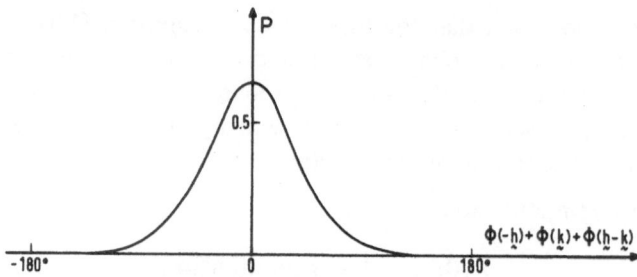

FIGURE 1.5. The probability $P(\phi)$ plotted against $\phi = \phi(-h) + \phi(k) + \phi(h - k)$ for $K(hk) = 2.9$.

case, if both its amplitude and that of $E(h)$ are large:

$$P_+(2h) = \tfrac{1}{2} + \tfrac{1}{2}\tanh\{\tfrac{1}{2}\sigma_3\sigma_2^{-3/2}\,|\,E(2h)\cdot[|\,E(h)\,|^2 - 1]\,|\} \qquad (1.19)$$

The value of $\sigma_3\sigma_2^{-3/2}$ equals $N^{-1/2}$ for identical atoms, which shows that the strength of a probability formula is bound to decrease with an increasing complexity of the structure. It should be pointed out that the expressions reported in (1.18) and (1.19) are strictly correct only when the number N of atoms in the cell is very large. In particular, there results in practice a slight underestimate of the probability information, as expressed by $|\,P_+(hk) - \tfrac{1}{2}\,|$ and by $|\,P_+(2h) - \tfrac{1}{2}\,|$, in the centrosymmetric case. Klug (1958) showed that the probability is expressed exactly as a power series, which can lead to any desired precision.

Karle and Hauptman (1958) and more recently Hauptman (1975), Hauptman and Green (1976) and Giacovazzo (1975), following different methods, were able to prove that the most probable value of a phase invariant $\phi = \sum_1^n \phi(h_i)$, where $\sum h_i = 0$, is not necessarily zero if a sufficient

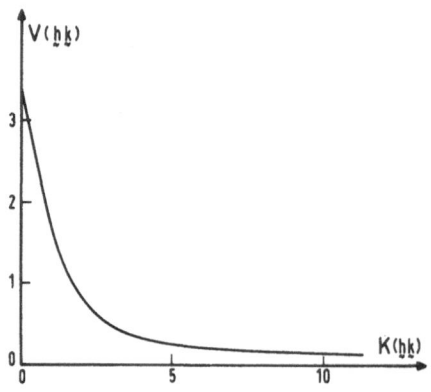

FIGURE 1.6. Variance $V(hk)$ of $P(\Phi(hk))$ as a function of $K(hk)$.

number of amplitudes are taken into account. Indeed, it appears that the most probable value of $|\phi|$ (remember that its sign cannot be obtained from the amplitudes, because it depends on the choice of enantiomorph) may be calculated with an accuracy that is better the larger the number of amplitudes considered. As an example, the probability distribution of the phase invariant $\phi = \phi(\mathbf{h}) + \phi(\mathbf{k}) + \phi(\mathbf{l}) + \phi(\mathbf{m})$, with $\mathbf{h} + \mathbf{k} + \mathbf{l} + \mathbf{m} = 0$, was obtained by Hauptman (1975) under two different hypotheses:

(a) The four amplitudes $R_1 = |E(\mathbf{h})|$, $R_2 = |E(\mathbf{k})|$, $R_3 = |E(\mathbf{l})|$, $R_4 = |E(\mathbf{m})|$ are known.

(b) In addition to the above, the amplitudes $R_{12} = |E(\mathbf{h} + \mathbf{k})|$, $R_{23} = |E(\mathbf{k} + \mathbf{l})|$, $R_{31} = |E(\mathbf{l} + \mathbf{h})|$ are also known, making seven amplitudes in all.

While the mathematical expressions are given in Chapter 4, Fig. 1.7 shows the probability distribution calculated in each case for a fictitious structure with 29 equal atoms in space group $P1$, the structure amplitudes being specified in the legend. In the case of four amplitudes, the probability has one maximum always centered at $\phi = 0$; with seven magnitudes, the maximum may be anywhere, and, in the actual case, it is displaced by only 9° from the true value. In particular, the maximum of the probability computed with seven magnitudes corresponds to $\phi = 180°$ if $R_{12} \simeq R_{23} \simeq R_{31} \simeq 0$.

A completely different approach has been followed (see Chapter 9) by Tsoucaris (1970), who studied the probability properties of the Karle–Hauptman determinants (1.11). He showed that the most probable phases

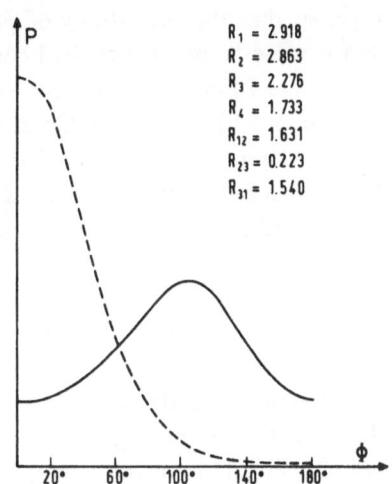

$R_1 = 2.918$
$R_2 = 2.863$
$R_3 = 2.276$
$R_4 = 1.733$
$R_{12} = 1.631$
$R_{23} = 0.223$
$R_{31} = 1.540$

FIGURE 1.7. Probability distribution of $\phi = \phi(\mathbf{h}) + \phi(\mathbf{k}) + \phi(\mathbf{l}) + \phi(\mathbf{m})$, with $\mathbf{h} + \mathbf{k} + \mathbf{l} + \mathbf{m} = 0$. The dashed and the solid lines correspond to considering four and seven amplitudes, respectively (after Hauptman, 1975).

of the reflections contained in any Karle–Hauptman determinant are those
that maximize the determinant itself, and he obtained a formal expression
for the probability distribution of m structure factors $E_1, E_2, \ldots, E_m$,
constituting the last row and column of the determinant, once all the other
structure factors are known. This "maximum determinant rule" is a particularly significant example of a bridge between the inequality and the
probability methods.

1.7. Solving the Phase Problem for a Real Structure

A method that has proved to be very effective in solving complicated
structures, in spite of its relative simplicity, is the *symbolic addition procedure*
(Karle and Karle, 1966). It is based essentially on a stepwise application
of the triple-phase relationships given in (1.8), which, in turn, belong to the
more general class of the so-called Σ_2 relationships (see Chapter 3). In
the particular case of a centrosymmetric structure, the probability formula
(1.19) (Σ_1 relationship), may also be exploited at the beginning of the
sign determinating process; it may give indications for some even-indices
reflection of being positive, although the corresponding probability is
seldom large enough to fulfill the corresponding inequality and so make the
sign acceptable with full confidence. Personal skill and experience still
play an important role in this type of approach. In the general case, once
the origin- and enantiomorph-fixing phases have been assigned, symbols
are assigned to the phases of a few properly chosen reflections involved in
strong triples, that is, triples where the product $| E(-\mathbf{h})E(\mathbf{k})E(\mathbf{h} - \mathbf{k}) |$ is
large, so that the probability of $\phi(-\mathbf{h}) + \phi(\mathbf{k}) + \phi(\mathbf{h} - \mathbf{k})$ being near to
zero is close to unity. Let $\phi(\mathbf{h})$ correspond to a symbol a, and $\phi(\mathbf{k})$ to b;
then $\phi(\mathbf{h} - \mathbf{k})$ may be taken as $a - b$ if the triple is strong enough [remem-
ber that $\phi(-\mathbf{h}) = -\phi(\mathbf{h})$]. Consequently, we have derived the phase of a
new reflection as a function of the starting set of symbols. Alternatively,
if $\phi(\mathbf{h}) = a$ but neither $\phi(\mathbf{k})$ nor $\phi(\mathbf{h} - \mathbf{k})$ is contained in the starting set,
we may add a new symbol (u) to the set, putting $\phi(\mathbf{k}) = u$, so that $\phi(\mathbf{h} - \mathbf{k})$
$\simeq a - u$. Subsequently, we may generate new phases in terms of symbols,
although this process may require from time to time the addition of new
symbols to the initial set. Fortunately, in this chain process the phase of
some particular reflection is often generated in more than one way, which
leads to equations among symbols. As an example, suppose the amplitudes
of the reflections belonging to the two triples with indices $1\bar{3}\bar{4}$, $\bar{2}21$, 113
and $1\bar{2}0$, $\bar{3}1\bar{3}$, 113 are very large; if $\phi(1\bar{3}\bar{4}) = -a$, $\phi(\bar{2}21) = b$, $\phi(1\bar{2}0) = 0$,

and $\phi(\bar{2}1\bar{3}) = c$, then $\phi(113)$ is given simultaneously by $a - b$ and by $-c$, so that $b - a \simeq c$. If the probability relations leading to an equality among symbols are all reliable enough, or if that equality occurs repeatedly, we may drop one of the symbols by expressing it in terms of the others; referring to the above example, we would put $b - a$ instead of c. If we are lucky (and clever) enough, we may eventually obtain either the phases of most of the strongest reflections in terms of very few symbols, or even the numerical values of all the phases. In the former case, we shall explore the different combinations of angular values for the remaining symbols through an appropriately limited sampling within the range 0 to 2π; of course, in centrosymmetric cases the possible phases are only 0 and π. After the numerical specification of as many phases as possible, they are usually refined through the tangent formula (see Section 1.8). If the phase angles chosen for the symbols are essentially correct, a Fourier map of the electron density, usually calculated with E instead of F, reveals the basic features of the structure.

A logically analogous procedure has been translated into a very powerful computing routine written in FORTRAN. This is the MULTAN program (Germain and Woolfson, 1968; Germain, Main, and Woolfson, 1971) resulting from the joint activity of the York and Louvain crystallographic groups. The program first selects as the best starting reflections (base reflections) those whose phase definition permits the phases of the majority of the strong reflections to be derived most effectively. After assigning the arbitrary phases for origin and enantiomorph definition, the phases of the remaining base reflections are permuted, and in each case other phases are derived through the Σ_2 relationships (1.8). Each phase set is tested according to different criteria, one of which consists of evaluating the following figure of merit (FOM):

$$\text{FOM} = \sum_{\mathbf{h}} \sum_{\mathbf{k}} |E(-\mathbf{h})E(\mathbf{k})E(\mathbf{h}-\mathbf{k})| \cos[-\phi(\mathbf{h})+\phi(\mathbf{k})+\phi(\mathbf{h}-\mathbf{k})] \quad (1.20)$$

Since each phase invariant on the right-hand side of (1.20) is more probably close to zero the larger the three-factor product, it is obvious that the above sum must tend to its largest possible value for the correct phase set.

The calculation of the most probable values of the cosines of phase invariants (Karle and Hauptman, 1958; Hauptman, 1975; Hauptman and Green, 1976; Giacovazzo, 1975) has also proved to be very effective in the solution of the phase problem, and it should provide even better opportunities in the future, after more accurate expressions involving larger numbers of structure-factor amplitudes have been developed. In the limiting hypothesis that a sufficiently large number of phase invariants can be evaluated so accurately as to be virtually exact, the phase problem would be implicitly solved, since each single phase can always be expressed as a linear combination of the origin- and enantiomorph-fixing phases with an appropriate set of phase invariants. As already pointed out, one important advantage implicit in the calculation of the cosines of phase invariants (cosine invariants) is that they may lead to nonzero values of $\sum_i \phi(\mathbf{h}_i)$ with $\sum_i \mathbf{h}_i = 0$, for any space group. This feature is especially important because, unless the space group contains screw axes or glide planes, the Σ_2 relationships are necessarily bound to lead to $\sum_i \phi(\mathbf{h}_i) \approx 0$, from (1.8).

1.8. Refinement of Phases

The most widely used method to improve approximate phase values obtained from stepwise application of the Σ_2 and, possibly, of the Σ_1 relationships is the tangent formula:

$$\tan \phi(\mathbf{h}) = \sum_{\mathbf{k}} K(\mathbf{hk}) \sin[\phi(\mathbf{h} - \mathbf{k}) + \phi(\mathbf{k})] \Big/ \sum_{\mathbf{k}} K(\mathbf{hk}) \cos[\phi(\mathbf{h} - \mathbf{k}) + \phi(\mathbf{k})]$$

$$(1.21)$$

where $K(\mathbf{hk})$ is defined in (1.18). Usually, the above formula is adopted in a cyclic routine; the approximate phases are introduced on the right-hand side, thus producing refined values. The formula may be proved by obtaining the value of $\phi(\mathbf{h})$ which maximizes the probability product $\prod_{\mathbf{k}} P[\phi(\mathbf{hk})]$, but it may also be derived more easily by equating the real and imaginary components in the two sides of Sayre's equation, in the form (1.10), as obtained by Hughes (1953). A conceptual difference between the two viewpoints is that (1.10) is strictly valid only if the sum is extended over all $\mathbf{k}$ vectors, while the above product of probabilities need not be, since any subset involving large values of the triple products may be significant on probability grounds.

A more complicated tangent formula has been obtained by Allegra and Colombo (1974), who imposed the requirement that, for structures consisting of identical atoms, the integral

$$\int_V [\hat{\varrho}(\mathbf{r}) \cdot C_1 - \hat{\varrho}^2(\mathbf{r}) \cdot C_2]^2 \, dV$$

$$(1.22)$$

is a minimum for the correct set of phases; C_1 and C_2 are normalizing constants and $\hat{\varrho}(\mathbf{r})$ is the point-atom electron density. The basic concept is still that which leads to Sayre's equation, namely, the normal and the squared structure must strictly resemble each other if the atoms are identical. The resulting tangent formula contains additional terms with sin(cos) $\times [\phi(\mathbf{h} - \mathbf{k} - \mathbf{l}) + \phi(\mathbf{k}) + \phi(\mathbf{l})]$ in the numerator and denominator of (1.21). The same formula may be obtained by calculating the value of $\phi(\mathbf{h})$ that minimizes the quantity

$$\sum_{\mathbf{h}} \{[E(\mathbf{h}) - N^{1/2}\langle E(\mathbf{k})E(\mathbf{h} - \mathbf{k})\rangle_{\mathbf{k}}] \times (\text{complex conjugate})\} \qquad (1.23)$$

which is simply the sum of the squares of the moduli of the differences between the two sides of (1.10). In fact, except possibly for a constant factor, (1.22) and (1.23) may be shown to reduce to the same expression. The new tangent formula should give better results inasmuch as it appears to be based on more rigid requirements than (1.21).

Instead of the tangent formula, Sayre's equation may sometimes be used directly in the refinement stage. This procedure has been followed quite successfully by Sayre (1972, 1974) to refine the phases of some crystalline proteins obtained through the usual isomorphous substitution method.

1.9. Possible Future Developments of Direct Methods

Although it is difficult, and to some extent arbitrary, to look into the future of this scientific field, which is undergoing rapid development along several lines, it may perhaps be useful to point out the basic shortcomings of current direct methods when applied to difficult structures, containing > 60–80 crystallographically independent atoms.

(a) The probability formulas generally lead to rather small levels of confidence, owing to the large number of atoms within the unit cell; as an example, $P_+(\mathbf{hk})$, from (1.18) may be rarely larger than 0.90 or smaller than 0.10.

(b) The reflections tend to cluster into islands, the phases being relatively well correlated, in a probabilistic sense, within each island while correlation is lacking between islands.

(c) As a consequence of (a) and (b), sequential procedures of phase determination, such as the symbolic addition procedure, may fail because some of the decisions taken along the sequence are likely to be wrong, entailing overall failure, especially if the faults happen at the early stages.

Also, no matter how the base symbols are chosen, a relatively high number of them are likely to remain unspecified at the end of the procedure, for lack of sufficient probability information.

It seems reasonable to predict that exploitation of *all* the possible probability distributions of phase invariants will be of considerable assistance in the solution of complex structures. It is fortunate that there are many more phase invariants and seminvariants than unknown phases, especially if we extend our treatment beyond the triples, so as to include quartets, and possibly quintets, of reflections. Each phase invariant may be evaluated approximately, and the degree of the approximation appears to be capable of substantial improvement in the future. If many more phase invariants than individual phases are calculated, and if the probable error of each of them is low enough—both hypotheses appearing reasonably realistic—any structure should be solved, no matter how complex it is, unless the experimental data are too poor. Even at present, an appropriate combination of the approach based on the calculation of the phase invariants and the sequential methods adopting the Σ_2 relationships (symbolic addition procedure, MULTAN, and so on) may lead to significant success with a difficult structure.

An interesting recent approach which has been designed specifically to reduce, at least in part, the probability of failure inherent with the sequential procedures, is the so-called "magic integers method" (White and Woolfson, 1975; Declercq, Germain, and Woolfson, 1975). This method maximizes the figure of merit (FOM) reported in (1.20) over a relatively large number of unknown phases. It is made possible through the adoption of the following intelligent strategy. Three, or even more, different phases, say ϕ_1, ϕ_2, ϕ_3, are expressed in terms of multiples of a single variable, say x, with values lying in the range 0 to 2π; in other words, $\phi_1 = n_1 x$, $\phi_2 = n_2 x$, $\phi_3 = n_3 x$ (modulo 2π). If the integers n_1, n_2, and n_3 are appropriately chosen (*magic integers*), the probable error involved in the above assumption may be reduced to a' sufficiently small value. As an example of magic integers, $n_1 = 5$, $n_2 = 6$, $n_3 = 7$, and the reader may check that any set of angles ϕ_1, ϕ_2, and ϕ_3 may be reproduced to a good approximation by a suitable choice of x. If two more variables, say, y and z, are considered, six more phases may be created with the same set of magic integers. Using the strongest Σ_2 relationships, other phases may be generated as a function of the above nine phases, plus those chosen for origin and enantiomorph definition. At this point, the figure of merit given in (1.20) is expressed in a three-dimensional Fourier series in terms of

x, y, z, whose maxima should correspond to the most probable phase sets. The method appears to be quite promising. Incidentally, appropriately weighted terms including cosine invariants depending on four or more phases might also be included in the sum, which could make the method even more general.

Another promising approach seems to consist of the automatic use of the Patterson function to extract information on phases [Anzenhofer and Hoppe (1962), Main and Woolfson (1963), and Allegra (1979)]. This should be of no surprise, since the Patterson function is the Fourier-transform of the intensities; therefore it contains essentially the same amount of information.

As a final remark, we may say that phase-determining methods based on sequential procedures should be avoided as much as possible in the case of a difficult structure. In this case, each new phase is likely to be predicted with a large margin of error, especially in the early stages when only a few phases are known, thus producing additional and larger errors for the remaining reflections. However, the large number of $|F(\mathbf{h})|$ values observable in such cases provides a huge body of probability information on the phase invariants, although each separate piece of information is generally weak, that is, it is determined with a small probability. Mathematical routines will be needed that are capable of handling simultaneously the probability distributions pertaining to several reflections. The magic integers approach may be a good example of a strategy pointing in this direction.

References

Allegra, G., and Colombo, A. (1974). *Acta Crystallogr. Sect. A* **30**, 727.

Allegra, G. (1979). *Acta Crystallogr. Sect. A* **35**, 213.

Anzenhofer, K. and Hoppe, W. (1962). *Phys. Verh.* **13**, 119.

Cochran, W. (1955). *Acta Crystallogr.* **8**, 473.

Cochran, W., and Woolfson, M. M. (1955). *Acta Crystallogr.* **8**, 1.

Declercq, J. P., Germain, G., and Woolfson, M. M. (1975). *Acta Crystallogr. Sect. A* **31**, 367.

Germain, G., and Woolfson, M. M. (1968). *Acta Crystallogr. Sect. B* 24, 91.

Germain, G., Main, P., and Woolfson, M. M. (1971). *Acta Crystallogr. Sect. A* **27**, 368.

Giacovazzo, C. (1975). *Acta Crystallogr. Sect. A* **31**, 252.

Harker, D., and Kasper, J. S. (1948). *Acta Crystallogr.* **3**, 374.

Hauptman, H. (1975). *Acta Crystallogr. Sect. A* **31**, 680.

Hauptman, H., and Green, E. A. (1976). *Acta Crystallogr. Sect. A* **32**, 43.

Hauptman, H., and Karle, J. (1953). *Solution of the Phase Problem. I. The Centrosymmetric Crystal*, American Crystallographic Association Monograph No. 3, Polycrystal Book Service, Pittsburgh, Pennsylvania.

Hauptman, H., and Karle, J. (1956). *Acta Crystallogr.* **9**, 45.

Hauptman, H., and Karle, J. (1959). *Acta Crystallogr.* **12**, 93.

Hughes, E. W. (1953). *Acta Crystallogr.* **6**, 871.

Karle, J., and Hauptman, H. (1950). *Acta Crystallogr.* **3**, 181.

Karle, J., and Hauptman, H. (1958). *Acta Crystallogr.* **11**, 264.

Karle, J., and Hauptman, H. (1961). *Acta Crystallogr.* **14**, 217.

Karle, J., and Karle, I. L. (1966). *Acta Crystallogr.* **21**, 849.

Klug, A. (1958). *Acta Crystallogr.* **11**, 515.

Main, P. and Woolfson, M. M. (1963). *Acta Crystallogr.* **16**, 1046.

Sayre, D. (1952). *Acta Crystallogr.* **5**, 60.

Sayre, D. (1972). *Acta Crystallogr. Sect. A* **28**, 210.

Sayre, D. (1974). *Acta Crystallogr. Sect. A* **30**, 180.

Tsoucaris, G. (1970). *Acta Crystallogr. Sect. A* **26**, 492.

White, P. S., and Woolfson, M. M. (1975). *Acta Crystallogr. Sect. A* **31**. 53.

Definition of Origin and Enantiomorph and Calculation of $|E|$ Values

D. ROGERS

PART I:
DEFINITION OF ORIGIN AND ENANTIOMORPH

2.1. Introduction

The historic approach to this aspect of direct methods was analytical (Hauptman and Karle, 1953, 1956, 1959; Karle and Hauptman, 1961) and has been authoritatively reviewed by Hauptman (1972). Here the approach relies on spatial concepts and deals directly with phases and phase planes. It was developed by the author for teaching and to provide more penetrating insight into certain problems that behaved unexpectedly. It is hoped it will help those who wish to have a better understanding of what they are doing but have had difficulty with the analytical papers.

Section 2.2 introduces the concepts of a structure (S) and its inverse (I); the *reference point* to which all atomic vectors are referred; the fundamentally important *shift theorem*, which relates the phase shift of each reflection to a shift of reference point; and the topics of *primitivity* and

---

D. ROGERS • Imperial College of Science and Technology, South Kensington, London SW7 2AY, United Kingdom.

accessibility. It also introduces a novel notation that may initially look unnecessarily pedantic, but the need for it and its capabilities will emerge later.

Section 2.3 is concerned with the topics of *invariance* and the *universal invariance theorem*; *seminvariance* and the corresponding *equivalence classes*; and seminvariant phases or phase sums.

Section 2.4 uses the Bertaut–Waser structure-factor algebra to discuss the variation of phase among the Laue-symmetric reflections. The variations are referred to as a *phase pattern* and this proves to be characteristic of both the space group and the equivalence class of the preferred origins. Such patterns play a vital role in the subsequent discussion and are a distinctive feature of the approach adopted in this chapter.

Section 2.5 discusses the number and nature of the arbitrary phase assignments needed to define a unique origin or reference point, and how to proceed if they cannot be met. Section 2.5.1 deals with the primitive, and 2.5.2 with the nonprimitive centrosymmetric space groups. Section 2.5.3 introduces the noncentrosymmetric space groups through the unique space group $P1$, and the need for a further arbitrary phase assignment to distinguish between a structure (S) and its inverse (I). Sections 2.5.4 and 2.5.5 deal, respectively, with primitive and nonprimitive noncentrosymmetric space groups. A feature of the treatment is that nonprimitive cells are not reindexed in terms of their primitive equivalents. It is shown by means of the new notation how to analyze the S/I ambiguity left after defining the origin, and thus how to choose the S/I-selector term and give it an effective phase. Section 2.5.4 shows that most noncentrosymmetric space groups can accommodate either the S or I structure; the exceptions are the 11 pairs of enantiomorphic space groups where one member of each pair accommodates S and the other I (see Table 2A.3 in Appendix 1). In every other case it is possible by superimposing S and I to identify a centrosymmetric ($S \& I$) space group, for which centers of symmetry usually coincide with the origin O of the corresponding noncentrosymmetric (S or I) space group. Certain exceptions are dealt with in Section 2.6.

2.2. Some Preliminaries

Consider a *known* noncentrosymmetric structure on which is superimposed the outline of the unit cell whose axes are those chosen, using standard conventions, for the collection of diffracted intensity data (see

Fig. 2.1a). The origin O of this cell is selected arbitrarily from among those that are conventionally preferred for the space group and have been adopted in the *International Tables for X-Ray Crystallography*, Vol. 1. We designate this structure S_O and its constituent atomic vectors $\mathbf{r}_{O,j}$. For every non-centrosymmetric structure there exists a second structure I_O which is the inverse of S_O: it has the same unit cell and origin as S_O. Except for the 11 enantiomorphic pairs of space groups mentioned above, every noncentro-symmetric space group can accommodate either the S or I structures, so for brevity we shall refer to them as the S *or* I space groups. In every such case it is possible in principle to superimpose S_O and I_O to yield a centro-symmetric structure whose space group will be designated S & I, any one of whose centers of symmetry C constitutes an inversion center between S_O and I_O. In all but nine space groups, C coincides with O to give the usual situation depicted in Fig. 2.1(a), for which the vectors $\mathbf{r}'_{O,j}$ in I_O are related to those in S_O by

$$\mathbf{r}'_{O,j} = -\mathbf{r}_{O,j} \tag{2.1a}$$

In the nine abnormal space groups, the preferred origin O of the S *or* I space group does not coincide with C in the S & I space group (see Figs. 2.1b and 2.9), and in this case

$$\mathbf{r}'_{O,j} = -\mathbf{r}_{O,j} + 2 \cdot \overrightarrow{OC} \tag{2.1b}$$

We shall defer discussion of these nine space groups to Section 2.6, and until then confine our attention to the commoner situation of Fig. 2.1a.

The structure factor for reflection $\mathbf{h}$ (i.e., $h\mathbf{a}^* + k\mathbf{b}^* + l\mathbf{c}^*$) and structure S_O is defined by

$$F(S_O, \mathbf{h}) = \int \varrho(\mathbf{r}_O) \exp[2\pi i(\mathbf{h} \cdot \mathbf{r}_O)] \, dV \tag{2.2Sa}$$

which can be expressed with sufficient accuracy for the purposes of direct methods as

$$F(S_O, \mathbf{h}) = \sum_{j=1}^{N} g_j \exp[2\pi i(\mathbf{h} \cdot \mathbf{r}_{O,j})] = |F(S_O, \mathbf{h})| \exp[i\phi(S_O, \mathbf{h})] \tag{2.2Sb}$$

where g_j is the thermally attenuated atomic scattering factor for the jth atom in S_O. We assume throughout this book that anomalous dispersion effects are negligible in the structures studied, so that all g_j are real. If

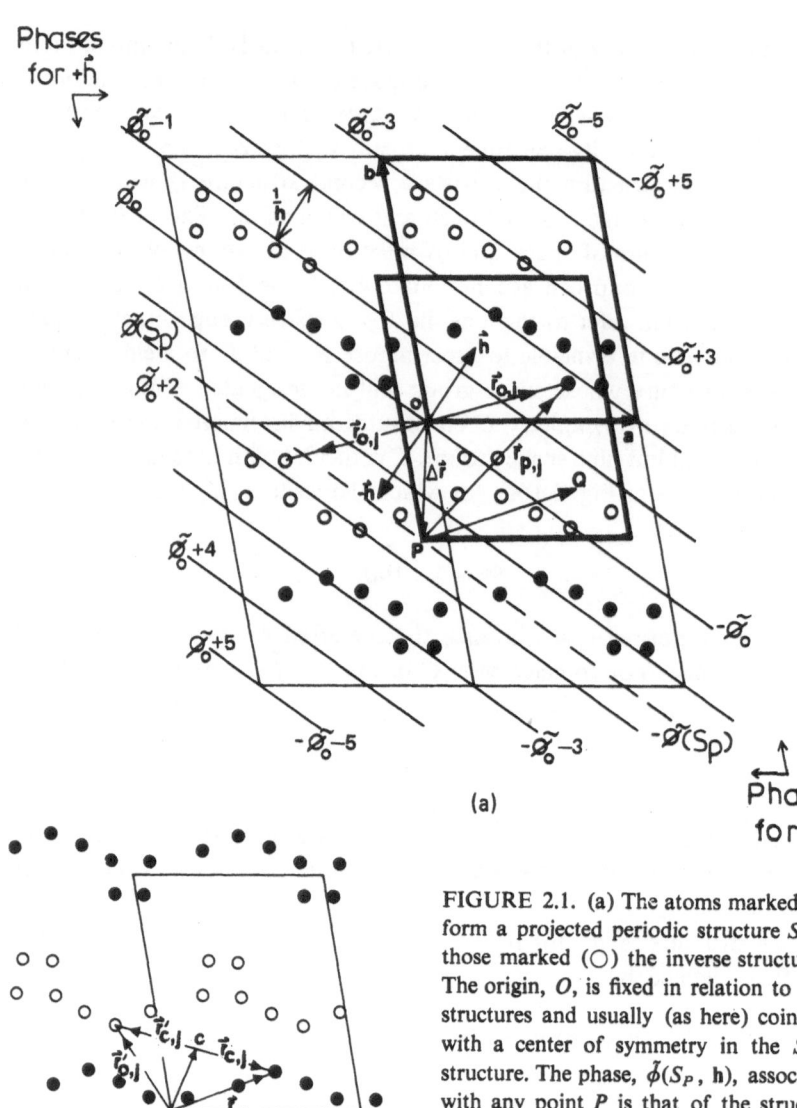

(a)

Phases
for -h̄

(b)

FIGURE 2.1. (a) The atoms marked (●)
form a projected periodic structure S and
those marked (○) the inverse structure I.
The origin, O, is fixed in relation to both
structures and usually (as here) coincides
with a center of symmetry in the S & I
structure. The phase, $\overset{\smile}{\phi}(S_P, \mathbf{h})$, associated
with any point P is that of the structure
factor for reflection, $\mathbf{h}$, for the unit cell of
S having its origin at P. The phase planes
(for the 230 reflection in this example) are
loci of constant phase as P moves through-
out the cell. The labels along the top and left-hand edges indicate the numerical values
of $\overset{\smile}{\phi}(S_P, \mathbf{h})$ and $\overset{\smile}{\phi}(I_P, -\mathbf{h})$, while those on the other two edges denote $\overset{\smile}{\phi}(S_P, -\mathbf{h})$
and $\overset{\smile}{\phi}(I_P, \mathbf{h})$.

(b) The relation between the cells of the S or I space group and the S & I structure for the nine exceptional space groups starred in Table 2A.1 of Appendix 1. See
also Fig. 2.9 for a specific example.

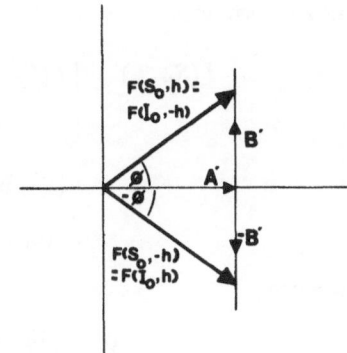

FIGURE 2.2. Complex-plane representations of the structure factors for reflections h and −h for both enantiomeric structures S and I referred to the fixed origin, O, in Fig. 2.1a.

they are not, excellent alternative ways of solving structures are available which would probably be used in preference to direct methods. It follows from (2.2Sb) and the corresponding equation for $F(I_0, \mathbf{h})$ that

$$F(S_0, \mathbf{h}) = A' + iB' = F(I_0, -\mathbf{h})$$

and   (2.3)

$$F(S_0, -\mathbf{h}) = A' - iB' = F(I_0, \mathbf{h})$$

All four structure factors have the same amplitude, but their phases vary as shown in Fig. 2.2 and Table 2.1. For the structure (S & I), if it is centrosymmetric about O as in Fig. 2.1a, the phases become

$$\phi[(S \& I)_0, \mathbf{h}] = \phi[(S \& I)_0, -\mathbf{h}] = 0 \text{ or } \pi$$

so that $F(\mathbf{h})$ is always *real.* Such reflections have a distinctive distribution of $|F(\mathbf{h})|$ which is called *centric*, whereas that from a noncentrosymmetric structure is different and called *acentric*.

If the outline of the unit cell is translated, the origin moving $\Delta \mathbf{r}$ $(= \overrightarrow{OP} = \mathbf{a}\, \Delta x + \mathbf{b}\, \Delta y + \mathbf{c}\, \Delta z)$ from O to any point P as in Fig. 2.1a,

TABLE 2.1. Phase Relationships for Structures S_0 and I_0

Reflection	$\phi(S_0)$	$\phi(I_0)$
h	ϕ	$-\phi$
−h	$-\phi$	ϕ

(2.2*S*) becomes

$$F(S_P, \mathbf{h}) = |F(S_P, \mathbf{h})| \exp[i\phi(S_P, \mathbf{h})] \tag{2.4Sa}$$

$$= \sum_j g_j \exp[2\pi i\mathbf{h} \cdot (\mathbf{r}_{0,j} - \Delta\mathbf{r})]$$

$$= F(S_0, \mathbf{h}) \exp[2\pi i(-\mathbf{h} \cdot \Delta\mathbf{r})] \tag{2.4Sb}$$

$$= |F(S_0, \mathbf{h})| \exp[i\phi(S_0, \mathbf{h}) - 2\pi i\mathbf{h} \cdot \Delta\mathbf{r}] \tag{2.4Sc}$$

whence

$$\phi(S_P, \mathbf{h}) = \phi(S_0, \mathbf{h}) - 2\pi\mathbf{h} \cdot \overrightarrow{OP} \tag{2.5S}$$

Similarly,

$$F(I_P, \mathbf{h}) = |F(I_0, \mathbf{h})| \exp[i\phi(I_0, \mathbf{h}) - 2\pi i\mathbf{h} \cdot \Delta\mathbf{r}] \tag{2.4Ic}$$

$$= |F(S_0, \mathbf{h})| \exp[-i\phi(S_0, \mathbf{h}) - 2\pi i\mathbf{h} \cdot \Delta\mathbf{r}] \tag{2.4Id}$$

So, for every reflection, $|F|$ is invariant with respect to a shift of origin, but the phase is shifted by

$$\boxed{\Delta\phi = -2\pi\mathbf{h} \cdot \Delta\mathbf{r}} \tag{2.6}$$

This is the *shift theorem*, which is fundamental to this chapter. It will be used repeatedly, but usually in the more convenient form

$$\boxed{\Delta\tilde{\phi} = -\mathbf{h} \cdot \Delta\mathbf{r} = -(h\,\Delta x + k\,\Delta y + l\,\Delta z)} \tag{2.6a}$$

The introduction of $\tilde{\phi}$ ($= \phi/2\pi$, read as "frac ϕ," short for fractional ϕ) affords valuable simplifications, and, because the periodicity of $\tilde{\phi}$ is unity, permits the use of a phase algebra modulo 1.

Equation (2.5*S*) shows that the variation of $\tilde{\phi}(S_P, \mathbf{h})$ as P ranges through the structure corresponds to a three-dimensional system of planar fringes in which each plane normal to $\mathbf{h}$ is characterized by a specific value of the phase and is, therefore, termed a *phase plane*. Thus the plane characterized by the value $\tilde{\phi}$ is defined by

$$\mathbf{h} \cdot \mathbf{r}_0 = \tilde{\phi}(S_0, \mathbf{h}) - \tilde{\phi} \tag{2.7S}$$

and its position is fixed in relation to the structure S and to the known origin O. The shift theorem shows that the *form* of this equation is invariant with respect to shift of origin to any other point P, thus

$$\mathbf{h} \cdot \mathbf{r}_P = \tilde{\phi}(S_P, \mathbf{h}) - \tilde{\phi} \qquad (2.7Sa)$$

provided all the real-space vectors ($\mathbf{r}$ and those in S) are referred to a common point.

The same fringe system also represents the contribution made by reflection $\mathbf{h}$ to the electron-density sum at P in structure S, viz.,

$$\varrho(P, S) = V^{-1} \sum |F(S_0, \mathbf{h})| \exp[i\phi(S_0, \mathbf{h}) - 2\pi i \mathbf{h} \cdot \overrightarrow{OP}] \qquad (2.8S)$$

$$= V^{-1} \sum F(S_P, \mathbf{h}) \qquad (2.8Sa)$$

A similar relation exists for the inverse structure.

Such fringe systems were introduced by Bragg and Lipson (1936), and were the basis of several subsequent analog Fourier-summation devices such as Huggins' masks (1945) and von Eller's photosommateur (1955).

For a *known* structure the phase can be calculated for every reflection and any point P, and with such phases the electron-density summation at any other point, Q, which is

$$\varrho(Q, S) = V^{-1} \sum F(S_P, \mathbf{h}) \exp(-2\pi i \mathbf{h} \cdot \overrightarrow{PQ})$$

depicts that part of the periodic structure (S) lying within the unit cell having P as its origin (see Fig. 2.1a). It is in this sense that the term *origin* is used in the literature, in the title of this chapter, and elsewhere in this book; but, to avoid confusion, we shall throughout this chapter refer to the point P as a *reference point* and shall reserve the term *origin* for the point O which is fixed in relation to the structure.

For an *unknown* structure we have no knowledge of $\tilde{\phi}$ for any $\mathbf{h}$ at any reference point. The best one can do is to assign an arbitrary value $\tilde{\phi}_1$ to a selected reflection $\mathbf{h}_1$. This action is tantamount to restricting the reference point to any position on the family of equispaced phase planes defined by

$$\mathbf{h}_1 \cdot \mathbf{r}_O = \tilde{\phi}(S_0, \mathbf{h}_1) - \tilde{\phi}_1 \text{ (mod 1)} \qquad (2.9S)$$

which are fixed in relation to the structure and whose interplanar spacing is $|h_1|^{-1}$ (see Fig. 2.3).

If two other terms, $\mathbf{h}_2$ and $\mathbf{h}_3$, are assigned phases, $\tilde{\phi}_2$ and $\tilde{\phi}_3$, two further sets of equispaced phase planes are introduced, which, if the three vectors are noncoplanar, divide the crystal volume into a series of identical, con-

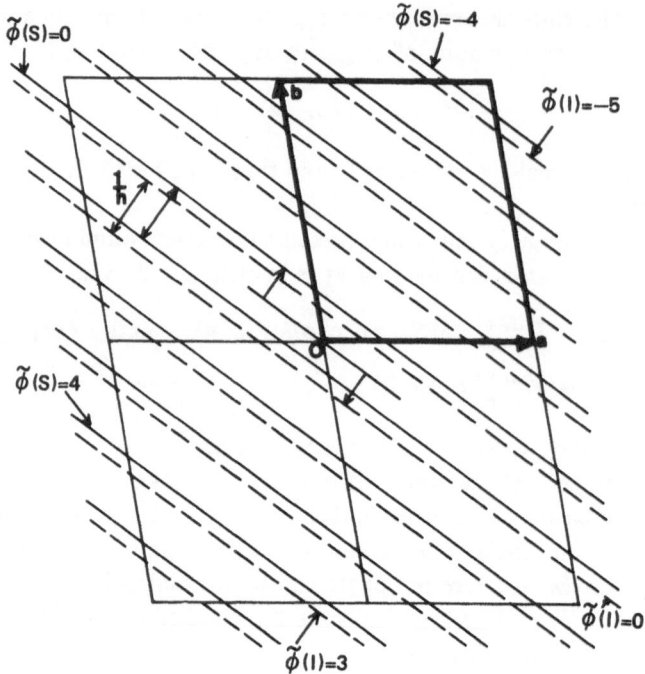

FIGURE 2.3. The interleaved sets of phase planes for integer values of $\tilde{\phi}(h)$ in a non-centrosymmetric structure: ——— for $\tilde{\phi}(S_P, h)$; – – – – for $\tilde{\phi}(I_P, h)$. The signs of the numerical values should be reversed for $-h$. The two sets of planes are displaced from O by $\pm \tilde{\phi}(S_0, h)/h$. For a centrosymmetric structure the two sets coincide, either running through O for $\tilde{\phi} = 0$ (mod 1), or straddling O with intercepts $\pm 1/2h$ for $\tilde{\phi} = \frac{1}{2}$ (mod 1).

tiguous parallelepipeds, and the reference point is now restricted to their vertices. If we express the trio of vectors in the form

$$\begin{pmatrix} h_1 \\ h_2 \\ h_3 \end{pmatrix} = \begin{pmatrix} h_1 & k_1 & l_1 \\ h_2 & k_2 & l_2 \\ h_3 & k_3 & l_3 \end{pmatrix} \begin{pmatrix} a^* \\ b^* \\ c^* \end{pmatrix} = H \begin{pmatrix} a^* \\ b^* \\ c^* \end{pmatrix} \tag{2.10}$$

and thereby define the transformation, H, we can conveniently compare the parallelepiped defined by h_1, h_2, h_3 with the conventional cell which is similarly bounded by the (100), (010), and (001) phase planes. Thus, if the reference point is to be uniquely defined, there can only be one parallelepiped vertex per unit cell, so the volumes of the two alternative cells, $[a \cdot b \times c]$ and that of the parallelepiped, must be equal (see Fig. 2.4), and so too must the volumes of the corresponding "cells" in reciprocal space; thus

$$| [\mathbf{h}_1 \cdot \mathbf{h}_2 \times \mathbf{h}_3] | = | [\mathbf{a}^* \cdot \mathbf{b}^* \times \mathbf{c}^*] |$$

which reduces to

$$\det(\mathbf{H}) = \varDelta = \pm 1 \tag{2.11}$$

This is known as the *primitivity condition.*

Furthermore, the two sets of reciprocal-lattice vectors constitute alternative bases for indexing any reflection:

$$(h'k'l') \begin{pmatrix} \mathbf{h}_1 \\ \mathbf{h}_2 \\ \mathbf{h}_3 \end{pmatrix} = (h'k'l')\mathbf{H} \begin{pmatrix} \mathbf{a}^* \\ \mathbf{b}^* \\ \mathbf{c}^* \end{pmatrix} = (hkl) \begin{pmatrix} \mathbf{a}^* \\ \mathbf{b}^* \\ \mathbf{c}^* \end{pmatrix} \tag{2.12}$$

whence

$$(h'k'l') = (hkl)\mathbf{H}^{-1} \tag{2.13}$$

When the primitivity condition is fulfilled (so that all the elements of $\mathbf{H}^{-1}$ are integral), it follows that every reciprocal-lattice vector can be

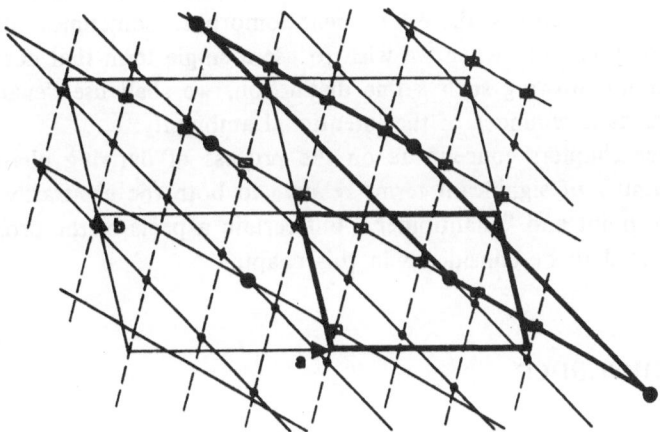

FIGURE 2.4. The intersections of the (10), (01), (11), (21), and ($\bar{3}$1) phase planes in a two-dimensional structure. The number of intersections per unit cell, $\varDelta = (h_1k_2 - h_2k_1)$, is 1 (●) for (21) and (11), 4 (□) for (11) and ($\bar{3}$1), and (5) (•) for (21) and ($\bar{3}$1). The conventional cell is bounded by the phase planes for (10) and (01) which thereby define its shape, orientation, and origin. An alternative primitive cell, also heavily outlined, is likewise defined by the phase planes for (21) and (11).

equally well expressed as a linear combination of either set of basis vectors with *integral* indices, hkl or $h'k'l'$, respectively. The two sets of basis vectors are then said to *span* the reciprocal lattice. The condition, $\Delta = \pm 1$, ensures, therefore, both the *primitivity* of the alternative real-space cells, and the *accessibility* of reciprocal space to both sets of basis vectors.

The treatment so far relates strictly only to the space group $P1$, for we have assumed that any point may serve as a reference point, but in every other space group the set of points that can serve as permissible or preferred origins is limited. So, in what follows, we have to consider the restrictions on the defining reflections, $\mathbf{h}_1$, $\mathbf{h}_2$, $\mathbf{h}_3$ appropriate to each space group. They are often less restrictive than the primitivity condition, and in the higher-symmetry space groups can take a variety of forms.

Two other considerations must be faced. First, if the primitivity condition cannot be met from among the strong, well-connected terms in a particular problem, the choice of three terms [for which $\det(\mathbf{H}) = n$] leaves an n-fold ambiguity in the definition of the origin (see Fig. 2.4). This makes it impossible for the considerable fraction, $(n - 1)/n$, of all reflections either to express all their indices as integers or to determine their phases, and we must consider what steps to take when this situation arises. Usually further terms are co-opted into the starting set in some way.

Secondly, in all noncentrosymmetric structures one must impose a further arbitrary condition to remove the ambiguity between the S and I structures. Sometimes these are enantiomorphs, sometimes alternative polar structures, but when we wish to use a single term that describes S, say, without drawing such a fine distinction, we shall use "enantiomer" in quotes as a reminder of the intentional ambiguity.

Later chapters concentrate on the process of deriving phases for a large number of significant terms relative to both the arbitrarily selected reference point and "enantiomer," but certain aspects of the propagation process need to be considered in this chapter.

2.3. Invariance

It is essential in all direct methods work to distinguish between entities that vary with respect to reference point and those that are invariant to its position. Examples are:

Variable: atomic vectors, $\mathbf{r}_j$;
            phases of reflections, $\phi(\mathbf{h})$ or $\tilde{\phi}(\mathbf{h})$.

Invariant: interatomic vectors, $\mathbf{r}_{ij} = \mathbf{r}_i - \mathbf{r}_j$;

the position of phase planes;

the associated amplitude, $|F|$ or $|E|$ and all functions of them, such as $|F|^2, |E|^2 - 1$, and the many functions used in later chapters to calculate the probabilities of various phase relationships.

Further distinctions need to be drawn, all of which will be illustrated later:

1. Certain entities are invariant in all space groups and for a reference point *anywhere* in the cell: they are called *universal invariants*.

2. Some entities are invariant only if the reference point ranges over a limited set of points known as permissible origins which are characteristic of the space group. Such points fall into one or more subsets known as *equivalence classes*, each of which is characterized by a *distinctive functional form* of the structure-factor expression, which applies only when the reference point coincides with the points constituting the class. The entity is then said to be a *structure seminvariant* with respect to that equivalence class.

3. Many space groups contain only one equivalence class, but in those that contain more than one, it is sometimes useful to distinguish certain entities whose *numerical value* is invariant over more than one equivalence class: for these entities we coin the word *multiseminvariant*.

Consider the invariance of the linear sum of phases

$$\sum_i A_i \bar{\phi}(S_0, \mathbf{h}_i) \text{ (mod 1)} \tag{2.14}$$

As fractions of a periodic phase are not uniquely definable, all the coefficients, A_i, must be integral. When the reference point is shifted to any point P the function becomes $\sum_i A_i [\bar{\phi}(S_0, \mathbf{h}_i) - \mathbf{h}_i \cdot \varDelta \mathbf{r}]$ (mod 1), which is invariant if

$$\sum_i (A_i \mathbf{h}_i) \cdot \varDelta \mathbf{r} = 0 \tag{2.15}$$

This equation can be fulfilled in two ways: In the first case, if

$$\sum_i A_i \mathbf{h}_i = 0 \tag{2.16}$$

the phase sum is *universally invariant provided all the $\bar{\phi}(S_0, \mathbf{h}_i)$ are referred to a common reference point and relate to the same structure, S.* This result is so important and so frequently used that it will be termed the *universal* *invariant theorem*. It is true for all space groups and any shift $\Delta \mathbf{r}$. Hauptman and Karle refer to this phase sum as a *structure invariant*. The simplest example, $\tilde{\phi}(\mathbf{h}) + \tilde{\phi}(-\mathbf{h}) = 0$, is trivial, and the next,

$$2\tilde{\phi}(\mathbf{h}) + \tilde{\phi}(-2\mathbf{h}) = C(\mathbf{h})$$

where $C(\mathbf{h})$ is a constant different for each $\mathbf{h}$, is of limited value. The commonest, and hitherto the most widely used, is the triple relation,

$$\tilde{\phi}(\mathbf{h}_1) + \tilde{\phi}(\mathbf{h}_2) + \tilde{\phi}(\mathbf{h}_3) = \tilde{\Phi}_{123} \qquad (2.17)$$

where $\tilde{\Phi}_{123}$ is invariant and is uniquely associated with the three reciprocal-lattice vectors if $\mathbf{h}_1 + \mathbf{h}_2 + \mathbf{h}_3 = 0$. It is shown elsewhere that the probability of finding a given value of $\tilde{\Phi}_{123}$ can be calculated from the invariant values of $|E_i|$ $(i = 1, 2, 3)$; so, when (2.17) is expressed complete with its probability [as in (1.18)] it is known as a Σ_2 relation. The general forms of the distributions of $\tilde{\Phi}_{123}$ for centric and acentric structure factors are shown in Fig. 2.5: the larger $|E_1 E_2 E_3|$, the greater is the probability that $\tilde{\Phi}_{123} = 0$ (rather than 1/2) for centric reflections, or the more closely it approximates to zero for acentric reflections. This is the basis of the commonest process of propagating phases, that of inferring $\tilde{\phi}(\mathbf{h}_3)$ when $\tilde{\phi}(\mathbf{h}_1)$ and $\tilde{\phi}(\mathbf{h}_2)$ are known. *Any* three reflections in *any* space group may be combined so long

(a)

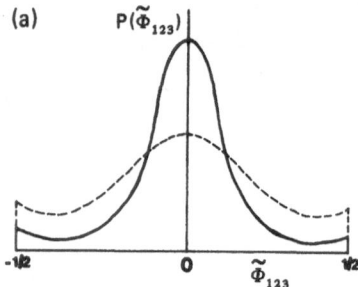

(b)

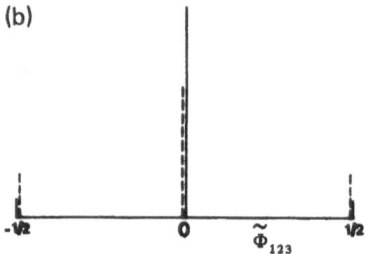

FIGURE 2.5. Typical probability distributions for finding a specified value of the seminvariant triple-phase sum, $\tilde{\Phi}_{123} = \tilde{\phi}(\mathbf{h}_1) + \tilde{\phi}(\mathbf{h}_2) + \tilde{\phi}(\mathbf{h}_3)$, where $\mathbf{h}_1 + \mathbf{h}_2 + \mathbf{h}_3 = 0$: (a) for acentric reflections, (b) for centric reflections. (———) large $|E_1 E_2 E_3|$; (— —) medium values of $|E_1 E_2 E_3|$.

as $\mathbf{h}_1 + \mathbf{h}_2 + \mathbf{h}_3 = 0$, and the error in assuming $\bar{\Phi}_{123} = 0$ decreases with increasing values of their $|E|$'s. An alternative way of expressing the triple relation is

$$\bar{\phi}(\mathbf{h}_1 \pm \mathbf{h}_2) \approx \bar{\phi}(\mathbf{h}_1) \pm \bar{\phi}(\mathbf{h}_2) \tag{2.18}$$

where $\approx$ signifies a probability-weighted approximation to equality.

In recent years the usefulness of quadruple and even higher phase sums has been investigated: they are discussed in Chapters 4 and 5.

The second way in which (2.15) can be fulfilled is when the right-hand side is zero, modulo 1, for *all* the Δr's associated with the permissible origins in a given equivalence class, and only for such points. As an example, consider the space group $P\bar{1}$ for which there are eight alternative origins at centers of symmetry, 000, $\frac{1}{2}$00, etc., which can be represented by $\varepsilon_j/2$, $j = 1, 2, 3$; $\varepsilon = 0$ or 1. Equation (2.15) then becomes, by use of the shift theorem,

$$\sum_i A_i(h_i\varepsilon_1 + k_i\varepsilon_2 + l_i\varepsilon_3) = 0 \pmod 2 \tag{2.19}$$

for all combinations of ε. In this case, unlike the first solution, (2.15) can be fulfilled *for a single reflection*. Thus, in this particular space group, (2.19) is fulfilled for all eight alternative origins by any reflection whose indices are all even (*ggg*). The structure factor, and more significantly the phase, of such a reflection is a structure seminvariant, that is, its value is uniquely determined by the structure alone and is independent of the choice of permissible origin. Such reflections cannot, therefore, be assigned arbitrary phases and so cannot be used to define origins. But they can be used in the propagation process based on universally invariant sums and can often be used as enantiomorph-selecting terms. They play a very important role in all direct methods work, but the parity rules specifying seminvariant reflections differ from one space group to another and will be considered later; they are not always *ggg* reflections. However, the *ggg* rule applies to all those space groups, whether centrosymmetric or not, having permissible origins at the eight *points* indicated above.

When more than one term is included in the sum of (2.15), it follows that the parity of $\sum_i A_i\mathbf{h}_i$ must be *ggg* for $P\bar{1}$ and similar space groups, and that $\sum_i A_i\bar{\phi}(\mathbf{h}_i)$ is likewise a structure seminvariant. This is an extension of the earlier rule [see (2.17)], and a much less restrictive one too. Thus for two terms, $\bar{\phi}(\mathbf{h}_1) + \bar{\phi}(\mathbf{h}_2)$ is seminvariant if both terms come from the same parity group. For three terms, $\bar{\phi}(\mathbf{h}_1) + \bar{\phi}(\mathbf{h}_2) + \bar{\phi}(\mathbf{h}_4)$ is seminvariant if $\mathbf{h}_4$ is any term in the parity group of $\mathbf{h}_1 + \mathbf{h}_2$. This wider triplet result is not used as often as it should be. It is potentially capable of propagating phases faster than the Σ_2 triple, because there are many more triples of type (2.21) than there are of type (2.20) and it can even give values of $\tilde{\Phi}_{124}$ diverging widely from zero. It can be compared with the Σ_2 triple

$$\mathbf{h}_1 + \mathbf{h}_2 + \mathbf{h}_3 = 0 \qquad\qquad (2.20)$$

$$\mathbf{h}_1 + \mathbf{h}_2 + \mathbf{h}_4 = \mathbf{h}_4 - \mathbf{h}_3 = \mathbf{h}_5 \qquad (\text{a } ggg \text{ term}) \qquad (2.21)$$

Thus, $\tilde{\phi}(\mathbf{h}_1) + \tilde{\phi}(\mathbf{h}_2) + \tilde{\phi}(\mathbf{h}_4) - \tilde{\phi}(\mathbf{h}_5) \approx 0$, and this becomes accessible through the ordinary triples procedure,

$$\tilde{\phi}(\mathbf{h}_1) + \tilde{\phi}(\mathbf{h}_2) \approx \tilde{\phi}(-\mathbf{h}_3) \approx \tilde{\phi}(\mathbf{h}_5) - \tilde{\phi}(\mathbf{h}_4)$$

This is a valuable generalization of the "Coincidence Method" introduced by Grant, Howells, and Rogers (1957) and is of wider applicability. The joint probability of such a coincidence was evaluated in the appendix of the same paper.

2.4. Variation of Phase among the Laue-Related Reflections

As a prelude to the consideration of invariance for individual space groups we must consider the variation of phase among the Laue-symmetry-related reflections. Every one of the m crystallographically independent atoms in the asymmetric unit is repeated n times (or a submultiple of n) by the space-group symmetry operations. Equation (2.2S) can then be rewritten as

$$F(S_O, \mathbf{h}) = \sum_{q=1}^{m} g_q \left\{ \sum_{j}^{n} \exp[2\pi i \mathbf{h} \cdot (\mathbf{r}_{O,j})_q] \right\} = \sum_{q}^{m} g_q T_q \qquad (2.22)$$

where the $(\mathbf{r}_O, j)_q$ are the n equivalent positions (equipoints) related to the qth atom in the coordinate list at x_q, y_q, z_q. The sum (T) is usually written as $A + iB$, though Hauptmann and Karle used $\xi + i\eta$ in their early papers. The distinctive and often complicated trigonometrical formulas for A and B for each space group are given in the *International Tables*, Vol. 1. Here we prefer to use the related property of the variations of phase among the Laue-symmetric reflections, and these we refer to as the *phase pattern*. It is readily obtained using a generalization of Bertaut's structure-factor algebra (1956), and each pattern is uniquely associated with and characteristic of the space group and equivalence class. A few examples are given here; others appear later.

Space Group $P2_12_12_1$

The equipoints, referred to the conventional axes and origin, are listed in Table 2.2 and give rise to

$$A = 4 \cos 2\pi\left(hx - \frac{h-k}{4}\right) \cos 2\pi\left(ky - \frac{k-l}{4}\right) \cos 2\pi\left(lz - \frac{l-h}{4}\right)$$

and

$$B = -4 \sin 2\pi\left(hx - \frac{h-k}{4}\right) \sin 2\pi\left(ky - \frac{k-l}{4}\right) \sin 2\pi\left(lz - \frac{l-h}{4}\right)$$

These particular functional forms are found at all eight points, 000, $\frac{1}{2}$00, $0\frac{1}{2}0$, $00\frac{1}{2}$, $\frac{1}{2}\frac{1}{2}0$, $\frac{1}{2}0\frac{1}{2}$, $0\frac{1}{2}\frac{1}{2}$, $\frac{1}{2}\frac{1}{2}\frac{1}{2}$, which, therefore, constitute the equivalence class.

In Table 2.2 each of the reflections in the Laue set is paired off with that equipoint that gives an expression for $\tilde{\phi} (= \mathbf{h} \cdot \mathbf{r})$ of the form $\pm(hx + ky + lz)$. More detail in justification of this simple procedure appears in Appendix 2A.2.

It is easy in this way to write the phase pattern for any space group merely by inspection of the list of equipoints: indeed, in some, such as trigonal or hexagonal, it is very much easier to derive patterns this way than from the trigonometrical formulas for A and B, which, especially for the higher-symmetry space groups, can take on varied functional forms whose equivalence is far from obvious.

The pattern in Table 2.2 is typical of those found for any equivalence class in any space group. It contains two characteristic components.

1. The set of Laue-symmetric reflections falls into two antipodal groups. Those that are point-group related to hkl, designated $\{hkl\}$, all contain the term $+\tilde{\phi}$, while the antipodal set, $\{\bar{h}\bar{k}\bar{l}\}$, all contain $-\tilde{\phi}$. This point-group

TABLE 2.2. Phase Relationships for Laue-Related Reflections in $P2_12_12_1$

$\mathbf{r}$	$\mathbf{h}$	$\mathbf{h} \cdot \mathbf{r}$	$-\mathbf{h}$	$-\mathbf{h} \cdot \mathbf{r}$
x, y, z	hkl	$+\tilde{\phi}$	$\bar{h}\bar{k}\bar{l}$	$-\tilde{\phi}$
$\frac{1}{2} + x, \frac{1}{2} - y, \bar{z}$	$h\bar{k}\bar{l}$	$+\tilde{\phi} + (h+k)/2$	$\bar{h}kl$	$-\tilde{\phi} - (h+k)/2$
$\bar{x}, \frac{1}{2} + y, \frac{1}{2} - z$	$\bar{h}kl$	$+\tilde{\phi} + (k+l)/2$	$h\bar{k}\bar{l}$	$-\tilde{\phi} - (k+l)/2$
$\frac{1}{2} - x, \bar{y}, \frac{1}{2} + z$	$\bar{h}\bar{k}l$	$+\tilde{\phi} + (l+h)/2$	$hk\bar{l}$	$-\tilde{\phi} - (l+h)/2$

dependence was pointed out by Waser (1955) and is discussed further by Rogers (1975) in connection with the observation and use of Bijvoet anomalies. The numerical value of $\breve{\phi}$ depends, of course, on the indices, hkl, and the structure, but it must be *uniform* (mod 1) throughout the Laue set. In view of the antisymmetry of $\breve{\phi}$, we shall henceforth only list the pattern for the $\{hkl\}$ set.

2. Some reflections have additional terms $[(h + k)/2$, *etc.* in this case], which we shall call *incremental terms*. They are not functions of the structure and *they alone give rise to the variations in the phase throughout the Laue set*, which, together with the point group, uniquely define the space group. They may contain:

(a) *translatory terms* which are derived from the translatory components of the symmetry elements of the space group (2_1 screws in this example) and are independent of the choice of reference point; and/or

(b) *displacement terms* due to the displacement of the corresponding symmetry element from the chosen reference point. These can, therefore, vary and serve to specify the chosen equivalence class.

Incremental terms are specially significant in what follows as they characterize the space group and the chosen equivalence class. This is demonstrated in Table 2.3, which gives the patterns for the primitive space groups in Class 222 and shows why in later sections one can often discuss together all the space groups isomorphous with a given point group. In $P222$ there are no incremental terms as all three symmetry elements are diads which concur in the origin. The terms for $P2_12_12_1$ are easily memorized: the indices required are the positive one and the next that follows it cyclically.

TABLE 2.3. Phase Patterns for the Primitive Space Groups in Crystal Class 222 [a]

Point-group related reflections	$P222$	$P222_1$	$P2_12_12$	$P2_12_12_1$
hkl	$+\breve{\phi}$	$+\breve{\phi}$	$+\breve{\phi}$	$+\breve{\phi}$
$h\bar{k}\bar{l}$	$+\breve{\phi}$	$+\breve{\phi}$	$+\breve{\phi} + (h_t + k_d)/2$	$+\breve{\phi} + (h_t + k_d)/2$
$\bar{h}k\bar{l}$	$+\breve{\phi}$	$+\breve{\phi} + l_d/2$	$+\breve{\phi} + (k_t + h_d)/2$	$+\breve{\phi} + (k_t + l_d)/2$
$\bar{h}\bar{k}l$	$+\breve{\phi}$	$+\breve{\phi} + l_t/2$	$+\breve{\phi}$	$+\breve{\phi} + (l_t + h_d)/2$

[a] The subscripts t and d identify the translatory and displacement terms.

When tackling any structure, the appropriate phase pattern is built in and applied to every reflection. If, for example, we assign a phase to hkl in $P2_12_12_1$ we thereby automatically define the phases of the seven other members of the Laue set. Some computer programs store only the unique set of indices and phases, and generate others by subroutine as needed. Others generate lists of indices and phases for all the reflections in a hemisphere of reciprocal space and store them for immediate use. However it be achieved, the imposition of a phase pattern on all the reflections in a data set ensures certain simplifications characteristic of the space group and the chosen equivalence class, but it leaves three important kinds of ambiguity.

1. The imposed pattern of incremental terms ensures that every Laue set is related to a reference point that is restricted to sites lying within the chosen equivalence class, but it cannot, of itself, distinguish between the alternatives and cannot, therefore, ensure that all Laue sets relate to a common reference point—except in those few space groups where the equivalence class consists of a single point.

2. The association of a uniform value, $+\tilde{\phi}$, with $\{hkl\}$ and $-\tilde{\phi}$ with $\{\bar{h}\bar{k}\bar{l}\}$ ensures that all reflections within a given Laue set relate to a common "enantiomer," but it cannot ensure that different Laue sets relate to the same "enantiomer," that is, all to S or all to I.

3. It is often possible, as here, to relabel the cell axes, so we must consider the invariance of the phase pattern under this process. $P2_12_12_1$ offers an instructive and nontrivial example. If we shift the reference point by an arbitrary vector from O, the conventional origin, the phase pattern becomes

$$\tilde{\phi}(S_P, hkl) = +\tilde{\phi}_0 - (h\,\Delta x + k\,\Delta y + l\,\Delta z)$$
$$\tilde{\phi}(S_P, h\bar{k}\bar{l}) = +\tilde{\phi}_0 - (h\,\Delta x - k\,\Delta y - l\,\Delta z) + (h + k)/2$$
$$\tilde{\phi}(S_P, \bar{h}k\bar{l}) = +\tilde{\phi}_0 - (-h\,\Delta x + k\,\Delta y - l\,\Delta z) + (k + l)/2 \tag{2.23}$$
$$\tilde{\phi}(S_P, \bar{h}\bar{k}l) = +\tilde{\phi}_0 - (-h\,\Delta x - k\,\Delta y + l\,\Delta z) + (l + h)/2$$

which may be rewritten as

$$\tilde{\phi}(S_P, hkl) = +\tilde{\phi}_P$$
$$\tilde{\phi}(S_P, h\bar{k}\bar{l}) = +\tilde{\phi}_P + [\qquad\quad 2k\,\Delta y + 2l\,\Delta z + (h + k)/2]$$
$$\tilde{\phi}(S_P, \bar{h}k\bar{l}) = +\tilde{\phi}_P + [2h\,\Delta x \qquad\quad + 2l\,\Delta z + (k + l)/2] \tag{2.24}$$
$$\tilde{\phi}(S_P, \bar{h}\bar{k}l) = +\tilde{\phi}_P + [2h\,\Delta x + 2k\,\Delta y \qquad\quad + (l + h)/2]$$

There are two ways in which the incremental terms in square brackets can be transformed into simple patterns suitable for imposition:

1. The pattern reduces to the conventional form when the shifts $\Delta x, \Delta y, \Delta z$ are all the combinations of 0 or $\frac{1}{2}$. This constitutes the conventional equivalence class of eight points, which we shall denote briefly as [000].

2. There is a second set of eight points displaced $\frac{1}{4}, \frac{1}{4}, \frac{1}{4}$ from the conventional set, written as $[\frac{1}{4}\frac{1}{4}\frac{1}{4}]$, and for these the pattern alters, by changes of the displacement terms only, to that given in Fig. 2.6. It follows that imposition of the conventional pattern eliminates this alternative, unconventional equivalence class. Figure 2.6 depicts the symmetry environment around representative points from each set. They are identical *but differently oriented*, and it is easy to see how the displacement terms arise in each case. Around [000] each screw axis is displaced a quarter of the axis that *follows* cyclically, whereas around $[\frac{1}{4}\frac{1}{4}\frac{1}{4}]$ each is displaced a quarter of the *preceding* axis. It follows that the conventional pattern applies also if the cell is relabeled cyclically, *bca* or *cab*. If, however, the cell is relabeled anticyclically (*bāc, aċb, ċba*), the displacement terms change so that the conventional pattern is now associated with the points that were formerly

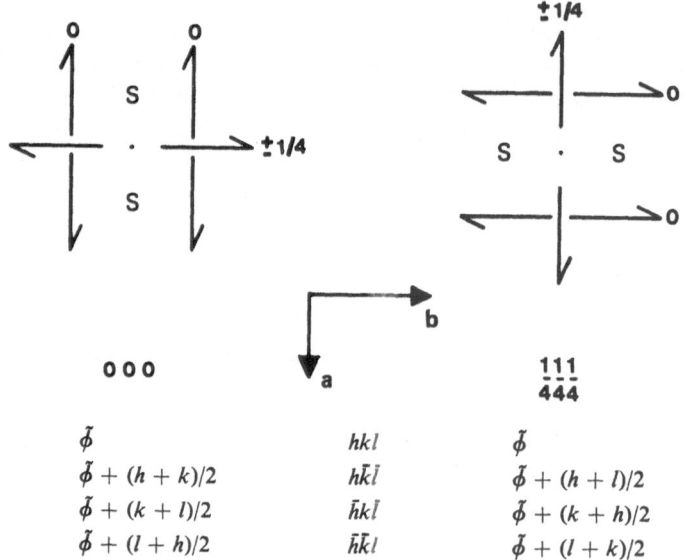

$$
\begin{array}{ccc}
\stackrel{\circ}{\phi} & hkl & \stackrel{\circ}{\phi} \\
\stackrel{\circ}{\phi} + (h+k)/2 & h\bar{k}\bar{l} & \stackrel{\circ}{\phi} + (h+l)/2 \\
\stackrel{\circ}{\phi} + (k+l)/2 & \bar{h}k\bar{l} & \stackrel{\circ}{\phi} + (k+h)/2 \\
\stackrel{\circ}{\phi} + (l+h)/2 & \bar{h}\bar{k}l & \stackrel{\circ}{\phi} + (l+k)/2
\end{array}
$$

FIGURE 2.6. Comparison of the symmetry environments of the points 000 and $\frac{1}{4}\frac{1}{4}\frac{1}{4}$ in $P2_12_12_1$, and the corresponding phase patterns.

TABLE 2.4. Alternative Labeling of the Unit-Cell of 2,2′-Cyclocytidine Hydrochloride Adopted by Different Authors

Cell edges (Å)	RVSH	KP	BS
16.94	*a*	*b*	*c*
11.08	*b*	*a*	*b*
6.04	*c*	*c*	*a*

$[\frac{1}{4}\frac{1}{4}\frac{1}{4}]$. These now become the conventional equivalence class [000] for the anticyclically relabeled cell.

So, if two workers independently collect data for the same compound in $P2_12_12_1$ but use anticyclically labeled cells, the same conventional phase pattern which they both employ demands the same symmetry environment at their respective origins, which are, therefore, $\frac{1}{4}\frac{1}{4}\frac{1}{4}$ (or some related vector) apart in the structure. An example of this was encountered a few years ago in this laboratory when a comparison was made of the coordinates obtained in three independent direct-method determinations of the structure of 2,2′-cyclocytidine hydrochloride. Rogers, Venkatasubramanian, Sørum, and Hjortås (RVSH) (1972) and Kartha and Phillips (KP) (1973, 1975) studied the L form, while Brennan and Sundaralingam (BS) (1973) studied the D form. The cell edges were labeled as in Table 2.4, so that both the KP and BS structures were labeled anticyclically compared with RVSH. Consequently the relationships between the coordinates (omitting increments of $\frac{1}{2}$) took the form shown in Table 2.5, which reveals both the displacement of the origin by $\frac{1}{4}\frac{1}{4}\frac{1}{4}$ and the enantiomorphism of the BS structure.

This situation occurs in other contexts too, namely, that the choice of axis labeling (or frame of reference) can of itself discriminate against certain equivalence classes through the imposed corresponding phase pattern.

TABLE 2.5. Relations between the Coordinates for the Different Labeling Schemes Shown in Table 2.4

RVSH		KP		BS
x	$=$	$y - \frac{1}{4}$	$=$	$\frac{1}{4} - z$
y	$=$	$x - \frac{1}{4}$	$=$	$\frac{1}{4} - y$
z	$=$	$z - \frac{1}{4}$	$=$	$\frac{1}{4} - x$

TABLE 2.6. Phase Shifts for Alternative Reference Points in Class 4

h	$\tilde{\phi}(000)$ [a]	$\Delta\tilde{\phi}(00z)$	$\Delta\tilde{\phi}(\frac{1}{2}\frac{1}{2}z)$	$\Delta\tilde{\phi}(\frac{1}{2}0z)$	$\Delta\tilde{\phi}(0\frac{1}{2}z)$
hkl	$\tilde{\phi}_0$	$-lz$	$(h+k)/2 - lz$	$h/2 - lz$	$k/2 - lz$
$\bar{k}hl$	$\tilde{\phi}_0 + (nl/4)$	$-lz$	$(h+k)/2 - lz$	$k/2 - lz$	$h/2 - lz$
$\bar{h}\bar{k}l$	$\tilde{\phi}_0 + (nl/2)$	$-lz$	$(h+k)/2 - lz$	$h/2 - lz$	$k/2 - lz$
$k\bar{h}l$	$\tilde{\phi}_0 + (3nl/4)$	$-lz$	$(h+k)/2 - lz$	$k/2 - lz$	$h/2 - lz$

[a] In the last three terms in this column use $n = 0$ for $P4$, $+1$ for $P4_1$, $+2$ for $P4_2$, $+3$ for $P4_3$.

In the simpler space groups the choice of equivalence class is limited and straightforward. Thus, in Class 2 there is but one equivalence class comprising all points on the four diad axes. In Class 4, however, the origin could be located either on one of the two tetrad axes ($00z$; $\frac{1}{2}\frac{1}{2}z$) or on one of the diads ($\frac{1}{2}0z$; $0\frac{1}{2}z$). The phase pattern and phase shifts (mod 1) on shifting the reference point about on these four axes are given in Table 2.6. After referring the coordinates to the new reference point these entries become those shown in Table 2.7. There are two phase patterns here, one for the tetrads and the other for the diads, so there are two equivalence classes for these space groups. In addition, reference to the patterns of Table 2.6 shows that $\Delta\tilde{\phi}$ is zero for all points on all four axes for $gg0$ reflections, whereas for the $uu0$ it is 0 on the tetrads but $\frac{1}{2}$ on the diads. Both parity groups are, therefore, structure seminvariants, that is, $hk0$ is seminvariant if $h + k = g$, but the $gg0$ reflections are also multiseminvariants.

Other high-symmetry space groups can be analyzed equally neatly and multiple equivalence classes are the rule in these space groups, but

TABLE 2.7. The Phase Patterns of Table 2.6 after Shifting the Reference Point to the Sites Indicated

h	$\tilde{\phi}(000)$	$\Delta\tilde{\phi}(00z)$	$\Delta\tilde{\phi}(\frac{1}{2}\frac{1}{2}z)$	$\Delta\tilde{\phi}(\frac{1}{2}0z)$	$\Delta\tilde{\phi}(0\frac{1}{2}z)$
hkl	$\tilde{\phi}_P$	0	0	0	0
$\bar{k}hl$	$\tilde{\phi}_P + (nl/4)$	0	0	$(h+k)/2$	$(h+k)/2$
$\bar{h}\bar{k}l$	$\tilde{\phi}_P + (nl/2)$	0	0	0	0
$k\bar{h}l$	$\tilde{\phi}_P + (3nl/4)$	0	0	$(h+k)/2$	$(h+k)/2$

they also occur in nonprimitive low-symmetry space groups. Nevertheless, the imposition of a specific phase pattern establishes the corresponding equivalence class and excludes all others from consideration.

2.5. Defining the Origin

2.5.1. Primitive Centrosymmetric Space Groups

The definition of a unique reference point in the cell is achieved by arbitrarily specifying the phases of a few reflections specially chosen to break the ambiguities existing in the equivalence class. We shall deal first with the simpler problem of the primitive centrosymmetric space groups. Their treatment is summarized in Table 2.8.

The centers of symmetry are the only points in the cell through which the phase planes of *all* reflections are restricted to $\tilde{\phi} = 0$ or $\frac{1}{2}$ and for which $\tilde{\phi}(-\mathbf{h}) = \tilde{\phi}(\mathbf{h})$. They constitute, therefore, the preferred equivalence class(es) in centrosymmetric space groups. As soon as the reference point moves away from these points the phases become general and $\tilde{\phi}(-\mathbf{h}) \neq -\tilde{\phi}(\mathbf{h})$.

For all the space groups in category 1 there is a single equivalence class consisting of the eight points listed in Table 2.8, the set [000], which, as before, we can express in the form: $\varepsilon_i/2$, $i = 1, 2, 3$; $\varepsilon = 0$ or 1. The phase shifts, $\Delta\tilde{\phi}$, corresponding to these alternative origins are given by

$$\delta = 2\Delta\tilde{\phi} = (h\varepsilon_1 + k\varepsilon_2 + l\varepsilon_3) \pmod{2} \tag{2.25}$$

where $\delta = 0$ or 1. So, as in (2.19) the ggg reflections are the seminvariants for all these space groups: their signs cannot be assigned arbitrarily nor can they be used to differentiate between alternative origins. Three of the other parity groups make $\delta = 0$ and four make $\delta = 1$. To remove the site ambiguity an arbitrary assignment of $\tilde{\phi}$ (0 or $\frac{1}{2}$) or sign ($+$ or $-$) is made for *three* reflections: each halves the ambiguity. This can be represented in the form

$$\begin{pmatrix} h_1 & k_1 & l_1 \\ h_2 & k_2 & l_2 \\ h_3 & k_3 & l_3 \end{pmatrix} \begin{pmatrix} \varepsilon_{11} & \cdots & \varepsilon_{18} \\ \varepsilon_{21} & \cdots & \varepsilon_{28} \\ \varepsilon_{31} & \cdots & \varepsilon_{38} \end{pmatrix} = \begin{pmatrix} \delta_{11} & \cdots & \delta_{18} \\ \delta_{21} & \cdots & \delta_{28} \\ \delta_{31} & \cdots & \delta_{38} \end{pmatrix} \pmod{2} \tag{2.26}$$

where all the ε's and δ's are 0 or 1. In view of the mod 2 character of this equation it follows that all the elements in the matrix of indices, $\mathbf{H}$, can be

TABLE 2.8. The Primitive Centrosymmetric Space Groups

Category	1	2	3	4
Space groups	$P\bar{1}$ $P2/m$ $P2_1/m$ $P2/c$ $P2_1/c$ $Pmmm$ $Pnnn$ $Pccm$ $Pban$ $Pmma$ $Pnna$ $Pmna$ $Pcca$ $Pbam$ $Pccn$ $Pbcm$ $Pnnm$ $Pmmn$ $Pbcn$ $Pbca$ $Pnma$	$P4/m$ $P4_2/m$ $P4/n$ $P4_2/n$ $P4/mmm$ $P4/mcc$ $P4/nbm$ $P4/nnc$ $P4/mbm$ $P4/mnc$ $P4/nmm$ $P4/ncc$ $P4_2/mmc$ $P4_2/mcm$ $P4_2/nbc$ $P4_2/nnm$ $P4_2/mbc$ $P4_2/mnm$ $P4_2/nmc$ $P4_2/ncm$	$P\bar{3}$ $P\bar{3}1m$ $P\bar{3}1c$ $P\bar{3}m1$ $P\bar{3}c1$ $P6/m$ $P6_3/m$ $P6/mmm$ $P6/mcc$ $P6_3/mcm$ $P6_3/mmc$	$R\bar{3}$ $R\bar{3}m$ $R\bar{3}c$ $Pm\bar{3}$ $Pn\bar{3}$ $Pa\bar{3}$ $Pm\bar{3}m$ $Pn\bar{3}n$ $Pm\bar{3}n$ $Pn\bar{3}m$
Preferred equivalence class	000, $\tfrac{1}{2}00$, $0\tfrac{1}{2}0$, $00\tfrac{1}{2}$, $\tfrac{1}{2}\tfrac{1}{2}\tfrac{1}{2}$, $0\tfrac{1}{2}\tfrac{1}{2}$, $\tfrac{1}{2}0\tfrac{1}{2}$, $\tfrac{1}{2}\tfrac{1}{2}0$	000, $00\tfrac{1}{2}$, $\tfrac{1}{2}\tfrac{1}{2}0$, $\tfrac{1}{2}\tfrac{1}{2}\tfrac{1}{2}$	000, $00\tfrac{1}{2}$	000, $\tfrac{1}{2}\tfrac{1}{2}\tfrac{1}{2}$
Seminvariant reflections for the preferred equivalence class	ggg	ggg and uug, i.e., hkg with $h + k = g$	$hkg \begin{cases} ggg \\ uug \\ gug \\ ugg \end{cases}$	ggg, guu, ugu, uug, i.e., hkl with $h + k + l = g$
Number of phases to be assigned arbitrarily: terms to be chosen from different nonseminvariant parity groups	3	2	1	1 (The 3R cells are referred to rhombohedral axes.)

replaced by their parities. Then, if each alternative reference point (a column of ε's) is to be uniquely defined by a trio of assigned signs or phases (a column of δ's) the determinant of the matrix of *index parities* must be ± 1:

$$\begin{vmatrix} h_1 & k_1 & l_1 \\ h_2 & k_2 & l_2 \\ h_3 & k_3 & l_3 \end{vmatrix} (\text{mod } 2, 2, 2) = \pm 1 \qquad (2.27)$$

where the modulus expression requires $h, k,$ and l on every row to be expressed modulo 2. This condition requires that

1. no one row or column may have parity *ggg*, i.e., a row or column of zeroes;
2. no two rows or columns may have the same parity, i.e., all three must be drawn from different parity groups; and
3. the sum of all three rows or columns may not be *ggg*, i.e., all three parity groups must be linearly independent.

These conditions are equivalent to the exclusion of one-, two-, and three-term seminvariant phase sums among the terms of the starting set. In $P\bar{1}$ the specification of signs for three such terms means one starts with known signs for six reciprocal-lattice points ($\pm \mathbf{h}$ for each), but in Class $2/m$ if they are general terms one starts with 12, and in mmm with 24. The more there are in the starting set the faster propagation gets under way. If, in addition, it is possible to determine signs for a few seminvariant *ggg* terms from inequalities or high-probability Σ_1 terms, they are well worth including as they accelerate propagation, especially if they fulfil the requirements for co-option (see p. 47). The use of triple or higher-structure invariants to deduce new phases takes place first between the terms in the starting set, then with their progeny, and later between their progeny, and leads in principle to known signs in all parity groups.

However, any trio that satisfies the modular equation (2.27), which we shall rewrite more compactly as

$$\det[\mathbf{H} (\text{mod } 2, 2, 2)] = \pm 1 \qquad (2.27a)$$

will give

$$\det(\mathbf{H}) = \pm n \qquad (2.28)$$

where n is an odd integer. Reference to Fig. 2.4 shows that there are then n reference points per cell strung along lines joining lattice-related points in adjacent cells, and, as n is odd, only one of the n points can coincide with a center of symmetry as required by the imposed phase pattern; all the others are suppressed during phase propagation. So the less demanding "uniqueness" condition (2.27) is reconcilable with the primitivity condition (2.11). The latter, which related to $P1$, had to assume the whole responsibility for selecting one reference point from the totality in the cell, whereas in $P\bar{1}$ the phase pattern has already narrowed the choice to eight points and (2.29) has merely to identify one of the eight. So here, as in $P1$, *three* terms are needed, but the reason this time is quite different.

Once the reference point is defined the signs of all reflections are determin*ate*, but they are determin*able* only for those reflections that are accessible. Equations (2.12) showed that any reflection **h** can be reindexed in terms of the three chosen basis vectors as

$$\mathbf{h} = h'\mathbf{h_1} + k'\mathbf{h_2} + l'\mathbf{h_3} \tag{2.29}$$

where h', k', and l' are given by (2.13). As shown earlier, when $\det(\mathbf{H}) = \pm 1$ the indices $h'k'l'$ are integral for all reflections so that both (2.29) and the structure invariant (2.30), corresponding to (2.14) and (2.16), are *linearly dependent* on the chosen trio and their assigned phases. Thus,

$$\tilde{\phi}(\mathbf{h}) - h'\tilde{\phi}(\mathbf{h_1}) - k'\tilde{\phi}(\mathbf{h_2}) - l'\tilde{\phi}(\mathbf{h_3}) = \tilde{\Phi}(\mathbf{h}) \; (\mathrm{mod}\; 1) \tag{2.30}$$

where $\tilde{\Phi}(\mathbf{h})$ has a distinctive invariant value for each reciprocal-lattice vector. This equation applies to all space groups, but if we consider centrosymmetric space groups only for the moment, $\tilde{\Phi}(\mathbf{h}) = 0$ or $\frac{1}{2}$. This route to phase propagation is direct but hardly a practical procedure, unfortunately, for experience shows that the probability bias in favor of $\tilde{\Phi}(\mathbf{h}) = 0$, rather than $\frac{1}{2}$, is usually inadequate. So propagation has to be done indirectly through structure invariants with a stronger bias.

However, when $|\det(\mathbf{H})| = n > 1$, some or all of the elements of $\mathbf{H}^{-1}$ and, therefore, also of the indices $h'k'l'$ will be fractions (multiples of $1/n$). Such terms are inaccessible from structure invariant sums. Both (2.29) and (2.30) are then *rationally dependent* on the basis trio, but, as fractional coefficients are inadmissable in structure invariants [see (2.14) *et seq.*], the only way to convert (2.30) into an invariant is to multiply through by n. Thus,

$$n\tilde{\phi}(\mathbf{h}) - nh'\tilde{\phi}(\mathbf{h_1}) - nk'\tilde{\phi}(\mathbf{h_2}) - nl'\tilde{\phi}(\mathbf{h_3}) = n\tilde{\Phi}(\mathbf{h}) \; (\mathrm{mod}\; 1) \tag{2.31}$$

But it is also true, after multiplying (2.29) throughout by n, that

$$\tilde{\phi}(n\mathbf{h}) - nh'\tilde{\phi}(\mathbf{h_1}) - nk'\tilde{\phi}(\mathbf{h_2}) - nl'\tilde{\phi}(\mathbf{h_3}) = \tilde{\Phi}(n\mathbf{h}) \; (\mathrm{mod}\; 1) \tag{2.32}$$

which, because it is linearly dependent, allows $\tilde{\phi}(n\mathbf{h})$ to be determined given a knowledge of $\tilde{\Phi}(n\mathbf{h})$. On the other hand, although the value of $\tilde{\phi}(\mathbf{h})$ cannot normally be derived uniquely from (2.31), yet in the special case of centrosymmetric structures, for which $\tilde{\phi}(\mathbf{h}) = 0$ or $\frac{1}{2}$ and n is odd, we can replace $n\tilde{\phi}(\mathbf{h})$ by $\tilde{\phi}(\mathbf{h})$ in the modular equation (2.31). So, in centrosymmetric structures the phases (signs) are in principle determinable for *all* reflections. All we need is some means of rendering accessible the rationally dependent terms (those with fractional $h'k'l'$). In practice, this is done by co-opting one or more of them to extend the defining matrix $\mathbf{H}$. Usually the choice is made empirically, but if desired the suitability of a proposed term can be analyzed as in the following examples.

1. Consider a structure in $P\bar{1}$ for which the best defining trio appears to be

$$\mathbf{H} = \begin{pmatrix} 1 & 2 & 0 \\ 1 & 1 & \bar{4} \\ 1 & 3 & 1 \end{pmatrix}, \quad \det(\mathbf{H}) = n = 3, \quad \mathbf{H}^{-1} = \begin{pmatrix} \frac{13}{3} & -\frac{2}{3} & -\frac{8}{3} \\ -\frac{5}{3} & \frac{1}{3} & \frac{4}{3} \\ \frac{2}{3} & -\frac{1}{3} & -\frac{1}{3} \end{pmatrix} \quad (2.33)$$

Because n is prime no more than one term can be co-opted, and each of the following terms appears superficially to be suitable as they have strong $|E|$'s and are well connected:

	hkl	$h'k'l'$	$nh'\ nk'\ nl'$
(a)	011	$\bar{1}\ 0\ 1$	$\bar{3}03$
(b)	$01\bar{1}$	$-\frac{7}{3}\ \frac{2}{3}\ \frac{5}{3}$	$\bar{7}25$
(c)	$12\bar{1}$	$\frac{1}{3}\ \frac{1}{3}\ \frac{1}{3}$	111

The first, (a), when transformed to $h'k'l'$ has all-integral indices and is, therefore, one of the terms linearly related to the defining trio. We cannot use (a), therefore, for it cannot give us access to the rationally dependent terms. Both (b) and (c), however, have fractional indices and so are rationally dependent; either could be used. With $01\bar{1}$, for example, we can express *all* hkl reflections in the form

$$(hkl) = (h'k'l'm') \begin{pmatrix} 1 & 2 & 0 \\ 1 & 1 & \bar{4} \\ 1 & 3 & 1 \\ 0 & 1 & \bar{1} \end{pmatrix} \quad (2.34)$$

where h', k', l', m' are all integral, though the combination is not unique.

For example, the 345 reflection has $h'k'l'$ indices $\frac{29}{3}$ $-\frac{7}{3}$ $-\frac{13}{3}$ and is, therefore, inaccessible until combined with $01\bar{1}$ $(-\frac{7}{3}\,\frac{2}{3}\,\frac{5}{3})$ to give $(h'k'l'm') = (12\,\bar{3}\,\bar{6}\,1)$. Another way of deriving nh' nk' nl' for a proposed co-opted term is to evaluate the three determinants formed by replacing each row of H in turn by the proposed term. The reader can show that, of the seminvariant terms, 204 can and $20\bar{4}$ cannot be co-opted.

2. Consider now

$$\mathbf{H} = \begin{pmatrix} 3 & 1 & \bar{2} \\ 1 & 5 & 1 \\ 1 & 2 & 1 \end{pmatrix}, \quad \det(\mathbf{H}) = n = 15, \quad \mathbf{H}^{-1} = \begin{pmatrix} \frac{3}{15} & -\frac{5}{15} & \frac{11}{15} \\ 0 & \frac{1}{3} & -\frac{1}{3} \\ -\frac{3}{15} & -\frac{5}{15} & \frac{14}{15} \end{pmatrix} \quad (2.35)$$

Here n contains two factors so we can use either one co-opted term that resolves the 15-fold ambiguity, or two terms resolving three- and five-fold ambiguities, respectively. The following first-order terms appear as possible candidates:

	hkl	$h'k'l'$	$n'h'\,n'k'\,n'l'$	$n'h\,n'k\,n'l$
(a)	$11\bar{1}$	$\frac{2}{3}\,\frac{1}{3}\,-\frac{8}{15}$	$65\bar{8}$	$15\ 15\ \bar{13}$
(b)	$10\bar{1}$	$\frac{2}{3}\,0\,-\frac{1}{5}$	$20\bar{1}$	$5\ 0\ \bar{3}$
(c)	141	$0\,\frac{2}{3}\,\frac{1}{3}$	021	$3\ 12\ 3$

We can, therefore, use either (a) alone, or (b) *and* (c). The last two columns give alternative designations for the first term in the relevant direction that is linearly related to the terms in H.

The signs of co-opted terms are unknown but cannot be assigned arbitrarily, so each must be tried with both $+$ and $-$. It is usual, as will be demonstrated later, to combine the co-opted terms with several others in multiple-sign trials.

We can now dispose of the remaining centrosymmetric space groups more quickly. Thus, all space groups in category 2 (Table 2.8) contain eight centers of symmetry at the same points as in category 1, but their symmetry environments are not now identical. Half of them, set (a) at 000, $00\frac{1}{2}$, $\frac{1}{2}\frac{1}{2}0$, $\frac{1}{2}\frac{1}{2}\frac{1}{2}$, constitute one equivalence class, and the other four (b) constitute a second class. In some space groups the sites of set (a) occur on 4 or 4_2 axes, while those of set (b) lie on 2 or $\bar{4}$ axes. There is, however, one exception, $P4_2/mnm$, where (a) are on diads and (b) on 4_2 axes. The phase pattern offers a convenient and reliable way of shifting the origin in this or any other space group if desired. In the remaining space groups in category 2 the centers of symmetry lie equidistant from the fourfold and twofold axes and the symmetry environment at every site is identical, but sites (a) and (b) now differ in the relative orientation of their environments. The definition of reference point is done by (i) building in the phase pattern of the (a) set; (ii) avoiding the structure seminvariants (the ggg and uug reflections); and (iii) choosing *two* reflections from the remaining parity groups to distinguish between the four centers of symmetry in set (a). The selection rules can be written as

$$\begin{vmatrix} h_1 + k_1 & l_1 \\ h_2 + k_2 & l_2 \end{vmatrix} \ (\text{mod } 2, 2) = \pm 1 \qquad (2.36)$$

In category 3, six of the eight centers lie on diad axes, and only two (at $000, 00\frac{1}{2}$) lie on a triad axis. These two points constitute the preferred equivalence class, so only *one* arbitrary assignment is needed to define one of the two points. As the seminvariants are of the form hkg one can choose any reflection with $l = u$.

In category 4, two of the centers (at $000, \frac{1}{2}\frac{1}{2}\frac{1}{2}$) constitute the preferred equivalence class. The symmetry environment about the other centers is either different or differently oriented. Here again *one* term only is needed, for which, however, $h + k + l = u$ since those with $h + k + l = g$ are seminvariant. Observe the footnote to Table 2.8 concerning the rhombohedral cells.

2.5.2. Nonprimitive Centrosymmetric Space Groups

In such structures experience has shown that there is no need to transform all the data to an equivalent primitive cell as Hauptman and Karle (1959) originally did. Indeed, to do so is deliberately to ignore much of the symmetry characteristic of the structure, and this is clearly contrary to the ethos of the basic principles of direct methods.

For a C-face-centered lattice, say, reflections occur only in the parity groups ggg, ggu, uug, and uuu. In $C2/m$, for example, there are two distinct sets of eight symmetry centers, each of which is an equivalence class: (a) (at $000, 0\frac{1}{2}0, 00\frac{1}{2}, 0\frac{1}{2}\frac{1}{2}$, and the C-face-centered equivalents) on diads with site symmetry $2/m$, and (b) at $\frac{1}{4}\frac{1}{4}0$, etc. on 2_1 screw axes; only four points in each set are independent because of the C-faced lattice. The ggg reflections are seminvariant with respect to both sets (a) and (b), but they can be subdivided into two groups: those with $h + k = 4n + 2$ have a different sign in each set, whereas those with $h + k = 4n$ have the same sign in each and are, therefore, multiseminvariants. The definition of reference point in this space group proceeds by (1) selecting set (a) by imposing the relevant phase pattern on all the data, and (2) selecting one of the four points in set (a) by an arbitrary assignment of signs to *two* reflections, each chosen from a different parity group, *ggu*, *uug*, or *uuu*. We can express this by a truncated version of (2.26):

$$
\begin{pmatrix} h_1 & k_1 & l_1 \\ h_2 & k_2 & l_2 \end{pmatrix} \begin{pmatrix} \varepsilon_{11} & \cdots & \varepsilon_{14} \\ \varepsilon_{21} & \cdots & \varepsilon_{24} \\ \varepsilon_{31} & \cdots & \varepsilon_{34} \end{pmatrix} = \begin{pmatrix} \delta_{11} & \cdots & \delta_{14} \\ \delta_{21} & \cdots & \delta_{24} \end{pmatrix} \ (\text{mod } 2) \qquad (2.37)
$$

where again the terms in the 2×3 matrix $\mathbf{H}$ are expressed modulo (2, 2, 2).

Table 2.9 summarizes the characteristics and treatment of all the nonprimitive centrosymmetric space groups. Each can be analyzed quickly on the above lines. Thus, as the *I*-centered cells in category 6 are also doubly nonprimitive, they too need only *two* arbitrary assignments, but in the *F*-centered cells of category 7, which are quadruply nonprimitive, the equivalence class is reduced to two points so that *one* arbitrary assignment suffices.

It is important to remember that in the higher-symmetry space groups the equivalence classes often divide one or more groups of equipoints simply because the symmetry environment of each equipoint though identical is not in parallel orientation, with the result that the phase patterns are not identical but are interrelated by appropriate cyclic changes of indices. This applies especially to the space groups in categories 8 and 9.

2.5.3. Noncentrosymmetric Space Groups

Here again the treatment is simplified by discussing them a point group at a time, but the equivalence classes and corresponding seminvariants are now much more varied. In particular, the *ggg* reflections are no longer seminvariants for some noncentrosymmetric space groups, but they are nevertheless always to be avoided if possible in the choice of defining reflections for they are excluded by the primitivity condition, $\det(\mathbf{H}) = \pm 1$, (2.11). This condition imposes the same three restrictions on the choice of defining reflections that were inferred from (2.27), but this time the restrictions are *mandatory* only if required by the point-group seminvariants, otherwise they are merely *advisable* in the interests of primitivity and accessibility, and as we saw earlier the primitivity condition can be breached if suitable terms can be co-opted.

TABLE 2.9. The Nonprimitive Centrosymmetric Space Groups

Category	5	6	7	8	9
Space groups	C2/m C2/c Cmcm Cmca Cmmm Cccm Cmma Ccca	Immm Ibam Ibca Imma	Fmmm Fddd Fm3 Fd3 Fm3m Fm3c Fd3m Fd3c	I4/m I4₁/a I4/mmm I4/mcm I4₁/amd I4₁/acd	Im3 Ia3 Im3m Ia3d
Preferred equivalence class	000, ½00, 0½0, ½0½	000, 0½½, ½0½, ½½0	000, ½½½	000, 00½	000
Seminvariant reflections for the preferred equivalence class	ggg	ggg	ggg	ggg, uug	ggg, guu, ugu, uug
Number of signs to be assigned arbitrarily	2	2	1	1	0
To be chosen from	ggu, uug, uuu	guu, ugu, uug	uuu	guu, ugu	

Space Group $P1$

This space group is exceptional in several respects. First, every point in the cell is a permissible origin for none is repeated by symmetry, and, as all give the same function $T = \exp(2\pi i \mathbf{h} \cdot \mathbf{r})$, the only equivalence class consists of the totality of points in the cell, and not a proper subset as in every other space group. Secondly, it follows that there are no sets of reflections for which $\tilde{\phi}$ is invariant at every possible reference point in the equivalence class. So there are no seminvariants for $P1$, though the universal structure invariants described earlier are available. Thirdly, the only restrictions placed on the origin-defining reflections are those advised by the primitivity condition, but this now assumes a much more significant role than it did for the centrosymmetric space groups, for as we saw [(2.28) *et seq.*] it has to assume the entire responsibility for defining the reference point.

The Laue-related reflections in this case are $\mathbf{h}$ and $-\mathbf{h}$, so, by imposing the phase pattern $\tilde{\phi}(-\mathbf{h}) = -\tilde{\phi}(\mathbf{h})$ on all reflections, we ensure that the phases for any pair, $\pm\mathbf{h}$, are related to the same reference point and "enantiomorph." But we must look more closely at this space group than we did on pages 32 and 46.

If we select a reflection $(\mathbf{h}_1)$ and give it an arbitrary phase $(\tilde{\phi}_1)$ we restrict the reference point to any position on the two interleaved families of phase planes for S and I (see Fig. 2.3) which are uniquely located in the structure and defined by

$$\mathbf{h}_1 \cdot \mathbf{r}_0 = \pm\tilde{\phi}(S_0, \mathbf{h}_1) - \tilde{\phi}_1 \;(\text{mod } 1) \tag{2.38}$$

($+$ for S planes; $-$ for I planes).

If we choose a second reflection $(\mathbf{h}_2)$ and assign it an arbitrary phase $(\tilde{\phi}_2)$ the reference point is further restricted to lie on a second set of interleaved phase planes, and, provided $\mathbf{h}_1$ and $\mathbf{h}_2$ are not parallel, the two sets intersect in a set of parallel lines. All the four sets of lines formed, respectively, by intersecting S/S, S/I, I/S, I/I planes are regularly spaced but displaced from one another (see Fig. 2.7) and all constitute possible loci for the reference point.

If $\tilde{\phi}_3$ is assigned to $\mathbf{h}_3$ (not in the plane of $\mathbf{h}_1$ and $\mathbf{h}_2$) we complete the partitioning of real space. Eight identical but displaced sets of contiguous parallelepipeds are produced corresponding to the intersections $S/S/S$, $I/I/I$, $S/S/I$, etc. Their vertices are possible reference points consistent with the arbitrary choices of $\tilde{\phi}_1$, $\tilde{\phi}_2$, and $\tilde{\phi}_3$. The selection of one of the eight sets is achieved in two steps.

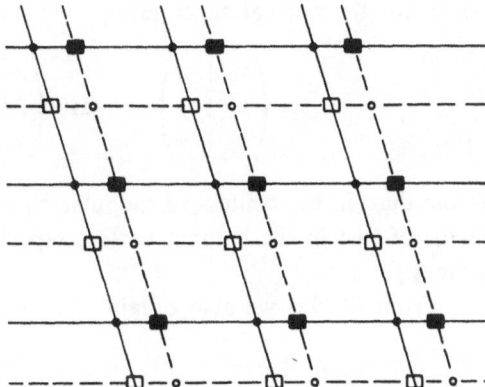

FIGURE 2.7. The intersections of the two families of S (——) and I (— —) phase planes. (●) S/S; (■) S/I; (□) I/S; (○) I/I.

1. The use of any form of universal invariant to extend the list of known phases ensures the elimination of all but the $S/S/S$ and $I/I/I$ sets because each phase sum (whether it contains three or more terms) is invariant only if all the participating phases relate to a common reference point *and to the same "enantiomorph"* [(2.16) *et seq.*].

This leaves a twofold ambiguity between (S, xyz), i.e., the structure S referred to one of the $S/S/S$ vertices, and $(I, x'y'z')$, the structure I referred to the point $x'y'z'$, one of the $I/I/I$ set. In most noncentrosymmetric space groups the inversion point, C, can coincide with O, the fixed origin, 000, but as we see there are nine space groups where this is not possible, so, to include these, we shall treat the problem generally and designate the vector

$$\overrightarrow{OC} = \mathbf{r}_C = x_C\mathbf{a} + y_C\mathbf{b} + z_C\mathbf{c}$$

Then for each defining phase,

$$\tilde{\phi}_i = \tilde{\phi}(S, xyz; \mathbf{h}_i) = \tilde{\phi}(I, x'y'z'; \mathbf{h}_i) \tag{2.39}$$

$$= \tilde{\phi}(S_C; \mathbf{h}_i) - \mathbf{h}_i \cdot (\mathbf{r} - \mathbf{r}_C) \tag{2.39a}$$

$$= -\tilde{\phi}(S_C; \mathbf{h}_i) - \mathbf{h}_i \cdot (\mathbf{r}' - \mathbf{r}_C)$$

whence

$$\mathbf{h}_i \cdot (\mathbf{r} + \mathbf{r}' - 2\mathbf{r}_C) = -2\tilde{\phi}_i \tag{2.40}$$

or

$$\begin{pmatrix} x' + x - 2x_C \\ y' + y - 2y_C \\ z' + z - 2z_C \end{pmatrix} = -\mathbf{H}^{-1} \begin{pmatrix} 2\tilde{\phi}_1 \\ 2\tilde{\phi}_2 \\ 2\tilde{\phi}_3 \end{pmatrix} \tag{2.40a}$$

which for the normal space groups becomes

$$\begin{pmatrix} x' + x \\ y' + y \\ z' + z \end{pmatrix} = -\mathbf{H}^{-1}\begin{pmatrix} 2\tilde{\phi}_1 \\ 2\tilde{\phi}_2 \\ 2\tilde{\phi}_3 \end{pmatrix} \text{ (mod 1)} \tag{2.40b}$$

(Note that, in the context of modular algebra, it is preferable to leave the factor of two in the column vector, especially if some of the phases are 0 or $\frac{1}{2}$.)

From (2.39a) we also obtain

$$\begin{pmatrix} x' - x \\ y' - y \\ z' - z \end{pmatrix} = -\mathbf{H}^{-1}\begin{pmatrix} 2\tilde{\phi}(S_C; \mathbf{h}_1) \\ 2\tilde{\phi}(S_C; \mathbf{h}_2) \\ 2\tilde{\phi}(S_C; \mathbf{h}_3) \end{pmatrix} \tag{2.41a}$$

which for the normal space groups becomes .

$$\begin{pmatrix} x' - x \\ y' - y \\ z' - z \end{pmatrix} = -\mathbf{H}^{-1}\begin{pmatrix} 2\tilde{\phi}(S_0; \mathbf{h}_1) \\ 2\tilde{\phi}(S_0; \mathbf{h}_2) \\ 2\tilde{\phi}(S_0; \mathbf{h}_3) \end{pmatrix} \text{ (mod 1)} \tag{2.41b}$$

For every (S or I), therefore, since O and C are fixed points, *the vector joining the alternative reference points, xyz and x'y'z', is invariant with respect to the numerical values arbitrarily assigned to $\tilde{\phi}_1, \tilde{\phi}_2, \tilde{\phi}_3$*, whereas the midpoint of the vector *is* a function of the assigned values.

Resolution of the ambiguity between (S, xyz) and (I, $x'y'z'$) is achieved by selecting a fourth reflection, $\mathbf{h}_4$, the S/I selector, whose phase takes different values for each of the ambiguous solutions. Thus,

$$\tilde{\phi}_4(S) = \tilde{\phi}(S, xyz; \mathbf{h}_4)$$
$$= \tilde{\phi}(S_C; \mathbf{h}_4) - \mathbf{h}_4 \cdot (\mathbf{r} - \mathbf{r}_C) \tag{2.42S}$$

$$\tilde{\phi}_4(I) = \tilde{\phi}(I, x'y'z'; \mathbf{h}_4)$$
$$= -\tilde{\phi}(S_C; \mathbf{h}_4) - \mathbf{h}_4 \cdot (\mathbf{r}' - \mathbf{r}_C) \tag{2.42I}$$

Similar relations exist for all other terms, so that every universal invariant in which $\tilde{\phi}_4$ enters takes the form

$$\tilde{\Phi}(S) = \tilde{\phi}_4(S) + \tilde{\phi}_5(S) + \tilde{\phi}_6(S) + \cdots$$
$$= \tilde{\phi}(S_C; \mathbf{h}_4) + \tilde{\phi}(S_C; \mathbf{h}_5) + \tilde{\phi}(S_C; \mathbf{h}_6) + \cdots$$
$$- (\mathbf{r} - \mathbf{r}_C) \cdot (\mathbf{h}_4 + \mathbf{h}_5 + \mathbf{h}_6 + \cdots)$$

which is invariant so long as all phases relate to the same enantiomorph and since $\Sigma\mathbf{h}_i = 0$ for a universal invariant.

Similarly,

$$
\begin{aligned}
\Phi(I) &= \tilde{\phi}_4(I) + \tilde{\phi}_5(I) + \tilde{\phi}_6(I) + \cdots \\
&= -\tilde{\phi}(S_C; \mathbf{h}_4) - \tilde{\phi}(S_C; \mathbf{h}_5) - \tilde{\phi}(S_C; \mathbf{h}_6) - \cdots \\
&\quad - (\mathbf{r}' - \mathbf{r}_C) \cdot (\mathbf{h}_4 + \mathbf{h}_5 + \mathbf{h}_6 + \cdots) \\
&= -\Phi(S)
\end{aligned}
$$

The invariant sums that are most effective in biasing new phases are those in which $\Phi(S)$ is well removed from 0 or $\frac{1}{2}$. The values of $\tilde{\phi}_4(S)$ and $\tilde{\phi}_4(I)$ are initially unknown and in $P1$ as in many other space groups are unrestricted, in which case the correct values must be found by tangent refinement. It is often assumed that one merely has to set $\tilde{\phi}_4(S) = +\frac{1}{4}$ (or $\frac{1}{8}$ and $\frac{3}{8}$) and refine over quadrants 1 and 2, i.e., $0 < \tilde{\phi}_4(S) < \frac{1}{2}$, but this is too naive. The problem is more complex even for normal space groups. Thus, subtracting and adding (2.42S) and (2.42I) give

$$
\begin{aligned}
\varDelta\tilde{\phi}_4 &= \tilde{\phi}_4(S) - \tilde{\phi}_4(I) \\
&= 2\tilde{\phi}(S_C; \mathbf{h}_4) + \mathbf{h}_4 \cdot (\mathbf{r}' - \mathbf{r}) \\
&= 2\tilde{\phi}(S_C; \mathbf{h}_4) - (h_4\,k_4\,l_4)\mathbf{H}^{-1}\begin{pmatrix} 2\tilde{\phi}(S_C; \mathbf{h}_1) \\ 2\tilde{\phi}(S_C; \mathbf{h}_2) \\ 2\tilde{\phi}(S_C; \mathbf{h}_3) \end{pmatrix} \quad\quad (2.43a)
\end{aligned}
$$

and (ii),

$$
\begin{aligned}
2\tilde{\phi}_4(S) - \varDelta\tilde{\phi}_4 &= -\mathbf{h}_4 \cdot (\mathbf{r}' + \mathbf{r} - 2\mathbf{r}_C) \\
&= (h_4\,k_4\,l_4)\mathbf{H}^{-1}\begin{pmatrix} 2\tilde{\phi}_1 \\ 2\tilde{\phi}_2 \\ 2\tilde{\phi}_3 \end{pmatrix} \quad\quad (2.43b)
\end{aligned}
$$

$\varDelta\tilde{\phi}_4$ is a convenient measure of the efficacy of the chosen term, $\mathbf{h}_4$, as S/I selector: it is totally ineffective if $\varDelta\tilde{\phi}_4 = 0$ and is optimally effective if $\varDelta\tilde{\phi}_4 = \frac{1}{2}$. Its value in any problem is initially unknown, but (2.43a) shows it to be invariant with respect to the values arbitrarily assigned to $\tilde{\phi}_1$, $\tilde{\phi}_2$, $\tilde{\phi}_3$. Equation (2.43b) shows that the S/I selector would be optimally efficient if

$$
\tilde{\phi}_4(S) = \tfrac{1}{2}(h_4\,k_4\,l_4)\mathbf{H}^{-1}\begin{pmatrix} 2\tilde{\phi}_1 \\ 2\tilde{\phi}_2 \\ 2\tilde{\phi}_3 \end{pmatrix} \pm \tfrac{1}{4} \quad\quad (2.44)
$$

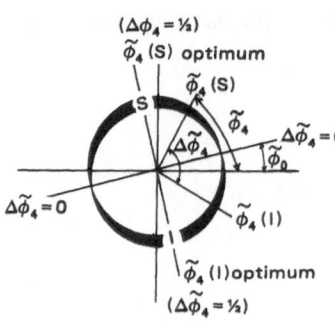

$(\Delta\phi_4 = \frac{1}{2})$
$\tilde{\phi}_4\,(S)$ optimum
$\tilde{\phi}_4\,(S)$
$\Delta\tilde{\phi}_4 = 0$
$\Delta\tilde{\phi}_4 = 0$
$\tilde{\phi}_4\,(I)$
$\tilde{\phi}_4\,(I)$ optimum
$(\Delta\tilde{\phi}_4 = \frac{1}{2})$

FIGURE 2.8. The ranges over which $\tilde{\phi}_4$, the phase of the S/I selector, distinguishes between S and I. For normal space groups an offset, $\tilde{\phi}_0$, occurs, defined by

$$\tilde{\phi}_0 = \tfrac{1}{2}(h_4\,k_4\,l_4)\mathbf{H}^{-1}\begin{pmatrix} 2\tilde{\phi}_1 \\ 2\tilde{\phi}_2 \\ 2\tilde{\phi}_3 \end{pmatrix}$$

and the optimum values of $\tilde{\phi}_4$ occur at $\tilde{\phi}_0 \pm \tfrac{1}{4}$, when $\Delta\tilde{\phi}_4 = \tfrac{1}{2}$. The offsets for the nine abnormal space groups are given in Table 2A.2, column 6, in Appendix 2A.1.

which we rewrite as

$$\tilde{\phi}_4(S) = \tilde{\phi}_0 \pm \tfrac{1}{4} \tag{2.44a}$$

The situation is depicted in Figure 2.8, which shows that *even for normal space groups the ranges of values of $\tilde{\phi}_4$ leading to S or to I are not necessarily $0 \to \tfrac{1}{2}$ (S) and $\tfrac{1}{2} \to 1$ (I), but they are advanced by an offset, $\tilde{\phi}_0$, which is calculable from the values assigned to $\tilde{\phi}_1, \tilde{\phi}_2, \tilde{\phi}_3$.* Tangent refinement to find the correct value of $\tilde{\phi}_4(S)$ should, therefore, cover the range $\tilde{\phi}_0 \to \tilde{\phi}_0 + \tfrac{1}{2}$, but it is, of course, permitted to assign values of $\tilde{\phi}_1, \tilde{\phi}_2, \tilde{\phi}_3$ to make the offset, $\tilde{\phi}_0$, zero and thus simplify the refinement. If $\tilde{\phi}_4(S)$ refines to a value that makes $\Delta\tilde{\phi}_4$ small, the S/I selector is badly chosen. Its efficacy cannot be improved by altering the numerical values of $\tilde{\phi}_1, \tilde{\phi}_2, \tilde{\phi}_3$, but only by changing one or more of the terms $\mathbf{h}_1, \ldots, \mathbf{h}_4$, and trying again. We shall have occasion several times in what follows to use this concept, but we shall not always follows this precise route.

If it is necessary to co-opt terms to achieve primitivity, their phases in this space group are quite general, so it is usual to try each one at starting values of $\tfrac{1}{8}, \tfrac{3}{8}, \tfrac{5}{8}, \tfrac{7}{8}$ and to refine them. Experience shows that tangent refinement is well capable of adjusting phases by $\pm 45°$, so one at least of the combinations should give better agreement factors than the others.

Nonprimitivity is more important in $P1$ than it was in $P\bar{1}$ because all n alternative reference points are now permissible and are not excluded by the phase pattern, but it is a trifle easier to achieve as there are now eight reciprocal-lattice vectors with known phases, and if any three of them fulfill the primitivity determinant condition co-option is unnecessary. In the higher-symmetry space groups, however, only one of the n ambiguous reference points has the correct symmetry environment, so the imposed phase pattern eliminates all but one of the alternatives. Co-option is then only needed to assist accessibility.

2.5.4. Primitive Noncentrosymmetric Space Groups

The remaining primitive noncentrosymmetric space groups are summarized in Table 2.10, which lists merely the minimum and simplest requirements. But before embarking on a discussion of this table it is necessary to refer to Table 2A.1 of Appendix 2A.1. This lists the 116 noncentrosymmetric space groups that can contain S or I structures, and gives for each the corresponding centrosymmetric space group that contains the $(S \& I)$ structure. In most cases the centers of symmetry of the latter coincide with points in the preferred equivalence class of the former; $\mathbf{r}_C$ in (2.39)–(2.43) is zero. In such cases the switch from S_0 to I_0 reverses the phase of every reflection (see Table 2.1). But there are some significant exceptions.

1. For nine pairs (marked with asterisks) the centers of symmetry of the $S \& I$ space group do not coincide with points in the preferred equivalence class of the S or I space group; $\mathbf{r}_C$ in (2.39)–(2.43) is nonzero. This introduces complications, for (2.39a) shows that a switch from S_0 to I_0 gives

$$\tilde{\phi}(I, 000; \mathbf{h}) = -\tilde{\phi}(S, 000; \mathbf{h}) + 2\mathbf{h} \cdot \mathbf{r}_C = -\tilde{\phi}(S, 000; \mathbf{h}) + t \quad (2.45)$$

and care must be taken to allow for the extra term, especially when assigning the phase of the S/I selector.

2. A similar complication arises in polar space groups if the phase chosen to define the reference point on the chosen axis is nonzero. It is discussed in some detail in connection with Class 2.

3. All the eleven pairs of enantiomorphic space groups ($P3_1, P3_2$; $P4_1, P4_3$; etc.) can contain only one structure (see Table 2A.3 of Appendix 2A.1). Thus if $P4_1$ contains the S structure, $P4_3$ contains I. In these cases, therefore, the arbitrary choice of space group and the imposition of the corresponding phase pattern eliminates the other space group and its mirror-image structure, so no S/I selector is needed nor can it be applied. It is noteworthy that there is no $S \& I$ space group corresponding to any one of these 22 space groups. The symmetry elements of $P3_1$, for example, occur in the hexagonal cell of $R\bar{3}$ along with those of $P3_2$, but $R\bar{3}$ contains 18 equipoints instead of the six which would arise in $S \& I$.

Class 2: Category 11

The only equivalence class consists of all points on the four diad axes, so the seminvariants are the $g0g$ reflections. The simplest choice of defining

TABLE 2.10. The Primitive

Category	10	11	12	13	14	15
Space groups (If starred, refer to Appendix 2A.1)	$P1$	$P2$ $P2_1$	Pm Pc	$P222$ $P222_1$ $P2_12_12$ $P2_12_12_1$	$Pmm2$ $Pmc2_1$ $Pcc2$ $Pma2$ $Pca2_1$ $Pnc2$ $Pmn2_1$ $Pba2$ $Pna2_1$ $Pnn2$	$P4$ $P4_1$ $P4_2$ $P4_3$ $P4mm$ $P4bm$ $P4_2cm$ $P4_2nm$ $P4cc$ $P4nc$ $P4_2mc$ $P4_2bc$
Preferred equivalence class	All points	$0y0, \frac{1}{2}y0,$ $0y\frac{1}{2}, \frac{1}{2}y\frac{1}{2}$	$x0z, x\frac{1}{2}z$	$000, \frac{1}{2}\frac{1}{2}\frac{1}{2},$ $\frac{1}{2}00, 0\frac{1}{2}\frac{1}{2},$ $0\frac{1}{2}0, \frac{1}{2}0\frac{1}{2},$ $00\frac{1}{2}, \frac{1}{2}\frac{1}{2}0$	$00z, \frac{1}{2}\frac{1}{2}z,$ $0\frac{1}{2}z, \frac{1}{2}0z$	$00z, \frac{1}{2}\frac{1}{2}z$
Seminvariant reflections	None	$g0g$	$0g0$	ggg	$gg0$	$gg0, uu0$
Number of phases needed to specify the origin	3	3	3	3	3	2
Selection rules (always excluding seminvariants)	Any 3 different parity groups except ggg	$\begin{vmatrix} h_1 & 0 & l_1 \\ h_2 & 0 & l_2 \\ h_3 & 1 & l_3 \end{vmatrix} = \pm 1$	$\begin{vmatrix} 0 & u & 0 \\ h_2 & k_2 & l_2 \\ h_3 & k_3 & l_3 \end{vmatrix} = \pm 1$	$\begin{vmatrix} h_1 & k_1 & l_1 \\ h_2 & k_2 & l_2 \\ h_3 & k_3 & l_3 \end{vmatrix} = \pm 1$	$\begin{vmatrix} h_1 & k_1 & 0 \\ h_2 & k_2 & 0 \\ h_3 & k_3 & 1 \end{vmatrix} = \pm 1$	$gu0$ or $ug0$ $+ hkl$
Modulus to be applied to rows in parentheses in determinant		(2, 0, 2)	(0, 2, 0)	(2, 2, 2)	(2, 2, 0)	

Noncentrosymmetric Space Groups

16	17	18	19	20	21	22
$P\bar{4}$	$P3$	$P312$	$P31m$	$P321$	$R3$	$R32$
$P422$	$P3_1$	$P3_112$	$P31c$	$P3_121$	$R3m$	$P23$
* $P42_12$	$P3_2$	$P3_212$	$P6$	$P3_221$	$R3c$	$P2_13$
$P4_122$	$P3m1$	$P\bar{6}$	$P6_1$	$P622$		$P432$
$P4_12_12$	$P3c1$	$P\bar{6}m2$	$P6_5$	$P6_122$		$P4_332$
$P4_222$		$P\bar{6}c2$	$P6_2$	$P6_522$		$P4_132$
$P4_22_12$			$P6_4$	$P6_222$		$P4_132$
$P4_322$			$P6_3$	$P6_422$		$P\bar{4}3m$
$P4_32_12$			$P6mm$	$P6_322$		$P\bar{4}3n$
$P\bar{4}2m$			$P6cc$	$P\bar{6}2m$		
* $P\bar{4}2c$			$P6_3cm$	$P\bar{6}2c$		
$P\bar{4}2_1m$			$P6_3mc$			
$P\bar{4}2_1c$						
$P\bar{4}m2$						
$P\bar{4}c2$						
$P\bar{4}b2$						
$P\bar{4}n2$						
$000, 00\frac{1}{2},$ $\frac{1}{2}\frac{1}{2}0, \frac{1}{2}\frac{1}{2}\frac{1}{2}$	$00z, \frac{1}{3}\frac{2}{3}z,$ $\frac{2}{3}\frac{1}{3}z$	$000, 00\frac{1}{2},$ $\frac{1}{3}\frac{2}{3}0, \frac{1}{3}\frac{2}{3}\frac{1}{2},$ $\frac{2}{3}\frac{1}{3}0, \frac{2}{3}\frac{1}{3}\frac{1}{2}$	$00z$	$000, 00\frac{1}{2}$	xxx	$000, \frac{1}{2}\frac{1}{2}\frac{1}{2}$
ggg, uug	$hk0$ with $h-k = 3n$	hkg with $h-k = 3n$	$hk0$	hkg	$h+k+l = 0$	$h+k+l = g$
2	2	2	1	1	1	1
$\begin{vmatrix} (h_1+k_1 \ l_1) \\ (h_2+k_2 \ l_2) \end{vmatrix}$ $= \pm 1$	$\begin{vmatrix} (h_1-k_1 \ 0) \\ h_2-k_2 \ 1 \end{vmatrix}$ $= \pm 1$	$\begin{vmatrix} (h_1-k_1 \ l_1) \\ (h_2-k_2 \ l_2) \end{vmatrix}$ $= \pm 1$	$hk1$	hku	$h+k+l = 3$	$h+k+l = u$
$(2, 2)$	$(3, 0)$	$(3, 2)$			See text for indexing on hexagonal cell	

reflections is governed by the determinant equation,

$$\begin{vmatrix} (h_1 & 0 & l_1) \\ (h_2 & 0 & l_2) \\ h_3 & 1 & l_3 \end{vmatrix} \bmod(2, 0, 2) = \pm 1 \tag{2.46}$$

in which the mod condition applies only to the rows in parentheses. The first two terms, whose $\tilde{\phi}$'s must be 0 or $\frac{1}{2}$, select a diad axis, and the third, whose phase can have any value, defines a unique reference point on it because $k_3 = 1$. However, this process needs looking at more closely.

The fixed origin, O, is set arbitrarily on the chosen diad and serves as the inversion center, C. Then,

$$\begin{aligned} \tilde{\phi}_3 &= \tilde{\phi}(S, 0y0; \mathbf{h}_3) = \tilde{\phi}(S_O; \mathbf{h}_3) - y \\ &= \tilde{\phi}(I, 0y'0; \mathbf{h}_3) = -\tilde{\phi}(S_O; \mathbf{h}_3) - y' \end{aligned} \tag{2.47}$$

whence

$$y + y' = -2\tilde{\phi}_3 \quad \text{and} \quad y - y' = 2\tilde{\phi}(S_O; \mathbf{h}_3) \tag{2.47a}$$

Again the interval between the alternative reference points is invariant, but the midpoint is dependent on the values assigned to $\tilde{\phi}_1, \tilde{\phi}_2, \tilde{\phi}_3$. Consequently,

$$\tilde{\phi}_4(S) = \tilde{\phi}(S, 0y0; \mathbf{h}_4) = \tilde{\phi}(S_O; \mathbf{h}_4) - k_4 y$$

and

$$\tilde{\phi}_4(I) = \tilde{\phi}(I, 0y'0; \mathbf{h}_4) = -\tilde{\phi}(S_O; \mathbf{h}_4) - k_4 y'$$

whence

$$\begin{aligned} \Delta\tilde{\phi}_4 &= 2\tilde{\phi}(S_O; \mathbf{h}_4) - k_4(y - y') \\ &= 2\tilde{\phi}(S_O; \mathbf{h}_4) - 2k_4\tilde{\phi}(S_O; \mathbf{h}_3) \tag{2.48} \\ &= 2[\tilde{\phi}_4(S) - k_4\tilde{\phi}_3] \tag{2.48a} \end{aligned}$$

which, like (2.43a), is invariant. The refinement range of $\tilde{\phi}_4(S)$ is offset by $\tilde{\phi}_0 = k_4\tilde{\phi}_3$.

If a suitable term with $k_3 = 1$ cannot be found, *two general* terms can be used instead, provided their indices, k_{3a} and k_{3b}, have no common factor. Each defines a string of k_{3a} or k_{3b} equispaced reference points per cell on the chosen axis exactly like the graduations on a Vernier scale. If both reflections are initially set at $\tilde{\phi} = 0$, one can pursue the analogy with the Vernier by keeping one phase constant, thereby fixing its k_{3a} reference points, and tangent-refining the other, i.e., sliding the k_{3b} points, until a match is found at a unique point at which value phase propagation proceeds efficiently. It is only necessary to refine the adjustable phase over the range $\pm(|k_{3a}| - |k_{3b}|)/2k_{3a}k_{3b}$, which can be made quite small by suitable choice of terms. However, there is a second Vernier corresponding to the I solutions, for there are two ambiguity equations instead of the one occurring previously [(2.47)]. Thus, by analogy,

$$\tilde{\phi}_{3a} = \tilde{\phi}(S_O; \mathbf{h}_{3a}) - k_{3a}y_a$$
$$= -\tilde{\phi}(S_0; \mathbf{h}_{3a}) - k_{3a}y_a'$$

and similarly for $\tilde{\phi}_{3b}$. Tangent refinement of $\tilde{\phi}_{3b}$ *versus* constant $\tilde{\phi}_{3a}$ will make *either* $y_a = y_b$ on the S Vernier *or* $y_a' = y_b'$ on the I Vernier. Normally these two equalities occur for different values of $\tilde{\phi}_{3b}$ and in that event one may not introduce an S/I selector: the selection is implicit in the four phases adopted. However, after refinement it is fortuitously possible for an ambiguity to exist between $(S, 0y0)$ and $(I, 0y'0)$ if

$$2\tilde{\phi}_{3a} = -k_{3a}(y + y') \ (\text{mod } 1)$$
$$2\tilde{\phi}_{3b} = -k_{3b}(y + y') \ (\text{mod } 1)$$

or

$$\frac{2\tilde{\phi}_{3a} + n_a}{k_{3a}} = \frac{2\tilde{\phi}_{3b} + n_b}{k_{3b}},$$

where n_a and n_b are integers less than k_{3a} and k_{3b}, respectively. In such case an S/I selector is needed.

In some instances, however, it may not prove possible to select $h0l$ reflections with which to define a unique diad axis. In that event one can specify a unique reference point by extending this notion of Vernier refinement to three reflections that fulfill

$$\begin{vmatrix} (h_1 & k_1 & l_1) \\ (h_2 & k_2 & l_2) \\ (h_3 & k_3 & l_3) \end{vmatrix} \text{mod}(2, 2, 2) = \pm 1$$

where the three k's contain no common factor. The phases of all three terms are initially set at zero, but the values of $\tilde{\phi}_2$ and $\tilde{\phi}_3$ are tangent refined. The phase pattern compels the reference point to lie on one of the diads; as the origin, O, can occur anywhere on a diad, let us take it at one of the reference points specified by $\tilde{\phi}_1$, and let us assume for the moment that the other two phases have been correctly refined till one of each of their reference points falls at 000. The condition that the k's contain no common factor ensures the uniqueness of the reference point on $0y0$. Each string of reference points on the other three diads has been shifted by some multiple of $\frac{1}{2}$, so uniqueness for the point 000 is achieved by inserting the *parities* of the k indices into the above determinantal equation, and this makes it much easier to satisfy than (2.46). Vernier refinement of $\tilde{\phi}_2$ is needed only over the range $\pm(|k_2| - |k_1|)/2k_1k_2$ and correspondingly for $\tilde{\phi}_3$. In this case there are three ambiguity relations analogous to (2.39) in which, however, x, x', z, z' are this time all 0 or $\frac{1}{2}$. But, as there are only three equations, Vernier refinement causes convergence on both (S, xyz) and $(I, x'y'z')$ simultaneously. An S/I selector is, therefore, needed, but it begins to exert its influence effectively only when $\tilde{\phi}_2$, $\tilde{\phi}_3$ are near their convergence values.

Class m: Category 12

In all the space groups in this Class the only equivalence class consists of all points on the planes $(x0z, x\frac{1}{2}z)$, so the seminvariants are $0g0$ reflections and the simplest choice of origin-defining terms is governed by the equation

$$\begin{vmatrix} (0 & k_1 & 0) \\ h_2 & k_2 & l_2 \\ h_3 & k_3 & l_3 \end{vmatrix} \bmod(0, 2, 0) = \pm 1$$

in which the modular condition applies only to the first row shown in parentheses. The first term, whose phase is 0 or $\frac{1}{2}$, selects one of the planes, and the other two define a unique point on it. An alternative form of this condition is $h_2l_3 - h_3l_2 = \pm 1$.

However, the likelihood of finding a suitable $0u0$ term to serve as $\mathbf{h}_1$ is rather poor, so it is often necessary to use three general reflections. These give

$$\tilde{\phi}_i = \tilde{\phi}(S, xyz; \mathbf{h}_i)$$
$$= \tilde{\phi}(S_0; \mathbf{h}_i) - (h_ix + k_iy + l_iz),$$

and

$$\tilde{\phi}_i = \tilde{\phi}(I, x'y'z'; \mathbf{h}_i)$$
$$= -\tilde{\phi}(S_0; \mathbf{h}_i) - (h_ix' + k_iy' + l_iz'),$$

where $y = \varepsilon/2$, $y' = \varepsilon'/2$ ($\varepsilon = 0$ or 1). Thus,

$$\begin{pmatrix} h_1 & (k_1) & l_1 \\ h_2 & (k_2) & l_2 \\ h_3 & (k_3) & l_3 \end{pmatrix} \begin{pmatrix} x \\ y \\ z \end{pmatrix} = \begin{pmatrix} \tilde{\phi}(S_0; \mathbf{h}_1) - \tilde{\phi}_1 \\ \tilde{\phi}(S_0; \mathbf{h}_2) - \tilde{\phi}_2 \\ \tilde{\phi}(S_0; \mathbf{h}_3) - \tilde{\phi}_3 \end{pmatrix} \qquad (2.49)$$

which can be rewritten as

$$\mathbf{H}\begin{pmatrix} x \\ \varepsilon/2 \\ z \end{pmatrix} = \begin{pmatrix} \delta\tilde{\phi}_1 \\ \delta\tilde{\phi}_2 \\ \delta\tilde{\phi}_3 \end{pmatrix}$$

If, as we may in this Class, we choose the origin, 000, at the point defined by the arbitrary choices $\tilde{\phi}_1$, $\tilde{\phi}_2$, $\tilde{\phi}_3$, we have as one solution, $x = \varepsilon = z = 0$. If there are to be no more on $x0z$ all the minors associated with the k_i terms in det(**H**) must not have a common factor, and if there are to be none on $x\frac{1}{2}z$, det(**H**) must be odd. In this case it is over-restrictive to require det(**H**) $= \pm 1$.

Another set of equations applies to the I structure:

$$\mathbf{H}\begin{pmatrix} x' \\ \varepsilon'/2 \\ z' \end{pmatrix} = -\begin{pmatrix} \tilde{\phi}(S_O; \mathbf{h}_1) + \tilde{\phi}_1 \\ \tilde{\phi}(S_O; \mathbf{h}_2) + \tilde{\phi}_2 \\ \tilde{\phi}(S_O; \mathbf{h}_3) + \tilde{\phi}_3 \end{pmatrix}$$

which for the above-chosen phases becomes

$$\begin{pmatrix} x' \\ \varepsilon'/2 \\ z' \end{pmatrix} = -\mathbf{H}^{-1}\begin{pmatrix} 2\tilde{\phi}(S_O; \mathbf{h}_1) \\ 2\tilde{\phi}(S_O; \mathbf{h}_2) \\ 2\tilde{\phi}(S_O; \mathbf{h}_3) \end{pmatrix} = -\mathbf{H}^{-1}\begin{pmatrix} 2\tilde{\phi}_1 \\ 2\tilde{\phi}_2 \\ 2\tilde{\phi}_3 \end{pmatrix} \qquad (2.50)$$

This equation defines a unique point, so there is need for an S/I selector. Whichever way one determines the reference point, the determination of the S and I ranges of $\tilde{\phi}_4$ follows the same route as for $P1$, and the offset, $\tilde{\phi}_0$, may again be calculated from (2.44).

In Classes such as this which contain rotation–inversion axes ($\bar{X}$) there are both D and L molecules present in S, so the structure I is not an enantiomorph. Instead, the assignment of a fourth phase fixes the orientation of the structure with respect to the chosen **a** and **c** axes; it removes a twofold ambiguity around **b**.

Class 222: Category 13

The four space groups in this Class were discussed in considerable detail in Section 2.4, where it was shown that for a given labeling of axes the equivalence class is the set of eight points [000] listed in Table 2.10. The phase patterns are given in Table 2.3. Anticyclic relabeling does not shift the reference point in $P222$ and $P2_12_12$, but shifts it to $00\frac{1}{4}$ in $P222_1$ and to $\frac{1}{4}\frac{1}{4}\frac{1}{4}$ in $P2_12_12_1$; the seminvariants are ggg.

The zonal reflections $0kl$, $h0l$, $hk0$ are centric and, when referred to a common reference point, have $\bar{\phi} = m/4$, the rules for m depending on the space group. For $P2_12_12_1$ m has the same parity as the index following the zero cyclically. It is convenient to make some use of zonal reflections as their phases can be assigned exact values that need no refinement, but it must be remembered that each occurs only four instead of eight times in reciprocal space, so the number of interactions tends to be smaller and the rate of propagation slower if one relies solely on zonal reflections. The trio of selected terms must satisfy the determinantal equation,

$$\begin{vmatrix} (h_1 & k_1 & l_1) \\ (h_2 & k_2 & l_2) \\ (h_3 & k_3 & l_3) \end{vmatrix} \mod(2, 2, 2) = \pm 1 \qquad (2.51)$$

in which each line is subject to the modular restriction.

The S/I selector term can be chosen from any parity group including the seminvariants ggg, but in this Class it is often possible to make an intelligent choice of an effective term, which need not necessarily be a general (acentric) reflection. Consider, for example, a situation recently encountered in the author's laboratory. The chosen terms were

$$\left. \begin{array}{llll} ugg & 9\ 4\ 0 & +\frac{1}{4} \\ guu & 0\ 1\ 1 & +\frac{1}{4} \\ uug & 31\ 7\ 0 & +\frac{1}{4} \end{array} \right\} \quad \text{define the reference point}$$

$$ugu \quad 31\ 0\ 3 \quad +\frac{1}{4} \qquad \text{defines the enantiomorph}$$

All four reflections can have $\bar{\phi} = \pm\frac{1}{4}$, but the above choices were made arbitrarily. The first three are commonly supposed to define the reference point, but, as usual, they leave an ambiguity which can be analyzed in either of the following ways.

	$S(xyz)$	$I(x'y'z')$
$\bar{\phi}(ugg)$ is invariant over	$000,\ 00\frac{1}{2},\ 0\frac{1}{2}0,\ 0\frac{1}{2}\frac{1}{2}$	$\frac{1}{2}00,\ \frac{1}{2}0\frac{1}{2},\ \frac{1}{2}\frac{1}{2}0,\ \frac{1}{2}\frac{1}{2}\frac{1}{2}$
$\bar{\phi}(guu)$ is invariant over	$000,\ \frac{1}{2}00,\ 0\frac{1}{2}\frac{1}{2},\ \frac{1}{2}\frac{1}{2}\frac{1}{2}$	$0\frac{1}{2}0,\ 00\frac{1}{2},\ \frac{1}{2}\frac{1}{2}0,\ \frac{1}{2}0\frac{1}{2}$
$\bar{\phi}(uug)$ is invariant over	$000,\ 00\frac{1}{2},\ \frac{1}{2}\frac{1}{2}0,\ \frac{1}{2}\frac{1}{2}\frac{1}{2}$	$0\frac{1}{2}0,\ \frac{1}{2}00,\ 0\frac{1}{2}\frac{1}{2},\ \frac{1}{2}0\frac{1}{2}$

The ambiguity is, therefore, between $S(000)$ and $I(\frac{1}{2}0\frac{1}{2})$, and this influences the choice of S/I selector. Table 2.11 shows which zonal reflections can and can not be used for this purpose. The table shows the value of $\bar{\phi}(I, \frac{1}{2}0\frac{1}{2})$

TABLE 2.11. Illustration of One Method of Identifying Valid Proposed S/I
Selector Terms

Proposed S/I selector	$\tilde{\phi}(S, 000)$	$\tilde{\phi}(I, \tfrac{1}{2}0\tfrac{1}{2})$
0gu	0	$0 + \tfrac{1}{2}$ ✓
0ug	$\tfrac{1}{4}$	$-\tfrac{1}{4}$ ✓
0uu	$\tfrac{1}{4}$	$-\tfrac{1}{4} + \tfrac{1}{2}$
g0u	$\tfrac{1}{4}$	$-\tfrac{1}{4} + \tfrac{1}{2}$
u0g	0	$0 + \tfrac{1}{2}$ ✓
u0u	$\tfrac{1}{4}$	$-\tfrac{1}{4}$ ✓
gu0	0	0
ug0	$\tfrac{1}{4}$	$-\tfrac{1}{4} + \tfrac{1}{2}$
uu0	$\tfrac{1}{4}$	$-\tfrac{1}{4} + \tfrac{1}{2}$

corresponding to each of the arbitrary assignments shown for $\tilde{\phi}(S, 000)$.
Each term with a check mark is both valid and optimally effective as an
S/I selector.

The same results are more neatly obtained by use of (2.40b), in which
H is the 3×3 matrix of *index parities* whose determinant appears in (2.51).
Thus, in this particular example,

$$\begin{pmatrix} x' \\ y' \\ z' \end{pmatrix} = -2 \begin{pmatrix} 1 & 0 & 0 \\ \bar{1} & 0 & 1 \\ 1 & 1 & \bar{1} \end{pmatrix} \begin{pmatrix} \tfrac{1}{4} \\ \tfrac{1}{4} \\ \tfrac{1}{4} \end{pmatrix} \quad (\text{mod } 1)$$

$$= \begin{pmatrix} \tfrac{1}{2} \\ 0 \\ \tfrac{1}{2} \end{pmatrix}$$

substitution of which into (2.44) gives

$$\tilde{\phi}_4 = (h_4 + l_4 \pm 1)/4 \pmod 1$$

This is fulfilled only for the four zonal reflections checked in Table 2.11,
and thus by the 31 0 3 term chosen.

Class $mm2$: Category 14

This corresponds closely to Class 2 (category 11): indeed, in the
monoclinic c-axis unique notation, there is no need to separate them.
The origin is conventionally on a diad or screw axis, except in $Pmn2_1$,

where it is on a mirror plane. The simplest determinant equation in this Class takes the form

$$\begin{vmatrix} (h_1 & k_1 & 0) \\ (h_2 & k_2 & 0) \\ h_3 & k_3 & 1 \end{vmatrix} \bmod(2, 2, 0) = \pm 1 \qquad (2.52)$$

in which the modular restriction applies strictly only to the first two rows. All the comments made concerning Class 2 apply here, except that the S/I selector now determines the polarity of the structure with respect to the c axis.

Classes 4 and 4mm: Category 15

The preferred equivalence class contains all the points on the two preferred axes, $00z$, $\frac{1}{2}\frac{1}{2}z$ (usually some form of fourfold axis), so the seminvariants are the $gg0$ and $uu0$ reflections, i.e., $hk0$ with $h + k = g$. Definition of the origin requires the selection of *two* reflections, one from $gu0$ or $ug0$ (with a phase of 0 or $\frac{1}{2}$) to select one of the axes, and an $hk1$ reflection with general phase that can conveniently be set at 0. The determinant equation is

$$\begin{vmatrix} (h_1 + k_1 & 0) \\ h_2 + k_2 & 1 \end{vmatrix} \bmod(2, 0) = \pm 1 \qquad (2.53)$$

The choice of S/I selector follows the method described for Class 2 and the same variants apply. In Class 4 it selects a true enantiomorph, but in 4mm it defines the polarity of the structure with respect to the c axis.

Classes $\bar{4}$, 422, and $\bar{4}2m$: Category 16

The preferred equivalence class consists of four points, so the seminvariants are the ggg and uug reflections. The sets of centric reflections differ in each Class and to some extent influence the choice of defining reflections. The determinant condition requires two reflections:

$$\begin{vmatrix} (h_1 + k_1 & l_1) \\ (h_2 + k_2 & l_2) \end{vmatrix} \bmod(2, 2) = \pm 1 \qquad (2.54)$$

The S/I selector selects an enantiomorph in Class 422, and defines the polarity of the structure with respect to the c axis in $\bar{4}2m$ or with respect to the a, b axes in $\bar{4}$.

Class 3 and Subclass $3m1$: Category 17

The preferred equivalence class consists of all points on the triad axes $00z$, $\frac{1}{3}\frac{2}{3}z$, $\frac{2}{3}\frac{1}{3}z$, all of which have indistinguishable symmetry environments. The seminvariants are reflections of the type $hk0$, where $h - k = 3n$, and there are no centric reflections (apart from $h00$ in $3m1$). Two reflections are needed, one a nonseminvariant $hk0$ to select one of the axes, and an $hk1$ to define the reference point on it. They must fulfil the determinant condition

$$\begin{vmatrix} (h_1 - k_1 & 0) \\ h_2 - k_2 & 1 \end{vmatrix} \bmod(3, 0) = \pm 1 \qquad (2.55)$$

The phase of the first term must be refinable over the range $\pm\frac{1}{6}$. The S/I selector can be any term and, although it is not possible to choose it to be efficient, the same sort of procedures can be used as in Class 2. It selects an enantiomorph in Class 3 and defines the polarity of the structure with respect to c in $3m1$.

Classes $\bar{6}$, $\bar{6}m2$ and Subclass 312: Category 18

The preferred equivalence class comprises the six points 000, $00\frac{1}{2}$, $\frac{1}{3}\frac{2}{3}0$, $\frac{1}{3}\frac{2}{3}\frac{1}{2}$, $\frac{2}{3}\frac{1}{3}0$, $\frac{2}{3}\frac{1}{3}\frac{1}{2}$, so the seminvariants are reflections of the type hkg where $h - k = 3n$. Two reflections are needed which fulfil

$$\begin{vmatrix} (h_1 - k_1 & l_1) \\ (h_2 - k_2 & l_2) \end{vmatrix} \bmod(3, 2) = \pm 1 \qquad (2.56)$$

and their phases must be refined over $\pm\frac{1}{6}$. The S/I term selects an enantiomorph in 312, but in $\bar{6}$ and $\bar{6}m2$ it reduces a sixfold symmetry axis to $\bar{6}$.

Classes 6, $6mm$ and Subclass $31m$: Category 19

The preferred equivalence class consists of all points on the axis $00z$, as the symmetry and symmetry environment of this axis differ from those of the adjacent axes, $\frac{1}{3}\frac{2}{3}z$ and $\frac{2}{3}\frac{1}{3}z$. The seminvariants are, therefore, *all* the $hk0$ reflections. Only *one* term of the form $hk1$ is needed to locate the origin on this axis, and its phase can have any value, conveniently 0. This may seem an impossibly slender basis on which to build, but it must be remembered that its Laue set contains at least 12 vectors, and co-option of terms with multiple alternative starting phases in tangent refinement is usual.

The S/I term selects an enantiomorph in Class 6, but in $6mm$ and $31m$ defines the polarity of the structure relative to the c axis.

Classes 622, $\bar{6}2m$, and Subclass 321: Category 20

The preferred equivalence class consists merely of the two points 000 and $00\frac{1}{2}$, so all the hkg reflections are seminvariants. Only *one* term of the form hku is needed to define the origin. It is advisable to set its phase initially and for several cycles at 0, but later to allow its phase to refine over the range $\pm 1/4l$. As there are three mutually orthogonal centric zones in 622 it is possible to approach the choice of S/I term on the lines of Class 222. It selects an enantiomorph in 622 and 321, but in $\bar{6}2m$ it converts a sixfold axis to $\bar{6}$.

Subclasses $R3$ and $R3m$: Category 21

If the structure is referred to rhombohedral axes, there are no systematic absences and the preferred equivalence class consists of all points on [111]. The seminvariants are, therefore, those reflections for which $h + k + l = 0$, and one reflection of the form $h + k + l = 1$ defines the origin.

If, however, the structure is referred to hexagonal axes the preferred equivalence class consists of all points on the axis $00z$ (those on the adjacent axes $\frac{1}{3}\frac{2}{3}z$ and $\frac{2}{3}\frac{1}{3}z$ being lattice equivalents), so the seminvariants are *all* the observed reflections[†] of the form $HK0$ (with $K - H = 3n$). Only *one* term of the form $HK1$ is needed to define the origin on this axis and its phase is general. Any reflection may serve as S/I selector. In $R3$ it selects an enantiomorph, but in $R3m$ and $R3c$ it defines the polarity of the structure with respect to the c axis.

Classes 23, 432, $\bar{4}3m$, and Subclass $R32$: Category 22

Here the distinction is merely between the points 000 and $\frac{1}{2}\frac{1}{2}\frac{1}{2}$, or two points separated by the vector $\frac{1}{2}\frac{1}{2}\frac{1}{2}$, so the seminvariants are all reflections of the form $h + k + l = g$. A single reflection of the form $h + k + l = u$ suffices to define the origin; it is preferable to use a centric reflection if possible. There are three orthogonal centric zones in 23 and 432 so it may prove possible in these classes to choose the S/I-selector term efficiently

---

[†] H, K, L are the Miller indices referred to the triply primitive hexagonal unit cell.

as in 222 and it selects an enantiomorph. In $\bar{4}3m$, however, it defines the polarity of the structure along [111].

If $R32$ is indexed in terms of the rhombohedral cell, the choice is between 000 and $\frac{1}{2}\frac{1}{2}\frac{1}{2}$: the seminvariants are hkl with $h + k + l = g$, so the defining term must have $h + k + l = u$. If, however, it is indexed in terms of the hexagonal cell the choice is between 000 and $00\frac{1}{2}$ so the seminvariants are all the HKg's. One term of the form HKu with $-H + K + L = 3n$ defines the origin, but this condition precludes any of the centric reflections. The term must, therefore, be input with starting phases of, say, $\pm\frac{1}{8}$ and refined within the range $\pm\frac{1}{4}$.

2.5.5. Nonprimitive Noncentrosymmetric Space Groups

The procedures for these are summarized in Table 2.12, which should by now be largely self-explanatory. Considerable simplifications have been introduced compared with Karle and Hauptman's original table (1961) by retaining the nonprimitive cell and indices, rather than converting them to the equivalent primitive cell. An example of a structure solved in Cc is given by Allen, Rogers, and Trotter (1972).

Every nonprimitive space group contains symmetry elements and equivalence classes extra to those possessed by the corresponding primitive space group in the same Class (cf. $C222$ with $P222$). Usually, they differ from those originally present and can, therefore, readily be eliminated by imposing the phase pattern of the preferred equivalence class. But in category 30, new points enter the preferred equivalence class. So, whereas nonprimitivity usually reduces the number of phases to be specified compared with that needed for the related primitive cell, this is not always true.

In category 30 there are four subsets of points within the preferred equivalence class: A_1 (000 plus the F-centered equivalents), A_2 ($\frac{1}{2}\frac{1}{2}\frac{1}{2}$, etc.), B_1 ($\frac{1}{4}\frac{1}{4}\frac{1}{4}$, etc.), B_2 ($\frac{3}{4}\frac{3}{4}\frac{3}{4}$, etc.). It is easy to show that all four points within any one of these groups give the same phase for all reflections in a Laue set, so only the four representative terms need be considered. Their $\Delta\bar{\phi}$'s relative to 000 are tabulated below for three types of reflection.

	$A_1(000)$	$A_2(\frac{1}{2}\frac{1}{2}\frac{1}{2})$	$B_1(\frac{1}{4}\frac{1}{4}\frac{1}{4})$	$B_2(\frac{3}{4}\frac{3}{4}\frac{3}{4})$
$ggg(h + k + l = 4n)$	0	0	0	0
$ggg(h + k + l = 4n + 2)$	0	0	$\frac{1}{2}$	$\frac{1}{2}$
$uuu(h + k + l = 4n + m)$	0	$\frac{1}{2}$	$+m/4$	$-m/4$

TABLE 2.12. The Nonprimitive Noncentrosymmetric Space Groups

Category	23	24	25	26	27	28	29	30	31	32	33
Space groups (If starred, refer to Appendix 2A.1)	$C2$	Cm Cc	$C222$ $C222_1$ $I222$ $I2_12_12_1$	$Cmm2$ $Cmc2_1$ $Ccc2$ $Cm2m$ $C2ma$ $C2cm$ $Cc2a$ $Imm2$ $Iba2$ $Ima2$	$Fmm2$ * $Fdd2$	$I4$ * $I4_1$ $I4mm$ $I4cm$ * $I4_1md$ * $I4_1cd$	$I\bar{4}$ $I\bar{4}m2$ $I\bar{4}c2$ $I\bar{4}2m$ * $I\bar{4}2d$	$F222$ $F23$ * $F4_132$ $F\bar{4}3m$ $F\bar{4}3c$	$I422$ * $I4_122$	$F432$	$I23$ $I2_13$ $I432$ $I4_132$ $I\bar{4}3m$ $I\bar{4}3d$
Preferred equivalence class	$0y0, 0y\tfrac{1}{2}$	$x0z$	$000, 00\tfrac{1}{2},$ $\tfrac{1}{2}00, \tfrac{1}{2}0\tfrac{1}{2}$	$00z, \tfrac{1}{2}0z$	$00z$	$00z$	$000, 00\tfrac{1}{2}$	$000, \tfrac{1}{4}\tfrac{1}{4}\tfrac{1}{4},$ $\tfrac{1}{2}\tfrac{1}{2}\tfrac{1}{2}, \tfrac{3}{4}\tfrac{3}{4}\tfrac{3}{4}$	$000, 00\tfrac{1}{2}$	$000, \tfrac{1}{4}\tfrac{1}{4}\tfrac{1}{4}$	000
Seminvariants	$g0g$	$0k0$	ggg	$gg0$	$gg0$	$hk0$	$gg, 4n$	$h+k+l$ $= 4n$	hkg	ggg	All
Number of phases needed to define origin	2	2	2	2	1	1	1	1	1	1	0
Selection rules	$u0u, h1l$	$2(hkl),$ $\begin{vmatrix} h_1 & l_1 \\ h_2 & l_2 \end{vmatrix}$ $= \pm 1$	from any 2 other parity groups	$uu0, hk1$	$uu1$	$hk1$	ugu or guu	uuu	hku	uuu	None

Only reflections of the first type are seminvariant, and it is evident that a single *uuu* reflection suffices to identify one of the four subsets by assigning it a phase which is some multiple of $\frac{1}{4}$.

2.6. Some Unusual Requirements of S/I Selectors

The phase of the S/I selector does not always have to be constrained to the range $0 \rightarrow \frac{1}{2}$ and excluded from $\frac{1}{2} \rightarrow 1$. First, because there are plenty of instances when it may enter into a universal invariant with defined terms to give a value of $\Phi = \pm\frac{1}{4}$, according to whether it is assigned a value of 0 or $\frac{1}{2}$. For example, two of the four zonal reflections checked in Table 2.11 as valid S/I terms have phases of 0 or $\frac{1}{2}$. Secondly, in $P1$ and polar space groups, as was shown in Class 2, the most efficient value of its phase depends on the phases assigned to define the reference point on the chosen axis, cf. (2.50).

But a rather more subtle situation was encountered by the author in the course of tackling a crystal structure in *Fdd*2. The work has recently been submitted for publication (Rogers, 1979), but appears of sufficient importance to justify devoting space to it here too. Indeed, it was primarily to understand this problem that the S/I notation was devised.

Figure 2.9, which compares the S or I space group, *Fdd*2, with the S & I space group, *Fddd*, leads to the following conclusions:

1. The symmetry environments of the $00z$ and $\frac{1}{4}\frac{1}{4}z$ axes in *Fdd*2 are the same but of opposite polarity, as is seen from the different orientations of the *d*-glide arrows or the dispositions of the $\bigcirc$ and $\odot$ equipoints.

2. The center of symmetry in *Fddd* cannot be inserted on either axis; it occurs midway between them on $\frac{1}{8}\frac{1}{8}z$.

3. The S structure shown in *Fdd*2 contains both D and L molecules. The I structure consists of all the *extra* points generated in *Fddd* (see Fig. 2.9).

4. Any relabeling of the cell of *Fdd*2 needs no shift of origin provided the orientation of $+\mathbf{c}$ is not reversed, e.g., $b\bar{a}c$. But $ba\bar{c}$ requires a shift to the $\frac{1}{4}\frac{1}{4}z$ axis. Hence the very act of labeling the axes and imposing the conventional phase pattern for *Fdd*2 on all Laue sets automatically preselects the $00z$ axis for the reference point. The $0\frac{1}{2}z$, $\frac{1}{2}0z$ and $\frac{1}{2}\frac{1}{2}z$ axes are symmetry related and need not be distinguished.

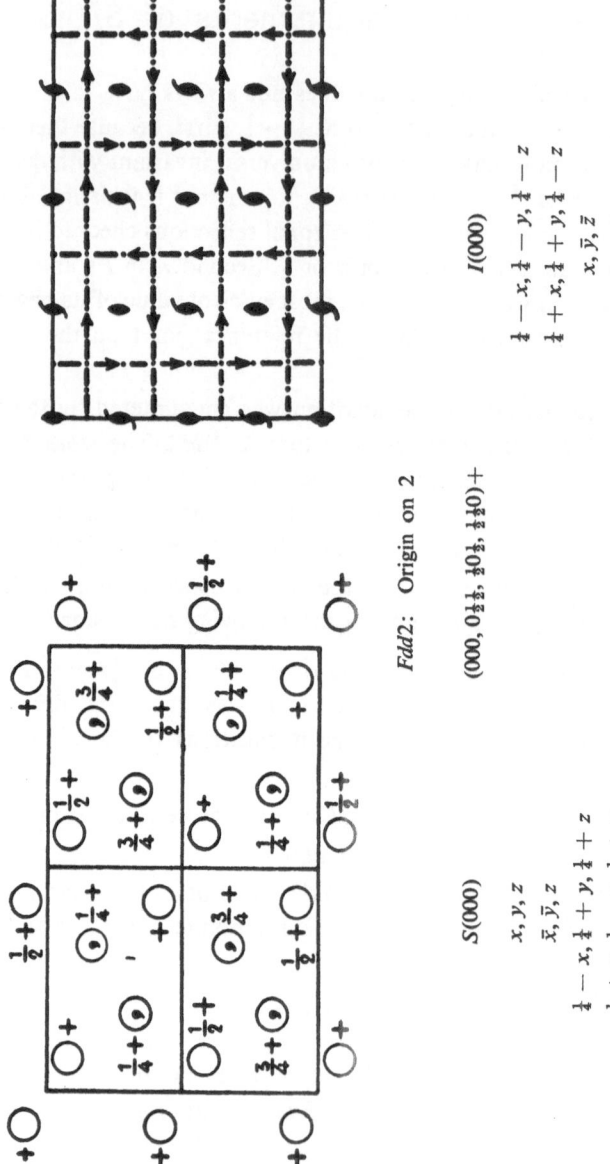

Fdd2: Origin on 2

$(000, 0\frac{1}{2}\frac{1}{2}, \frac{1}{2}0\frac{1}{2}, \frac{1}{2}\frac{1}{2}0)+$

$S(000)$

x, y, z

$\bar{x}, \bar{y}, z$

$\frac{1}{4} - x, \frac{1}{4} + y, \frac{1}{4} + z$

$\frac{1}{4} + x, \frac{1}{4} - y, \frac{1}{4} + z$

$I(000)$

$\frac{1}{4} - x, \frac{1}{4} - y, \frac{1}{4} - z$

$\frac{1}{4} + x, \frac{1}{4} + y, \frac{1}{4} - z$

$x, \bar{y}, \bar{z}$

$\bar{x}, y, \bar{z}$

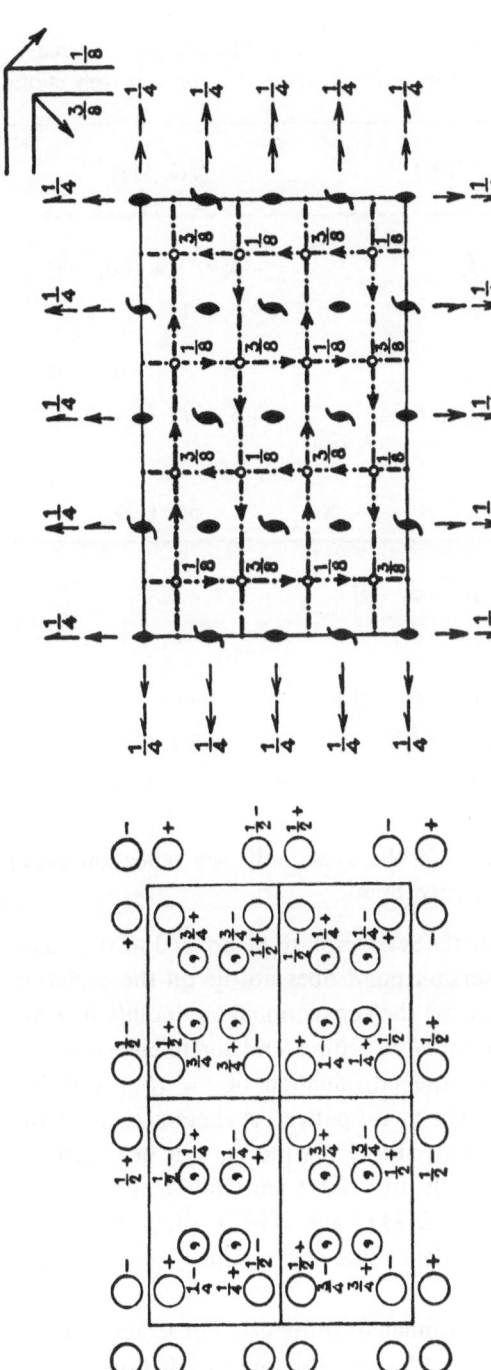

Fddd: Origin at 222, at $\bar{\tfrac{1}{8}}, \bar{\tfrac{1}{8}}, \bar{\tfrac{1}{8}}$ from $\bar{1}$

$I(\tfrac{1}{8}\tfrac{1}{8}\tfrac{1}{8})$

$\bar{x}', \bar{y}', \bar{z}'$

$\tfrac{1}{4} + x', \tfrac{1}{4} + y', \bar{z}'$

$x', \tfrac{1}{4} - y', \tfrac{1}{4} - z'$

$\tfrac{1}{4} - x', y', \tfrac{1}{4} - z'$

$S(\tfrac{1}{8}\tfrac{1}{8}\tfrac{1}{8})$

x', y', z'

$\tfrac{1}{4} - x', \tfrac{1}{4} - y', z'$

$\bar{x}', \tfrac{1}{4} + y', \tfrac{1}{4} + z'$

$\tfrac{1}{4} + x', \bar{y}', \tfrac{1}{4} + z'$

FIGURE 2.9. Comparison of the conventional cells for *Fdd2* and *Fddd*, and the coordinates of the *S* and *I* sets of equipoints.

TABLE 2.13. Phase Patterns for *Fdd*2 for the *S* and *I* Structures Referred to
000 and $\frac{1}{4}\frac{1}{4}\frac{1}{4}$; $t = (h + k + l)/4 = 0$ or $\frac{1}{2}$ for *ggg*, or $\pm\frac{1}{4}$ for *uuu* Reflections

Point-group related reflections	$\tilde{\phi}(S, 000)$	$\tilde{\phi}(S, \frac{1}{4}\frac{1}{4}\frac{1}{4})$
$hkl, \bar{h}\bar{k}l$	$+\tilde{\phi}_1$	$-\tilde{\phi}_2 \quad [= +\tilde{\phi}_1 - t]$
$\bar{h}kl, h\bar{k}l$	$+\tilde{\phi}_1 - t$	$-\tilde{\phi}_2 + t \, [= +\tilde{\phi}_1]$
$\bar{h}\bar{k}\bar{l}, hk\bar{l}$	$-\tilde{\phi}_1$	$+\tilde{\phi}_2 \quad [= -\tilde{\phi}_1 + t]$
$h\bar{k}\bar{l}, \bar{h}k\bar{l}$	$-\tilde{\phi}_1 + t$	$+\tilde{\phi}_2 - t \, [= -\tilde{\phi}_1]$

	$\tilde{\phi}(I, 000)$	$\tilde{\phi}(I, \frac{1}{4}\frac{1}{4}\frac{1}{4})$
$hkl, \bar{h}\bar{k}l$	$+\tilde{\phi}_2 \quad [= -\tilde{\phi}_1 + t]$	$-\tilde{\phi}_1$
$\bar{h}kl, h\bar{k}l$	$+\tilde{\phi}_2 - t \, [= -\tilde{\phi}_1]$	$-\tilde{\phi}_1 + t$
$\bar{h}\bar{k}\bar{l}, hk\bar{l}$	$-\tilde{\phi}_2 \quad [= +\tilde{\phi}_1 - t]$	$+\tilde{\phi}_1$
$h\bar{k}\bar{l}, \bar{h}k\bar{l}$	$-\tilde{\phi}_2 + t \, [= +\tilde{\phi}_1]$	$+\tilde{\phi}_1 - t$

5. Given a particular labeling of the axes, only *one* reflection (*uu*l)
is needed to define a unique reference point.

A special problem arises over the restriction to be applied to the phase
of the *S/I* selector, since the inversion point does not lie on the preferred
diads of *Fdd*2. The phase pattern for the conventional equipoints in *Fdd*2
is given in Table 2.13 for the *S* and *I* structures and for reference points
on both axes. It is simplified by the introduction of $t = (h + k + l)/4$
(mod 1). The table confirms that the phase pattern is characteristic of the
axis on which the reference point occurs. Thus the pattern in which $-t$
is associated with $+l$ applies equally to $S(000)$ and $I(000)$, whereas $-t$
associated with $-l$ applies to both $S(\frac{1}{4}\frac{1}{4}\frac{1}{4})$ and $I(\frac{1}{4}\frac{1}{4}\frac{1}{4})$. $\tilde{\phi}(S, 000; \mathbf{h})$ and
$\tilde{\phi}(I, 000; \mathbf{h})$ differ only in the numerical value of the point-group-invariant
component $\tilde{\phi}$.

We can now proceed on lines similar to those used for Class 2. Thus,
if the phase $\tilde{\phi}_1$ is assigned to the *uu*1 reflection, this value occurs once in the
interval $00z$ ($z = 0 \rightarrow 1$) for both the *S* and *I* structures, but at different

positions, 0 and z. If we insert the inversion point, C, at $\frac{1}{8}\frac{1}{8}\frac{1}{8}$ as in *Fddd*, then

$$\tilde{\phi}(S, 000; \mathbf{h}_1) = \tilde{\phi}_1 = \tilde{\phi}(I, 00z; \mathbf{h}_1)$$

$$= \tilde{\phi}(I, \tfrac{1}{8}\tfrac{1}{8}\tfrac{1}{8}; \mathbf{h}_1) + \frac{h_1 + k_1 + l_1}{8} - l_1 z$$

$$= -\tilde{\phi}(S, \tfrac{1}{8}\tfrac{1}{8}\tfrac{1}{8}; \mathbf{h}_1) + \frac{h_1 + k_1 + l_1}{8} - l_1 z$$

$$= -\tilde{\phi}(S, 000; \mathbf{h}_1) + \frac{h_1 + k_1 + l_1}{4} - l_1 z$$

$$= -\tilde{\phi}_1 + t_1 - l_1 z \pmod 1 \tag{2.57}$$

whence, since $l_1 = 1$,

$$z = t_1 - 2\tilde{\phi}_1 \tag{2.58}$$

The corresponding phases for the S/I selector ($\mathbf{h}_4$) are, say,

$$\tilde{\phi}(S, 000; \mathbf{h}_4) = \tilde{\phi}_4$$

and

$$\tilde{\phi}(I, 00z; \mathbf{h}_4) = -\tilde{\phi}_4 + t_4 + l_4(2\tilde{\phi}_1 - t_1)$$

from (2.57), so that the optimal value of $\tilde{\phi}_4$ is

$$\tilde{\phi}_4 = \pm \frac{1}{4} + \frac{t_4 - l_4 t_1}{2} + l_4 \tilde{\phi}_1 \tag{2.59}$$

which in certain circumstances can be 0 or $\frac{1}{2}$.

Before elucidating these considerations, we began tangent refinement with the following set of reflections and starting phases,

hkl	$\tilde{\phi}$	
17 7 1	0	to define origin
1 13 5	$\frac{1}{8}, \frac{3}{8}$	to select S or I
11 13 5	$\frac{1}{8}, \frac{3}{8}, \frac{5}{8}, \frac{7}{8}$	co-opted
16 6 6	$\frac{1}{8}, \frac{3}{8}, \frac{5}{8}, \frac{7}{8}$	

and these were applied to 136 more terms with $|E| \geq 1.38$. With this choice of starting set and value of $\tilde{\phi}_1$ it is now clear that $\tilde{\phi}_4$ should have

TABLE 2.14. A Selection of Reflections to Show How Their Phases Are Linked in Different Sets When the Enantiomorph Ambiguity Has Not Been Resolved

hkl	Phase in set 1	Phase in set 2
17 7 1	0	0
1 13 5	$\tilde{\phi}_2 (= \frac{1}{4})$	$\frac{1}{2} - \tilde{\phi}_2 (= \frac{1}{4})$
11 13 5	$\tilde{\phi}_3$	$- \tilde{\phi}_3$
16 6 6	$\tilde{\phi}_4$	$\frac{1}{2} - \tilde{\phi}_4$
12 0 0	$\tilde{\phi}_5$	$- \tilde{\phi}_5$
5 1 9	$\tilde{\phi}_6$	$\frac{1}{2} - \tilde{\phi}_6$
14 4 6	$\tilde{\phi}_7$	$\frac{1}{2} - \tilde{\phi}_7$
16 4 4	$\tilde{\phi}_8$	$- \tilde{\phi}_8$

been given starting phases of $\pm\frac{1}{8}$ (which must be refined in the range $-\frac{1}{4}$ to $+\frac{1}{4}$), or a value of 0 and the phase of 17 7 1 refined over $\pm\frac{1}{20}$. What we found can now be predicted, namely, that the 32 trials were not distinct but fell into two sets of 16 having identical figures of merit, etc. A few are listed in Table 2.14 to show how their phases were related. In every case the phases in sets 1 and 2 conform to (2.57) and indicate that the S/I ambiguity has not been resolved. In retrospect we can see that we covered only half the possible combinations that should have been examined for this starting set, and duplicated those we did examine. We were just lucky that the solution lay in the half we did examine. Had it not, and had we not succeeded in analyzing the problem, it might have been tried with other starting sets until we stumbled on a valid combination of phases (such as $\tilde{\phi}(17\ 7\ 1) = 0$, $\tilde{\phi}(2\ 2\ 4) = \pm\frac{1}{4}$), but such blind repetition is expensive and objectionable because it is inelegant and one learns little. Now, with the above theory and notation, it should not be necessary.

Formulas analogous to (2.57) and (2.59) are given in Table 2A.2 of Appendix 1 for each of the nine anomalous space groups.

2.7. In Conclusion

Examination of problems submitted because they had not turned out as expected or had failed to yield a solution has given me the following impressions.

1. Some workers were lucky to get solutions and sometimes indulged in needless computing; it was quicker and cheaper than thinking. The growing use of MULTAN and related programs in which the starting set is computer-derived has helped in this respect, but all too often they are used unquestioningly as "black boxes." Experience has shown that it is occasionally possible to improve on the proposed starting set.

2. Some failures were undoubtedly due to setting up the starting set and their phases incorrectly, especially in making the correct assignment of S/I selector, e.g., omitting nonzero values of $\bar{\phi}_3$ from (2.48a) or the like. The rules for $Fdd2$ in MULTAN were corrected after correspondence between the author and Dr. Main, and corrections are soon to be made for the other starred space groups in Table 2A.2 of Appendix 2A.1.

3. Some instances of "chicken-wire" effect in the E maps were traced to lack of primitivity, but in some structures nothing much could be done about it as the terms with strong E's occupied a sublattice in reciprocal space: the spurious peaks can only be eliminated by bringing in the numerous terms with weaker E's.

4. Unexpected results should always be thoroughly investigated, especially those that arise when two independent solutions of the same structure are compared. There are doubtless more lessons to be learned.

5. It is occasionally impracticable to tackle a space group by means of the terms specified in the foregoing tables, and in that event a nonstandard starting set must be used. The notation developed here has proved useful on several such occasions in analyzing the S/I ambiguity and in helping to select an effective starting set.

Appendix 2A.1

Table 2A.1 lists all the noncentrosymmetric (S or I) space groups (those that can accommodate S or I separately), together in each case with the corresponding centrosymmetric (S & I) space group (in which S and I coexist). In the nine pairs marked with an asterisk, the origins of the conventional (S or I) cells do not coincide with an inversion point in the (S & I) structure, so the phase assigned to the S/I selector needs special consideration as in Table 2A.2 and Section 2.6. Table 2A.3 gives a list of 11 pairs of enantiomorphic space groups in each of which one member accommodates S and the other I. There is no (S & I) counterpart to any of these pairs.

TABLE 2A.1. The 116 (*S or I*) and the Corresponding (*S & I*) [a]

(*S or I*)	(*S & I*)	(*S or I*)	(*S & I*)	(*S or I*)	(*S & I*)
$P1$	$P\bar{1}$	$P4$	$P4/m$	$P321$	$P\bar{3}m1$
$P2$	$P2/m$	$P4_2$	$P4_2/m$	$R32$	$R\bar{3}m$
$P2_1$	$P2_1/m$	$I4$	$I4/m$	$P3m1$	$P\bar{3}m1$
$C2$	$C2/m$	$I4_1$	* $I4_1/a$	$P31m$	$P\bar{3}1m$
Pm	$P2/m$	$P\bar{4}$	$P4/m$	$P3c1$	$P\bar{3}c1$
Pc	$P2/c$	$I\bar{4}$	$I4/m$	$P31c$	$P\bar{3}1c$
Cm	$C2/m$	$P422$	$P4/mmm$	$R3m$	$R\bar{3}m$
Cc	$C2/c$	$P42_12$	* $P4/nmm$	$R3c$	$R\bar{3}c$
$P222$	$Pmmm$	$P4_222$	$P4_2/mmc$	$P6$	$P6/m$
$P222_1$	$Pmcm$ (51)	$P4_22_12$	$P4_2/mnm$	$P6_3$	$P6_3/m$
$P2_12_12$	$Pbam$	$I422$	$I4/mmm$	$P\bar{6}$	$P6/m$
$P2_12_12_1$	$Pbca$	$I4_122$	* $I4_1/amd$	$P622$	$P6/mmm$
$C222_1$	$Cmcm$	$P4mm$	$P4/mmm$	$P6_322$	$P6_3/mmc$
$C222$	$Cmmm$	$P4bm$	$P4/mbm$	$P6mm$	$P6/mmm$
$F222$	$Fmmm$	$P4_2cm$	$P4_2/mcm$	$P6cc$	$P6/mcc$
$I222$	$Immm$	$P4_2nm$	$P4_2/mnm$	$P6_3cm$	$P6_3/mcm$
$I2_12_12_1$	$Ibca$	$P4cc$	$P4/mcc$	$P6_3mc$	$P6_3/mmc$
$Pmm2$	$Pmmm$	$P4nc$	$P4/mnc$	$P\bar{6}m2$	$P6/mmm$
$Pmc2_1$	$Pmcm$ (51)	$P4_2mc$	$P4_2/mmc$	$P\bar{6}c2$	$P6_3/mcm$
$Pcc2$	$Pccm$	$P4_2bc$	$P4_2/mbc$	$P\bar{6}2m$	$P6/mmm$
$Pma2$	$Pmam$ (51)	$I4mm$	$I4/mmm$	$P\bar{6}2c$	$P6_3/mmc$
$Pca2_1$	$Pcam$ (57)	$I4cm$	$I4/mcm$	$P23$	$Pm3$
$Pnc2$	$Pncm$ (53)	$I4_1md$	* $I4_1/amd$	$F23$	$Fm3$
$Pmn2_1$	$Pmna$	$I4_1cd$	* $I4_1/acd$	$I23$	$Im3$
$Pba2$	$Pbam$	$P\bar{4}2m$	$P4/mmm$	$P2_13$	$Pa3$
$Pna2_1$	$Pnam$ (62)	$P\bar{4}2c$	* $P4_2/mmc$	$I2_13$	$Ia3$
$Pnn2$	$Pnnm$	$P\bar{4}2_1m$	$P4/mbm$	$P432$	$Pm3m$
$Cmm2$	$Cmmm$	$P\bar{4}2_1c$	$P4/mnc$	$P4_232$	$Pm3n$
$Cmc2_1$	$Cmcm$	$P\bar{4}m2$	$P4/mmm$	$F432$	$Fm3m$
$Ccc2$	$Cccm$	$P\bar{4}c2$	$P4/mcc$	$F4_132$	* $Fd3m$
$Amm2$	$Ammm$ (65)	$P\bar{4}b2$	$P4/mbm$	$I432$	$Im3m$
$Abm2$	$Abmm$ (67)	$P\bar{4}n2$	$P4/mnc$	$I4_132$	$Ia3d$
$Ama2$	$Amam$ (63)	$I\bar{4}m2$	$I4/mmm$	$P\bar{4}3m$	$Pm3m$
$Aba2$	$Abam$ (64)	$I\bar{4}c2$	$I4/mcm$	$F\bar{4}3m$	$Fm3m$
$Fmm2$	$Fmmm$	$I\bar{4}2m$	$I4/mmm$	$I\bar{4}3m$	$Im3m$
$Fdd2$	* $Fddd$	$I\bar{4}2d$	* $I4_1/amd$	$P\bar{4}3n$	$Pm3n$
$Imm2$	$Immm$	$P3$	$P\bar{3}$	$F\bar{4}3c$	$Fm3c$
$Iba2$	$Ibam$	$R3$	$R\bar{3}$	$I\bar{4}3d$	$Ia3d$
$Ima2$	$Imam$ (74)	$P312$	$P\bar{3}1m$		

[a] Where the designation of (*S & I*) is nonstandard the space-group number is given. Asterisks indicate the nine abnormal pairs.

TABLE 2A.2. The Nine Abnormal Space Groups

Space groups (S or I)–(S & I)	$x_C y_C z_C$	t [a]	$x'y'z'$ coordinates in the ambiguity $\tilde{\phi}(S, 000; \mathbf{h}_1) = \tilde{\phi}(I, x'y'z'; \mathbf{h}_1)$ (mod 1) [b]	Possible values of $x'y'z'$	$\tilde{\phi}_0$ = offset of the optimum values of $\tilde{\phi}$, i.e., $\tilde{\phi}_4(\text{opt}) \pm \frac{1}{4}$ [cf. (2.44a)] or the value of $\tilde{\phi}_4$ for which $\Delta\tilde{\phi}_4 = 0$ [b],[c]	Preferred values of $\tilde{\phi}_1$ and $\tilde{\phi}_2$, and the corresponding value of $\tilde{\phi}_4$ for S [c]
*Fdd*2–*Fddd*	$\frac{1}{8}\frac{1}{8}\frac{1}{8}$	$(h + k + l)/4$	$x' = y' \quad =$	0	$l_4\tilde{\phi}_1 + (t_4 - l_4 t_1)/2$	$t_1/2$; —; $\frac{1}{4} + t_4/2$
$I4_1$–$I4_1/a$	$0\frac{1}{4}\frac{1}{8}$	$(2h - l)/4 = (2k + l)/4 = (h + k)/4$	$z' = t_1 - 2\tilde{\phi}_1 \quad =$	any value		
$I4_122$–$I4_1/amd$ $I4_1md$–$I4_1/amd$ $I\bar{4}2d$–$I4_1/amd$ $I4_1cd$–$I4_1/acd$ [d]			$x' = y' \quad =$ $z' = t_1 - 2\tilde{\phi}_1 \quad =$	0 0 or $\frac{1}{2}$		
$F4_132$–$Fd3m$	$\frac{1}{8}\frac{1}{8}\frac{1}{8}$	$(h + k + l)/4$	$x' = y' = z' = t_1 - 2\tilde{\phi}_1 \quad =$	0 or $\frac{1}{2}$	$(t_4/t_1)\tilde{\phi}_1$	
$P42_12$–$P4/nmm$	$\frac{1}{2}\frac{1}{2}0$	$(h + k)/2$	★ $x' = y' = \frac{1}{2} + 2[(l_1)\tilde{\phi}_2 - (l_2)\tilde{\phi}_1] \quad =$ $z' = 2[(h_1 + k_1)\tilde{\phi}_2 - (h_2 + k_2)\tilde{\phi}_1] =$	0 or $\frac{1}{2}$ 0 or $\frac{1}{2}$	★ $[(h_1 + k_1)l_4 - \{h_4 + k_4\}(l_1)]\tilde{\phi}_2 - [(h_2 + k_2)l_4 - \{h_4 + k_4\}(l_2)]\tilde{\phi}_1$	0; 0; $\frac{1}{4}$ [e]
$P\bar{4}2c$–$P4/mmc$	$00\frac{1}{2}$	$l/2$	★ $x' = y' = 2[(l_1)\tilde{\phi}_2 - (l_2)\tilde{\phi}_1] \quad =$ $z' = \frac{1}{2} + 2[(h_1 + k_1)\tilde{\phi}_2 - (h_2 + k_2)\tilde{\phi}_1] =$	0 or $\frac{1}{2}$ 0 or $\frac{1}{2}$		

[a] $t = 2\mathbf{h} \cdot \overrightarrow{OC} = 2(hx_C + ky_C + lz_C) = \tilde{\phi}(S, 000; \mathbf{h}) + \tilde{\phi}(I, 000; \mathbf{h})$ (mod 1) [cf. (2.45)].

[b] In boxes marked by a ★, $(h + k)$ and (l) denote the parities of the enclosed quantities, i.e., (mod 2); $\{h + k\}$ denotes the true value of $h + k$; all other functions are to be evaluated modulo 1.

[c] The *S*/*I* selector is treated as $\mathbf{h}_4$ with phase $\tilde{\phi}_4$ regardless of the number of origin-defining reflections needed.

[d] There is an error in the drawing of $I4_1/acd$ in the *International Tables*: the arrows on the *d* glides are drawn the wrong way round.

[e] With these particular values the space groups $P42_12$ and $P\bar{4}2c$ can be treated as normal.

TABLE 2A.3. The Eleven Enantiomorphic Pairs of Space Groups

$P4_1(S)$	$P3_112(S)$	$P6_122(S)$
$P4_3(I)$	$P3_212(I)$	$P6_522(I)$
$P4_122(S)$	$P3_121(S)$	$P6_222(S)$
$P4_322(I)$	$P3_221(I)$	$P6_422(I)$
$P4_12_12(S)$	$P6_1(S)$	$P4_132(S)$
$P4_32_12(I)$	$P6_5(I)$	$P4_332(I)$
$P3_1(S)$	$P6_2(S)$	
$P3_2(I)$	$P6_4(I)$	

Appendix 2A.2

Table 2A.4 uses $P2_12_12_1$ as an example to justify the simple rule given in Table 2.2 for writing a phase pattern by inspection. Each row relates to a different reflection in the Laue set $\{hkl\}$. The entries in columns B–E are the angular arguments $(\mathbf{h} \cdot \mathbf{r}_{0,j})$ of the exponential terms contributed by each of the four equipoints quoted in Table 2.2, the order of the scalar products having been rearranged to bring matching terms into columns. Column F gives the phase of T, the resultant of this one set of equipoints:

$$|T| \exp(2\pi i\psi_0) = \sum_{j}^{4} \exp(2\pi i\mathbf{h} \cdot \mathbf{r}_{0,j})$$

When the contributions of all the sets of equipoints are summed as in (2.22), the incremental terms in column F apply to every set and are thus attached to $\tilde{\phi}$, the phase of the reflection mentioned in column A, and so constitute the phase pattern. As the incremental terms in column F match those in column B it suffices merely to evaluate column B as explained in connection with Table 2.2.

TABLE 2A.4. Construction of Phase Pattern for $P2_12_12_1$

A	B	C	D	E	F
hkl	$(hx+ky+lz)_0$	$(hx-ky-lz)_0 + \dfrac{h+k}{2}$	$(-hx+ky-lz)_0 + \dfrac{k+l}{2}$	$(-hx-ky+lz)_0 + \dfrac{l+h}{2}$	$\bar{\psi}_0$
$h\bar{k}l$	$(hx+ky+lz)_0 + \dfrac{h+k}{2}$	$(hx-ky-lz)_0$	$(-hx+ky-lz)_0 + \dfrac{l+h}{2}$	$(-hx-ky+lz)_0 + \dfrac{k+l}{2}$	$\bar{\psi}_0 + \dfrac{h+k}{2}$
$\bar{h}kl$	$(hx+ky+lz)_0 + \dfrac{k+l}{2}$	$(hx-ky-lz)_0 + \dfrac{l+h}{2}$	$(-hx+ky-lz)_0$	$(-hx-ky+lz)_0 + \dfrac{h+k}{2}$	$\bar{\psi}_0 + \dfrac{k+l}{2}$
$\bar{h}\bar{k}l$	$(hx+ky+lz)_0 + \dfrac{l+h}{2}$	$(hx-ky-lz)_0 + \dfrac{k+l}{2}$	$(-hx+ky-lz)_0 + \dfrac{h+k}{2}$	$(-hx-ky+lz)_0$	$\bar{\psi}_0 + \dfrac{l+h}{2}$
$\bar{h}\bar{k}\bar{l}$	$-(hx+ky+lz)_0$	$-(hx-ky-lz)_0 + \dfrac{h+k}{2}$	$-(-hx+ky-lz)_0 + \dfrac{k+l}{2}$	$-(-hx-ky+lz)_0 + \dfrac{l+h}{2}$	$-\bar{\psi}_0$
$\bar{h}k\bar{l}$	$-(hx+ky+lz)_0 + \dfrac{h+k}{2}$	$-(hx-ky-lz)_0$	$-(-hx+ky-lz)_0 + \dfrac{l+h}{2}$	$-(-hx-ky+lz)_0 + \dfrac{k+l}{2}$	$-\bar{\psi}_0 + \dfrac{h+k}{2}$
$h\bar{k}\bar{l}$	$-(hx+ky+lz)_0 + \dfrac{k+l}{2}$	$-(hx-ky-lz)_0 + \dfrac{l+h}{2}$	$-(-hx+ky-lz)_0$	$-(-hx-ky+lz)_0 + \dfrac{h+k}{2}$	$-\bar{\psi}_0 + \dfrac{k+l}{2}$
$hk\bar{l}$	$-(hx+ky+lz)_0 + \dfrac{l+h}{2}$	$-(hx-ky-lz)_0 + \dfrac{k+l}{2}$	$-(-hx+ky-lz)_0 + \dfrac{h+k}{2}$	$-(-hx-ky+lz)_0$	$-\bar{\psi}_0 + \dfrac{l+h}{2}$

PART II:
CALCULATION OF $|E|$ VALUES

The normalized structure factor, $E(\mathbf{h})$, can be defined either by

$$E^2(\mathbf{h}) = \frac{|F(\mathbf{h})|^2/\mu(\mathbf{h})}{\langle|F(\mathbf{h})|^2/\mu(\mathbf{h})\rangle_\theta} \tag{2.60a}$$

or

$$= \frac{|\mathscr{I}(\mathbf{h})|/\mu(\mathbf{h})}{\langle|\mathscr{I}(\mathbf{h})|/\mu(\mathbf{h})\rangle_\theta} \tag{2.60b}$$

where $F(\mathbf{h})$ was defined in (2.2). $\mathscr{I}(\mathbf{h})$, which is the corresponding intensity on an arbitrary scale after correction for Lorentz and polarization factors (and absorption if necessary), can therefore be written as $K^2|F(\mathbf{h})|^2$: we shall assume that a single value of K, the scale factor, applies throughout the data set. The factor μ, which is a function of $\mathbf{h}$ but not of θ, the corresponding Bragg angle, can be either the *pwys*, $p(\mathbf{h})$, or the factor $\varepsilon(\mathbf{h})$, both of which are discussed below. They are defined in different ways, but in practice their numerical values are indistinguishable.

The averages in the denominators of (2.60a,b) are functions of θ, and the suffix θ indicates that the relevant local value must be used for evaluating each $E(\mathbf{h})$. On the other hand, it follows from these equations that $\langle E^2(\mathbf{h})\rangle_\theta = 1$, and therefore that

$$\langle E^2(\mathbf{h})\rangle = 1 \tag{2.61}$$

This relationship, which gives E its name, shows that the averaging of E^2 can be carried out regardless of θ. The same is not true, however, of either $|F(\mathbf{h})|^2$ or $|U(\mathbf{h})|^2$ and this simplification is one of the main reasons why E is preferred for use in direct methods. Other reasons are the simpler form of the probability expressions when written in terms of E, and the greater resolving power implicit in the use of E's since they correspond to diffraction from a structure composed of point atoms at rest.

The practical evaluation of E's is effected via (2.60b), and two related methods are in use: (i) Wilson-type plots, and (ii) the K curve.

Wilson-Type Plots. Wilson (1942) showed from the expansion of $|F(\mathbf{h})|^2$ in terms of g_j [cf. (2.2)] that

$$\langle|F(\mathbf{h})|^2\rangle_\theta = \left(\sum_{j=1}^{N} g_j^2\right)_\theta \tag{2.62}$$

This can usually be satisfactorily approximated by

$$\left(\sum_{j=1}^{N} f_j^2\right)_\theta \exp\left[\frac{-2\bar{B}\sin^2\theta}{\lambda^2}\right] \tag{2.63}$$

which we shall abbreviate to $(\sigma_2 \cdot T)_\theta$. Thus,

$$\langle\mathscr{I}(\mathbf{h})\rangle_\theta = K^2(\sigma_2 \cdot T)_\theta \tag{2.64}$$

or

$$\ln\left[\frac{\langle\mathscr{I}(\mathbf{h})\rangle}{\sigma_2}\right]_\theta = 2\ln K - (2\bar{B}/\lambda^2)\sin^2\theta \tag{2.65}$$

Values of $\langle\mathscr{I}(\mathbf{h})\rangle_\theta$ are obtained from a succession of adjacent or overlapping spherical shells in reciprocal space, excluding however a small region around the origin. The corresponding values of $\sin^2\theta$ are obtained by averaging $\sin^2\theta$ for all the reciprocal-lattice points within each shell, except the systematic absences, and making due allowance for multiplicity. Practical aspects of the process were discussed at some length by Rogers (1965), especially for data sets in which the averaging presents difficulty; little of the detail will be repeated here. Ladd (1978) has also critically assessed several details of the necessary procedures.

A Wilson plot of $\ln[\langle\mathscr{I}(\mathbf{h})\rangle/\sigma_2]_\theta$ versus $\sin^2\theta$ should ideally be a straight line whose intercept gives the scale factor, K, and whose slope gives $\bar{B}$. The plot is, however, often strongly curved at low θ and occasionally has perceptible "humps" partway along it, thus making the estimation of K and $\bar{B}$ unreliable. The origin of these nonlinearities was discussed by Rogers (1965), who also showed that the estimation could be improved by the simultaneous use of an auxiliary plot defined by

$$\ln\left[\langle\mathscr{I}(\mathbf{h})\rangle_\theta \bigg/ \sum_{j=1}^{N} Z_j^2\right] \quad \text{versus} \quad \sin^2\theta \tag{2.66}$$

which is often found to be the straighter of the two plots. Since $\sigma_2 \to \sum Z_j^2$ as $\theta \to 0$, the plots have the same intercept, so that K, the more important of the two parameters, can be estimated more accurately even when the curvature in Wilson's plot is marked.

It is important when considering "humps" to draw these two plots as in Fig. 2.10a, where the averages decrease with increase in θ. Wilson originally drew his plot as in Fig. 2.10b, and some authors have followed him, but such plots tend to be misleading as to the position and identity

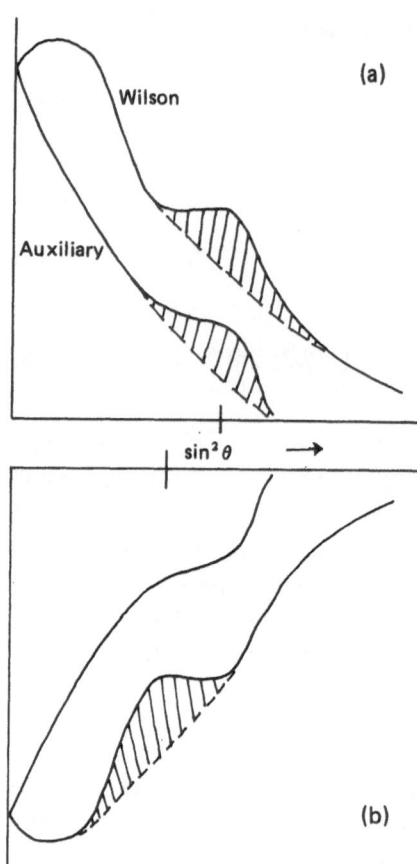

(a)

(b)

FIGURE 2.10. (a) Typical Wilson plot of $\ln(\langle\mathscr{I}(\mathbf{h})\rangle/\sigma_2)_\theta$ versus $\sin^2\theta$, and auxiliary plot of $\ln(\langle\mathscr{I}(\mathbf{h})\rangle_\theta/\Sigma Z_j^2)$ versus $\sin^2\theta$. (b) The same curves as in Fig. 2.10a plotted upside down and showing how the identity and position of a hump can be mistaken.

of the hump (cf. Figs. 2.10a,b). In some instances the hump can be related to known fragments of molecules, and in such cases allowance can be made via a Debye molecular-transform function:

$$\langle\mathscr{I}(\mathbf{h})\rangle_\theta = K^2 T(\theta)\left[\sigma_2 + \sum_{j\neq k}\sum f_j f_k\left(\frac{\sin 2\pi h r_{jk}}{2\pi h r_{jk}}\right)\right]_\theta \qquad (2.67)$$

where the sum σ_2 covers all the atoms in the cell, and that over j, k covers all atom pairs for which r_{jk} is known. Ladd (1978) has discussed the best way of handling data when humps re   in, and he concluded that it is better to use the minimum profile, by   aving off the hump as in Fig. 2.10a. This is not easy to do in a co   uter program, so it is common to find least-squares lines being fitted t   the humpy plot. He shows that such lines can produce appreciable err   s in the magnitudes of the E's, which in some circumstances make all the difference between success and failure in solving the structure. French and Wilson (1978) have also discussed this point. Severe thermal anisotropy can also make a considerable difference to the pecking order of the terms and thus alter those chosen for the starting set. It is not easy to allow for anisotropy, but even an approximation helps.

A more serious defect of several programs for Wilson plots lies in their handling of reflections with enhanced intensity averages. In 1950, Wilson, following a suggestion from the present author, demonstrated that certain zones and central rows have abnormally high intensity averages, and that for such sets of reflections (2.64) must be generalized to

$$\langle \mathscr{I}(\mathbf{h})\rangle_\theta = K^2 p(\mathbf{h})(\sigma_2 \cdot T)_\theta \qquad (2.68)$$

The modulation function, $p(\mathbf{h})$, takes on a distinct value for each different set of abnormal reflections, but was assumed to be unity for all other reflections. Rogers (1950) showed that the affected reflections are point-group (not space-group) dependent, and he listed all the corresponding values of p. The values, which in some point groups are as high as 12, were republished (Rogers, 1965) and named *pwysau* (singular *pwys*). The existence of such abnormally strong reflections poses a problem of the conservation of energy and of reconciling them with Wilson's equation (2.62). In a forthcoming article (Rogers, 1979) the present author derives a general theory to show that such abnormally high averages, which occur throughout well-defined sets of reflections, are compensated by regions of low average that are not defined until the structure is known. We can, therefore, no longer regard p as unity for all the unenhanced reflections; it fluctuates around unity, being somewhat below unity in a number of localities. The basic idea of the general treatment was touched upon by Rogers (1965). Wilson (1964) and Nigam (1965), using a less general theory, demonstrated the existence of compensation for the space groups *Pm* and *Pmm2*, respectively.

The general theory can be summarized as follows. The terms comprising the expansion of $|F(\mathbf{h})|^2 = F(\mathbf{h}) \cdot F^*(\mathbf{h})$ can be arranged in a square array (Rogers, 1965), each term of which is associated with a vector peak in the Patterson. Wilson's equation (2.62) corresponds to all the off-diagonal terms averaging to zero when summed over the whole of reciprocal space, and his equation is true for any space group. If, however, the individual terms in $F(\mathbf{h})$ are suitably sequenced, one can arrange all the Harker vectors into a succession of $n \times n$ blocks strung along the diagonal

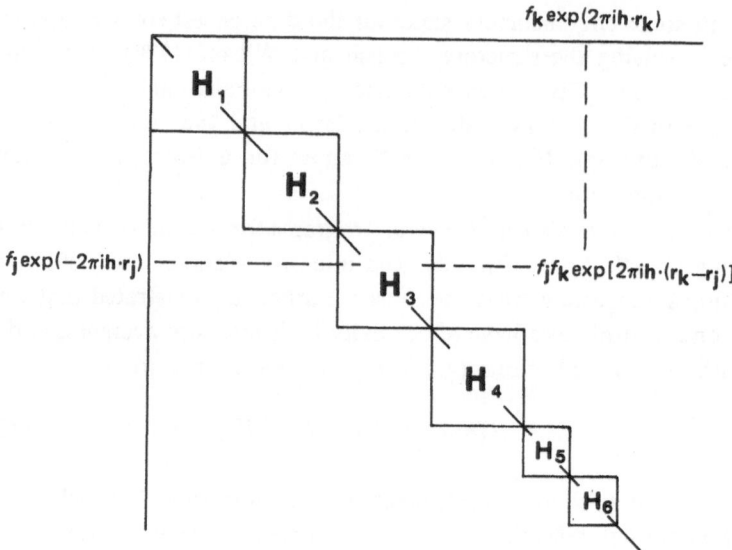

FIGURE 2.11. Schematic diagram of the array of products contributing to $F(\mathbf{h}) \cdot F^*(\mathbf{h})$. The atoms are arranged to bring all the Harker vectors for atom 1 into the $n \times n$ square $H1$, etc. Atoms 1–4 are in general positions, and atoms 5 and 6 are in special positions with fewer Harker vectors.

(see Fig. 2.11). Averaging now gives

$$\langle \mathscr{I}(\mathbf{h}) \rangle_\theta = K^2 \left\langle \sum_{j=1}^{N/n} g_j^2 \left(n + 2 \sum_{p>q}^{n} \sum^{n} \cos 2\pi \mathbf{h} \cdot \mathbf{r}_{pq} \right)_j \right\rangle_\theta \qquad (2.69)$$

where the vectors $(\mathbf{r}_{pq})_j$ constitute the Harker set for atom type j. If we make the usual approximation that all g_j are uniform, this becomes

$$\langle \mathscr{I}(\mathbf{h}) \rangle_\theta = K^2 (\sigma_2 T)_\theta \langle p(\mathbf{h}) \rangle \qquad (2.68)$$

where

$$p(\mathbf{h}) = 1 + \frac{2}{N} \sum_{j=1}^{N/n} \sum_{p>q}^{n} \sum^{n} (\cos 2\pi \mathbf{h} \cdot \mathbf{r}_{pq})_j \qquad (2.70)$$

For general reflections $p \approx 1$, and averaging over all $\mathbf{h}$ gives

$$\langle p \rangle = 1$$

but for certain sets of reflections some values of $\cos 2\pi \mathbf{h} \cdot \mathbf{r}_{pq}$ are systematically unity and so contribute to an enhancement of the corresponding

TABLE 2.15. The Association between the *Pwysau* (p) and the Harker
Vectors

Number and type of Harker vectors		Set of h with abnormal averages	h interacts with Harker subsets	ν	p
a	$00w$	8	$(00l)$	b, c, d, e, f	56	8
b	$u00$	8	$(h00)$	a, c, h	24	4
c	$0v0$	8	$(0k0)$	a, b, g	24	4
d	$uu0$	8	$(hh0)$	a, e, j	24	4
e	$u\bar{u}0$	8	$(\bar{h}h0)$	a, d, i	24	4
f	$uv0\ (u \neq v)$	24	$(hk0)$	a	8	2
g	$u0w$	8	$(h0l)$	c	8	2
h	$0vw$	8	$(0kl)$	b	8	2
i	uuw	8	(hhl)	e	8	2
j	$uvw\ (u{\neq}v{\neq}w{\neq}0)$	24	$(\bar{h}hl)$	d	8	2

intensity average, i.e.,

$$p = 1 + 2\nu/n \qquad (2.71)$$

where ν is the number of interacting Harker vectors in each set, and n is the symmetry number of the general positions in the relevant space group. As an example, consider space group $P4/mmm$. Its Harker vectors can be grouped as in Table 2.15, together with the corresponding sets of abnormal reflections and their values of p. These results were written compactly in my earlier tables as 8/2; 4/2; 4/2, the order being $(00l)/(hk0)$; $(h00)/(0kl)$; $(hh0)/(\bar{h}hl)$.

It should be noted in passing that the above definition of p makes it zero for all systematically absent reflections, and this gives further justification for ignoring all systematic absences when carrying out these conversions.

Equation (2.70) reveals the nature of the compensation. Thus, for the example of $P4/mmm$, if one averages over any layer hkl (L specified), excluding reflections common to enhanced zones, (2.70) becomes

$$p(L) = 1 + \frac{2}{N} \sum_{j=1}^{N/n} 8 \cos 2\pi L z_j \qquad (2.72)$$

The precise profile of this as a function of L depends on the z_j coordinates, but it is typically of the form shown in Fig. 2.12; it cannot be calculated till the coordinates are known. If atoms occur in special positions, they

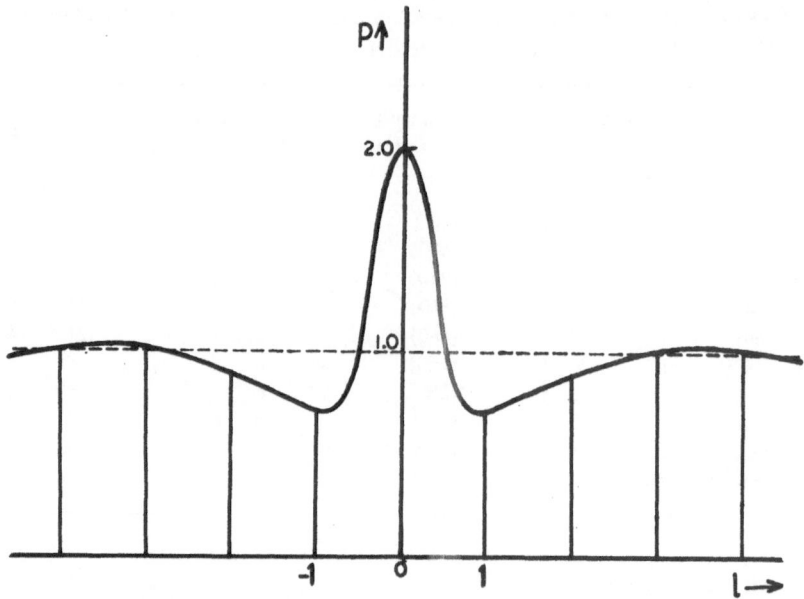

FIGURE 2.12. Variation of p for layers hkL for space group $P4/mmm$, showing $p = 2$ in $hk0$, due to the m plane parallel to (001), and the compensation for this in adjacent layers.

have smaller sets of Harker vectors which can be treated in the same way. Their contribution to p is smaller than that for atoms in general positions, and in such a situation p falls somewhat short of the ideal integer value.

In 1965, Rogers recommended that Wilson plots should be based on reduced intensities, $\mathcal{I}(\mathbf{h})/p(\mathbf{h})$, but a number of programs still do not appear to allow for p. In view of the large values of p in some point groups the consequential errors can be considerable. The above argument reinforces the earlier recommendation. Thus, (2.65) becomes

$$\ln\left[\frac{\langle\mathcal{I}(\mathbf{h})\rangle/p(\mathbf{h})}{\sigma_2}\right]_\theta = 2\ln K - (2\bar{B}/\lambda^2)\sin^2\theta \tag{2.73}$$

and (2.66) becomes

$$\ln\left[\frac{\langle\mathcal{I}(\mathbf{h})\rangle_\theta/p(\mathbf{h})}{\sum Z_j^2}\right] \quad \text{versus} \quad \sin^2\theta \tag{2.74}$$

Since in the low-average regions the values of p are unknown and do not differ greatly from unity, they are best approximated by putting $p = 1$, but it must be recognized that this is yet another approximation in the whole process, which prevents $\langle p \rangle$ from being unity as it should be. Allowing in this way for p has on occasion appreciably improved the linearity of Wilson plots.

As we shall see below, use of reduced intensities brings this scaling procedure fully into accord with the definition of E, whereas the omission of p corresponds to the definition of the quasinormalized structure factor $\mathscr{E}(\mathbf{h})$ (Karle and Hauptman, 1959).

K-Curve Plots. The K curve was introduced by Karle and Hauptman (1953b). In the notation used above they plotted

$$\left[\frac{\langle \mathscr{I}(\mathbf{h}) \rangle}{\sigma_2 \varepsilon(\mathbf{h})} \right]_\theta \quad \text{versus} \quad \sin \theta \qquad (2.75)$$

but, as this gives a very strongly curved plot, later authors used variants such as plotting the function against $\sin^n \theta$ $(n = 2$ or $3)$, or

$$\ln \left[\frac{\langle \mathscr{I}(\mathbf{h}) \rangle}{\sigma_2 \varepsilon(\mathbf{h})} \right]_\theta \quad \text{versus} \quad \sin^2 \theta \qquad (2.76)$$

The latter is a Wilson plot based on intensities reduced by the factor ε. Again, the earliest K curves, like the original Wilson plot, were drawn as ascending curves, but as explained above it is preferable to plot them as descending curves. The factor $\varepsilon(\mathbf{h})$ was introduced by Karle and Hauptman (1966) to denote a quantity they had used earlier (Hauptman and Karle, 1953; Karle and Hauptman, 1953a) in statistical arguments for calculating $P(|F|)\,dF$. Their definition can be written in more familiar notation as

$$\varepsilon(\mathbf{h}) = n^{-1} \int_0^1 \int_0^1 \int_0^1 (A^2 + B^2)\, dx\, dy\, dz$$

$$= n^{-1} \int_0^1 \int_0^1 \int_0^1 |F'(\mathbf{h})|^2\, dx\, dy\, dz \qquad (2.77)$$

where $F'(\mathbf{h}) = A + iB$ is the structure-factor expression for a single set of atoms in equivalent general positions, and n is the corresponding multiplicity. Equation (2.77) involves integration over all the vectors in a representative Harker set and can, therefore, be expressed in the notation of (2.70) in the form

$$\varepsilon(\mathbf{h}) = \frac{\langle \mathscr{I}(\mathbf{h}) \rangle_\theta}{K^2(\sigma_2 T)_\theta} = 1 + (2/n) \int_0^1 \int_0^1 \int_0^1 \sum_{p>q}^n \sum^n \cos 2\pi \mathbf{h} \cdot \mathbf{r}_{pq}\, dx\, dy\, dz \qquad (2.78)$$

Though this differs from (2.70), the two equations predict the same sets of enhanced reflections and the same numerical values for ε and p over those sets. But (2.78) is unable to predict the possibility of compensation [cf. (2.72)], so it has been the practice to regard ε^{-1} as a sort of three-dimensional δ function and hence $\langle \varepsilon \rangle \neq 1$. As was shown above in connection with p, this is an approximation, but one that cannot be improved upon. In this treatment it is possible to allow for atoms in special positions with results exactly matching those obtained for p, but there is no recognition that in many space groups atoms are excluded from certain volumes in the unit cell, contrary to the limits of integration in (2.77). This exclusion was, however, allowed for by Wilson (1964) in his treatment of compensation in *Pm*.

The statement "ε corrects for space-group extinctions" made by Karle and Hauptman themselves (1966) led to persistent misunderstanding, but it has recently been corrected by several authors, including Stewart and Karle (1976) and Stewart *et al.* (1977). The example adduced above of *P4/mmm* has plenty of enhanced sets of reflections but no absences. The properties of $\varepsilon(\mathbf{h})$ and $p(\mathbf{h})$ are indistinguishable in practice, so long as both are approximated to unity for all reflections not in enhanced sets. Likewise, both are point-group (not space-group) dependent, *provided systematic absences are excluded from the averaging*. The nature of the compensation in reciprocal space for space groups containing screws or glides is somewhat more complicated and will be elaborated in the forthcoming paper (Rogers, 1979), but it is of academic rather than practical significance. Other practical consequences will also be discussed.

Finally, it is often found after converting all the $\mathscr{I}(\mathbf{h})$ to $E(\mathbf{h})$ that some subsets of reflections do not give $\langle E^2 \rangle = 1$. This may be due to some atoms occupying special positions or to rationality between the coordinates of atoms in general positions. It has been discussed by Hauptman and Karle (1959), who showed that it is legitimate and very desirable to rescale the subsets independently, especially if the differences appear between the parity groups.

Acknowledgments

I am greatly indebted to Drs. Hauptman, Main, Wallwork, and Williams for correspondence and discussions during the development of the approach used in the first part of this chapter and for comments on earlier drafts.

References to Part I

Allen, F. H., Rogers, D., and Trotter, J. (1972). *J. Chem. Soc. B*, 166–171.

Bertaut, E. F. (1956). *Acta Crystallogr.* **9**, 769–770.

Bragg, W. L., and Lipson, H. (1936). *Z. Kristallogr.* **95**, 323.

Brennan, T., and Sundaralingam, M. (1973). *Biophys. Biochem. Res. Commun.* **52**, 1348.

Eller, G. von (1955). *Bull. Soc. Fr. Mineral. Cristallogr.* **78**, 157.

Grant, D. F., Howells, R. G., and Rogers, D. (1957). *Acta Crystallogr.* **10**, 489–497.

Hauptman, H. A. (1972). *Crystal Structure Determination*, Plenum, New York.

Hauptman, H. A., and Karle, J. (1953). "Solution of the Phase Problem I: The Centro-symmetric Crystal," ACA Monograph No. 3, Polycrystal Book Service, Pittsburgh, Pennsylvania.

Hauptman, H. A., and Karle, J. (1956). *Acta Crystallogr.* **9**, 45–55.

Hauptman, H. A., and Karle, J. (1959). *Acta Crystallogr.* **12**, 93–97.

Huggins, M. L. (1945). *Nature* **155**, 18–19.

International Tables for X-Ray Crystallography, Vol. 1 (1969). The Kynoch Press, Birmingham, England.

Karle, J., and Hauptman, H. A. (1961). *Acta Crystallogr.* **14**, 217–223.

Kartha, G., and Phillips, T. (1973). Stockholm Symposium on the Structure of Biological Molecules, (Abstracts p. 91).

Kartha, G., Phillips (II), T., and Ambady, G. (1975). *Acta Crystallogr. Sect. A* **31**, S52, 03.5-18.

Rogers, D. (1975). In *Anomalous Scattering*, Eds. S. Ramaseshan and S. C. Abrahams, International Union of Crystallography, Munksgaard, Copenhagen, pp. 231–250.

Rogers, D., Venkatasubramanian, K., Sørum, H., and Hjortås, J. A. (1972). Presented at Meeting of Chemical Crystallography Group, Nottingham, September 26.

Rogers, D. (1979). Submitted to *Acta Crystallogr.*

Waser, J. (1955). *Acta Crystallogr.* **8**, 595.

References to Part II

French, S., and Wilson, K. (1978). *Acta Crystallogr. Sect. A* **34**, 517–525.

Hauptman, H., and Karle, J. (1953). *Acta Crystallogr.* **6**, 136–141.

Hauptman, H., and Karle, J. (1959). *Acta Crystallogr.* **12**, 846–850.

Karle, J., and Hauptman, H. (1953a). *Acta Crystallogr.* **6**, 131–135.

Karle, J., and Hauptman, H. (1953b). *Acta Crystallogr.* **6**, 473–476.

Karle, J., and Hauptman, H. (1959). *Acta Crystallogr.* **12**, 404–410.

Karle, J., and Hauptman, H. (1966). *Acta Crystallogr.* **21**, 849–859.

Ladd, M. F. C. (1978). *Z. Kristallogr.*, **147**, 279–296.

Nigam, G. D. (1965). Personal communication.

Rogers, D. (1950). *Acta Crystallogr.* **2**, 455–464.

Rogers, D. (1965). *Computing Methods in Crystallography*, Ed. J. S. Rollet, Pergamon Press, Edinburgh, Chaps. 15 and 16, pp. 117–148.

Rogers, D. (1979). Submitted to *Acta Crystallogr.*

Stewart, J. M., and Karle, I. (1976). *Acta Crystallogr. Sect. A* **32**, 1005–1007.
Stewart, J. M., Karle, I., Iwasaki, H., and Ito, T. (1977). *Acta Crystallogr. Sect. A.* **33**, 519.
Wilson, A. J. C. (1942). *Nature* **150**, 151–152.
Wilson, A. J. C. (1950). *Acta Crystallogr.* **2**, 258–261.
Wilson, A. J. C. (1964). *Acta Crystallogr.* **17**, 1591–1592.

3

Symbolic Addition and Multisolution Methods

M. F. C. LADD and R. A. PALMER

3.1. Introduction

The symbolic addition technique and the multisolution method provide general approaches to crystallographic phase determination. Crystal structures with up to about 100 atoms in the asymmetric unit are solved routinely by these methods. They do not succeed in all cases, and other chapters in this book show new lines of approach both for difficult cases and for increasing the efficiency of phase determination generally. This article is written more from the point of view of a user than from one concerned actively in the development of these methods.

The positivity of electron density criterion leads to the phase addition formula known also as the triple phase relation (tpr), as proposed by Karle and Karle (1966):

$$\phi_h \approx \phi_k + \phi_{h-k} \tag{3.1}$$

where h, k, and $h - k$ form a vector triangle in reciprocal space with one apex at the origin. The sign $\approx$ indicates an approximation in the tpr which is better the larger the value of $|E_h|$. Where several triplets are involved with a given h, (3.1) becomes

$$\phi_h \approx \langle \phi_k + \phi_{h-k} \rangle_{k_r} \tag{3.2}$$

M. F. C. LADD • Department of Chemical Physics, University of Surrey, Guildford, Surrey GU2 5XH, United Kingdom.

R. A. PALMER • Department of Crystallography, Birkbeck College, Malet Street, London WC1E 7HX, United Kingdom.

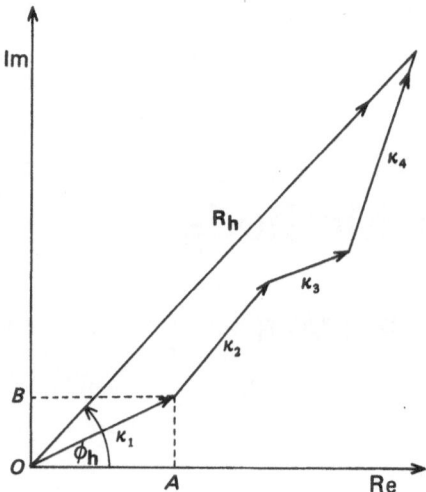

FIGURE 3.1. Combination of four tpr's to give an estimate of ϕ_h.

where $\langle \, \rangle_{k_r}$ implies a restricted set vector summation involving $r \, |E|$ values.

The $|F_o|$ data derived experimentally are converted to $|E|$ values by techniques described in Chapter 2. It is not uncommon to commence phase determination with $|E|$ values greater than about 1.5, in order to maintain acceptable probability limits. Equation (3.2) is illustrated by an Argand diagram in Fig. 3.1 for four values of $\mathbf{k}$; ϕ_h is the estimated phase angle associated with the resultant vector $\mathbf{R_h}$. Each vector labeled κ depends on a product $|E_k| \, |E_{h-k}|$ and may be resolved into components A and B along the real and imaginary axes, respectively, such that

$$A = |E_k| \, |E_{h-k}| \cos(\phi_k + \phi_{h-k}) \tag{3.3}$$

and

$$B = |E_k| \, |E_{h-k}| \sin(\phi_k + \phi_{h-k}) \tag{3.4}$$

It follows from (3.2)–(3.4) that

$$\tan \phi_h \approx \frac{\sum_{k_r} w_h |E_k| \, |E_{h-k}| \sin(\phi_k + \phi_{h-k})}{\sum_{k_r} w_h |E_k| \, |E_{h-k}| \cos(\phi_k + \phi_{h-k})} \tag{3.5}$$

Equation (3.5) is the weighted tangent formula, where weights w_h may be unity (Karle and Hauptman, 1956) or may be given values as explained on Section 3.5. Current phase-determining procedures are based largely on (3.5) as implemented in programs such as those of Karle and

Karle (1966) and Main *et al.* (1974). The reliability of (3.5) can be measured by the variance $V(\phi_h)$. Following Cochran (1955), the graph of Fig. 3.2 may be obtained for $V(\phi_h)$ as a function of α_h (Karle and Karle, 1966), where

$$\alpha_h{}^2 = \left[\sum_{k_r} \kappa_{hk} \cos(\phi_k + \phi_{h-k})\right]^2 + \left[\sum_{k_r} \kappa_{hk} \sin(\phi_k + \phi_{h-k})\right]^2 \quad (3.6)$$

where

$$\kappa_{hk} = 2\sigma_3\sigma_2{}^{-3/2} \, |E_h| \, |E_k| \, |E_{h-k}|$$

$$\sigma_n = \sum_{j=1}^{N} Z_j{}^n \quad (3.7)$$

Z_j being the atomic number of the jth atom in a unit cell containing a total of N atoms. The parameter α_h gives a measure of the reliability with which ϕ_h is determined by the tangent formula. When (3.6) contains only one term, as it may in the initial stages of phase determination, then $\alpha_h = \kappa_{hk}$ and is strongly dependent on the product $|E_h| \, |E_k| \, |E_{h-k}|$. Figure 3.2 shows clearly that $V(\phi_h)$ has acceptably small values when α_h is greater than about 4 or 5 ($<30°$) but increases rapidly for decreasing α_h less than about 3 ($>40°$): α_h depends also on $\sigma_3\sigma_2{}^{-3/2}$, which value depends on the number and types of atoms in the unit cell. This dependence may be illustrated by a hypothetical structure containing different numbers N of

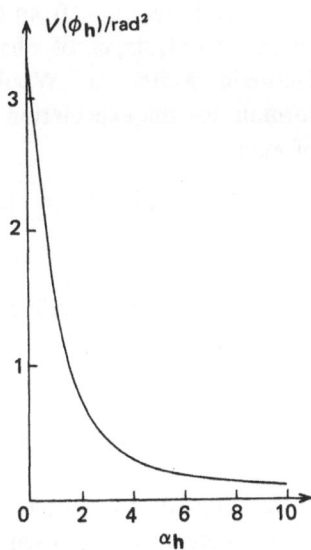

FIGURE 3.2. Variance $V(\phi_h)$ as a function of α_h.

TABLE 3.1. Values of $|E_{\min}|$ for $\alpha_h = 3.0$ in Structures Containing N Identical Atoms per Unit Cell

N	$|E_{\min}|$
25	1.96
36	2.08
49	2.19
64	2.29
81	2.38
100	2.47

identical atoms. $\alpha_h\ (= \kappa_{hk})$ is then given by

$$\alpha_h = \frac{2}{N^{1/2}}\,|E_h|\,|E_k|\,|E_{h-k}| \tag{3.8}$$

Table 3.1 lists the values of $|E_{\min}|$ needed to obtain $\alpha_h = 3$ for selected values of N from 25 to 100. The table illustrates clearly an important limitation of direct methods: the required $|E_{\min}|$ increases dramatically as a function of N whereas, as indicated in Chapter 2, the distribution of $|E|$ values is largely independent of structural complexity. Therefore it becomes more and more difficult to form a good starting set as N becomes larger and larger.

Calculation of α_h from (3.6) is possible only when phases are available. In the initial stages of phase determination this is not practicable, and Germain, Main, and Woolfson (1970a,b) have developed the following formula for the expectation value $(\alpha_E{}^2)$ of $\alpha_h{}^2$, which uses only the values of κ_{hk}:

$$\alpha_E{}^2 = \sum_{k_r} \kappa_{hk}^2 + \sum_{k_r}\sum_{\substack{k_{r'} \\ k \neq k'}} \kappa_{hk}\kappa_{hk'}\,\frac{I_1(\kappa_{hk})}{I_0(\kappa_{hk})}\,\frac{I_1(\kappa_{hk'})}{I_0(\kappa_{hk'})} \tag{3.9}$$

where I_0 and I_1 are modified Bessel functions of the zero and first orders, respectively. $I_1(\kappa)/I_0(\kappa)$ has the form shown in Fig. 3.3, but for computational purposes it may be expressed as the polynomial

$$I_1(\kappa)/I_0(\kappa) \approx 0.5658\kappa - 0.1304\kappa^2 + 0.0106\kappa^3$$

in the range $0 \leq \kappa \leq 6$; for $\kappa > 6$ the value of the function is essentially unity. These principles, used in conjunction with those evolved in Chapter 2

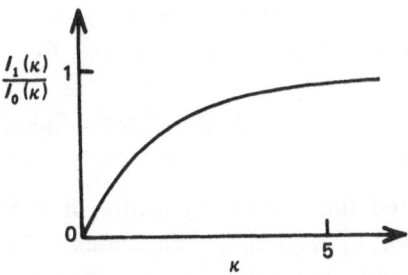

FIGURE 3.3. Variation of $I_1(\kappa)/I_0(\kappa)$ with κ.

for selecting the origin-determining reflections, may help a direct phase analysis to be established on a sound basis right from the beginning, and so lead to the production of a sufficient number of phases to give an interpretable E map. Experience shows, however, that even with the exercise of great care, the development of phases may not always be a successful process. In such an event the remedy is often to try again with a different starting set of reflections.

For a centrosymmetric structure with $\bar{1}$ at the origin and $w_h = 1$, (3.5) becomes (Hauptman and Karle, 1953)

$$s_h \approx s\left(\sum_{k_r} E_k E_{h-k}\right) \qquad (3.10)$$

where s means "the sign of," and the summation includes all pairs of reflections for which signs are known. A corresponding weighted sum function with $w_h \neq 1$ may also be formed, or, if all the $|E|$'s are large, (3.10) may be used in the form given by Zachariasen (1952):

$$s_h \approx s\left(\sum_{k_r} s_k s_{h-k}\right) \qquad (3.11)$$

Since $E = s\,|E|$, it is clear that s_k and s_{h-k} must both be known for application of (3.10), just as values of both ϕ_k and ϕ_{h-k} must be available for application of (3.5).

The probability associated with (3.10) as given by Woolfson (1954) and Woolfson and Cochran (1955) is

$$P_+(h) = \tfrac{1}{2}\left\{1 + \tanh\left[\sigma_3 \sigma_2^{-3/2}\,|E_h|\,\Big|\sum_{k_r} E_k E_{h-k}\Big|\right]\right\} \qquad (3.12)$$

If $P_+(h)$ approaches 1.0, it indicates strongly that $s_h = +1$, whereas $P_+(h)$ approaching 0.0 indicates strongly that $s_h = -1$. Intermediate values of

$P_+(\mathbf{h})$ are less reliable and must be treated with caution. For a structure containing N identical atoms, the probability becomes

$$P_+(\mathbf{h}) = \tfrac{1}{2}\left\{1 + \tanh\left[N^{-1/2}\,|E_\mathbf{h}|\sum_{\mathbf{k}_r} E_\mathbf{k}E_{\mathbf{h}-\mathbf{k}}\right]\right\} \qquad (3.13)$$

and for a single indication it depends on both $N^{-1/2}$ and the product $|E_\mathbf{h}|\,|E_\mathbf{k}|\,|E_{\mathbf{h}-\mathbf{k}}|$. Equations (3.2) and (3.10) are often called Σ_2 (sigma-two) relationships. When only a single triplet is considered, (3.10) may be written as the Cochran–Zachariasen–Sayre sign relationship (Cochran, 1952; Zachariasen, 1952; Sayre, 1952)

$$s_\mathbf{h}s_\mathbf{k}s_{\mathbf{h}-\mathbf{k}} \approx 1 \qquad (3.14)$$

known as a triple-product sign relationship (tpsr). If we take the special case that $\mathbf{h} = -\mathbf{k}$, in space group $P1$ (3.1) becomes

$$\phi_\mathbf{h} \approx \phi_{-\mathbf{h}} + \phi_{2\mathbf{h}} \approx \phi_{2\mathbf{h}}/2 \qquad (3.15)$$

Similarly in space group $P\bar{1}$, since $\phi_\mathbf{h} = 0$ or π, it follows that $s_\mathbf{h} = s_{\bar{\mathbf{h}}} = \pm 1$ and $s_\mathbf{h}s_{\bar{\mathbf{h}}} = +1$; hence

$$s(2h) \approx 1 \qquad (3.16)$$

Equations (3.15) and (3.16) are often referred to as Σ_1 relationships: they have different forms for different space groups or for different zones within a given space group, and different probability formulas (Karle and Karle, 1966; Main et al., 1974). The successful application of (3.15) or (3.16) also depends on the maintenance of high probability, which is associated with the use of reflections with high $|E|$ values.

   A listing of the triplets for each $\mathbf{h}$ and its several corresponding $\mathbf{k}$ and $\mathbf{h} - \mathbf{k}$ values, including all symmetry-related reflections, is a first stage in applying direct methods. Such a listing, known as a Σ_2 listing, may be produced either for a set of $|E|$ values limited by a value for $|E_{\min}|$, or up to a preset number of triplets to be produced, whichever is the most convenient for the problem in hand. The space group symmetry plays a large part in the development of phase sets from a starting set of reflections. For example, in the triclinic system, space group $P1$ or $P\bar{1}$, only interactions involving $\pm\mathbf{k}$, $\mathbf{h} \mp \mathbf{k}$ may be used to generate $\mathbf{h}$. In the monoclinic system, however, each $\mathbf{k}$ involves generally a total of four interactions, corresponding to $\mathbf{h}$ and $\mathbf{k}_1$, $\mathbf{h}$ and $\mathbf{k}_2$, $\mathbf{h}$ and $\mathbf{k}_3$, and $\mathbf{h}$ and $\mathbf{k}_4$, where $\mathbf{k}_1 - \mathbf{k}_4$ represent the four symmetry-related vectors from $\mathbf{k}$. This number increases as the number of independent symmetry-related $|E_h|$ values increases: in the trigonal system it is six, in the orthorhombic system eight, and so on for other crystal systems. From these considerations it is evident that structures with higher-symmetry space groups generally offer the greater chances of solution. The presence of centrosymmetric projection planes leads to special phases which may be 0 or π. In orthorhombic space groups, three-dimensional phases can be reached through direct vector interaction between two-dimensional, centric zone, phases.

A centrosymmetric structure presents the lesser problem, success being governed mainly by the molecular complexity. Both centrosymmetric and noncentrosymmetric structures may be treated by either symbolic addition (Karle and Karle, 1963, 1966; Karle, 1970) or multisolution techniques (Woolfson and Germain, 1968; Germain, Main, and Woolfson, 1970a,b). Both of these groups of workers use methods based on the tangent formula (3.5). Karle and Karle stress the desirability of determining by hand calculations some 40–50 phases for which $V(\phi_h)$ is less than about 0.5 rad^2 and scrutinizing the results of any relationship between sign or phase symbols which may have arisen. Germain *et al.* attempt to work from a smaller basis, the origin-defining phases, a small number of symbolic phases (having assigned values of $\pm\pi/4$, $\pm3\pi/4$ in the general noncentrosymmetric case), Σ_1 results and, where appropriate, an enantiomorph-restricting phase. The basic phase sets thus specified are expanded by the tangent formula to form possible phase sets, which are assessed initially for credibility by means of several figures of merit, and ultimately by their ability to produce interpretable E maps. If p variable phases are introduced in generating phase sets there will be 4^p possible sets of phases, assuming that each variable phase takes the values of $\pm\pi/4$, $\pm3\pi/4$. Half of the 4^p sets may be discounted by introducing, where applicable, an enantiomorph-specifying phase, but the problem is nevertheless much greater than for the centrosymmetric case, where there would be correspondingly only 2^p possible sign sets. In practice, a compromise must be struck between, on the one hand, too strict a probability limit, leading to the use of many symbols and a correspondingly large number of phases sets, and, on the other hand, probability limits that are too lax, leading to a small number of phase sets, but which include many inconsistencies and large phase errors, and perhaps no correct or nearly correct phase set at all.

We shall consider first symbolic addition for both centrosymmetric and noncentrosymmetric phase determination, and then look similarly at applications of the multisolution approach. Examples will include interesting results from our own and others' experiences.

3.2. Symbolic Addition: Centrosymmetric Case

For centrosymmetric structures a single phase relationship can be expressed as the tpsr (3.14), used in the form

$$s_\mathbf{h} \approx s_\mathbf{k} s_{\mathbf{h}-\mathbf{k}} \tag{3.17}$$

together with the associated probability given by (3.12). The symbolic addition procedure formulated by Karle and Karle (1963, 1966) may be summarized as follows:

(i) Obtain a Σ_2 listing of $|E| \geq |E_{\min}|$ in descending order of $|E|$, in parity groups if preferred.

(ii) Select from the list a valid set of $|E|$'s that specify the origin, according to the rules given in Chapter 2, for the space group in question. They should also be selected so as to give as many tpsr's with other $|E|$'s as possible in the Σ_2 listing.

(iii) Find new signs using tpsr's, each involving two known signs, accepting a new sign $s_\mathbf{h}$ only if $(1 - p) > P_+(\mathbf{h}) > p$, where p is a probability limit. The actual value for p may be chosen at one's discretion; 0.95 might be taken as a typical value.

(iv) When no further new signs can be generated under the above conditions, introduce a new reflection, also having a high $|E|$ value and as many Σ_2 interactions as possible, and allot it a symbolic sign $(a, b, \text{ or } c \cdots)$. It is desirable to keep the number of letter signs as small as possible, in order not to generate an unnecessarily large number of sign sets.

A single indication of sign for reflection $\mathbf{h}$ when the two other signs of a tpsr are known is given by (3.17), or if a consensus from several indications is to be formed then (3.11) may be used. The Σ_2 listing is invaluable in the application of this procedure.

3.2.1. Phase Determination in Space Group $P\bar{1}$: Pyridoxal Phosphate Oxime Dihydrate

Consider a structure in space group $P\bar{1}$; a sample of $|E|$ data is listed in Table 3.2. Symmetry considerations show that

$$|E_\mathbf{h}| = |E_{\bar{\mathbf{h}}}| \tag{3.18}$$

and

$$s_\mathbf{h} = s_{\bar{\mathbf{h}}} \tag{3.19}$$

TABLE 3.2. $|E|$ Data for a Crystal of Space Group $P\bar{1}$, Listed in Descending Order of Magnitude

hkl	$|E|$	hkl	$|E|$
$53\bar{2}$	3.5	311	3.1
$14\bar{1}$	3.4	$3\bar{3}2$	3.0
551	3.4	$26\bar{4}$	2.9
$51\bar{3}$	3.2	$5\bar{3}4$	2.7
$4\bar{1}\bar{1}$	3.1	162	2.6
023	3.1	$04\bar{1}$	2.6

From (3.18), it follows that a unique data set (hemisphere) contains the indices h, $\pm k$, $\pm l$, or an equivalent combination. We shall assume the starting set given in Table 3.3. A possible path for the application of symbolic addition to the data sample is shown by Table 3.4. Notice (i) the use of symmetry, and (ii) that a possible relationship

$$ab = c \qquad (3.20)$$

develops from the analysis. Relationships of this type may be strengthened in a full analysis by several such indications. The number of sign sets ultimately developed would be reduced by equations like (3.20), since the number of symbolic signs is reduced correspondingly.

TABLE 3.3. Assignments of Origin-Specifying Signs and Letter Symbols for Implementing the Σ_2 Relationship in Space Group $P\bar{1}$

$\mathbf{k}$ hkl	$s_\mathbf{k}$	$|E_\mathbf{k}|$	
023	$+$	3.1	Origin specification (see Chapter 2)
$4\bar{1}\bar{1}$	$+$	3.1	
$53\bar{2}$	$+$	3.5	
$26\bar{4}$	a	2.9	Symbolic signs: each letter may have only one of the possible values ± 1
311	b	3.1	
$5\bar{3}4$	c	2.7	

TABLE 3.4. Σ_2 Listing with Phase Assignments Developed from the Data in Tables 3.2 and 3.3

h	k	h − k	s_k	s_{h-k}	$s_h \approx s_k \cdot s_{h-k}$	$|E_h E_k E_{h-k}|$ [a]	Probable result
$14\bar{1}$	$\bar{4}11$ [b]	$53\bar{2}$	+ [b]	+	+	36.9	$s(14\bar{1}) = +$
551	023	$53\bar{2}$	+	+	+	36.9	$s(551) = +$
$51\bar{5}$	$53\bar{2}$	$0\bar{2}3$ [b]	+ [b]	+	+	34.7	$s(51\bar{5}) = +$
$\bar{3}32$	$\bar{2}\bar{6}4$ [b]	$53\bar{2}$	a [b]	+	a	30.5	$s(\bar{3}32) = a$
162	023	$14\bar{1}$	+	+	+	27.4 [c]	$s(162) = +$ [c]
	551	$\bar{4}11$ [b]	+	+ [b]	+	27.4 [c]	$s(162) = +$ [c]
$04\bar{1}$	311	$33\bar{2}$ [b]	b	a [b]	ab	24.2 [c]	$s(04\bar{1}) = ab$ [d]
	534	$51\bar{5}$	+	c [b]	c	22.5 [c]	$s(04\bar{1}) = c$ [d]
							$ab = c$ [d]

[a] Measure of probability according to (3.12).

[b] Uses the relation $s(hkl) = s(\bar{h}\bar{k}\bar{l})$.

[c] Weaker indications according to values in column 7 but reinforced by two separate indications.

[d] A probable relationship $ab = c$ emerges. Confidence in this relationship depends on the probability derived from (3.12).

To test the complete sign sets for plausibility, before calculating E maps, the following quantities may be determined:

$$M_1 = \sum s_h s_k s_{h-k} \tag{3.21}$$

$$M_2 = \sum |E_h E_k E_{h-k}| s_h s_k s_{h-k} \tag{3.22}$$

$$M_3 = \sum P_{hk} s_h s_k s_{h-k} \tag{3.23}$$

where P_{hk} is given by (3.12), written for a single tpsr. M_1, M_2, and M_3 would all be large for a set wherein consistent sign relationships had prevailed. The numerical values of these figures of merit indicate the order of use of sign sets in calculating E maps, a useful constraint, particularly in cases where many sign sets have emerged from the analysis.

A successful sign determination of this nature has been carried out for pyridoxal phosphate oxime dihydrate, $C_8H_{11}N_2O_6P \cdot 2H_2O$, space group $P\bar{1}$ (Barrett and Palmer, 1969). Among $165 |E|$ values greater than 1.5, it was found necessary to allot five symbols a–e in order to determine the signs. When the process was completed, several significant sign relationships were obtained:

$$ac = e$$
$$c = eb$$
$$b = ed$$
$$ad = e$$
$$ab = cd$$

Simple manipulation reduced this list to $a = b$, $c = d$, and $e = ac$, so that there were only two unknown symbols, a and c. The result $a = c = +1$ was rejected, and $a = c = -1$ was tried first on the grounds that it led to approximately equal numbers of $+$ and $-$ signs, a not unusual feature where there are no atoms in special positions. The E map based on this analysis revealed the nonhydrogen atoms of the molecule, including the water of crystallization, with only three small, spurious peaks.

3.3. Symbolic Addition: Noncentrosymmetric Case

In noncentrosymmetric crystals, the phase problem becomes one of finding general values between 0 and 2π for the phases ϕ_h for a sufficient number of reflections. The next example illustrates this somewhat different problem, usually far more complicated than that of the centrosymmetric case.

3.3.1. Phase Determination in Space Group $P2_1$: Tubercidin

The structure of tubercidin was determined by Stroud (1968, 1973), and we are indebted to him for permission to quote from his analysis and results. Table 3.5 lists the crystal data for this compound.

In space group $P2_1$, $|E(hkl)|$ has the following symmetry equivalents:

$$|E(hkl)| = |E(\bar{h}k\bar{l})| = |E(h\bar{k}l)| = |E(\bar{h}\bar{k}\bar{l})| \qquad (3.24)$$

The phases of the symmetry-related reflections in this space group are also linked, but in a different way, according to the parity of k:

$$k = 2n: \qquad \phi(hkl) = \phi(\bar{h}k\bar{l}) = -\phi(h\bar{k}l) = -\phi(\bar{h}\bar{k}\bar{l}) \qquad (3.25)$$

$$k = 2n+1: \qquad \phi(hkl) = \pi + \phi(\bar{h}k\bar{l}) = \pi - \phi(h\bar{k}l) = -\phi(\bar{h}\bar{k}\bar{l}) \qquad (3.26)$$

Although ϕ_h can, in general, have a value anywhere in the range 0–2π, the $h0l$ reflections are restricted to the values 0 or π in this space group; in other words the $h0l$ zone is centric.

The origin was specified by assigning phases to three reflections, according to the known rules (Chapter 2), as shown by Table 3.6.

Next, new phases were determined according to (3.1) or (3.2). In order to maintain an expected variance $V(\phi_h)$ (Fig. 3.2) of no more than 0.5 rad^2, the product $|E_h||E_k||E_{h-k}|$ must be greater than 8.5 for this structure. Two new phases $\phi(80\bar{2})$ and $\phi(612)$ were thus determined from

TABLE 3.5. Crystal Data for Tubercidin

Formula	$C_{11}H_{14}N_4O_4$
M_r	266.3
Space group	$P2_1$
a	9.724(9) Å
b	9.346(11)
c	6.762(10)
β	94.64(10)°
V_c	610.4 Å^3
D_m	1.449 g cm^{-3}
D_x	1.443
Z	2
$F(000)$	280

TABLE 3.6. Origin-Specifying Phases for Tubercidin

hkl	$|E_h|$	ϕ_h/deg
$10\bar{6}$	1.95	0
$40\bar{1}$	2.09	0
$71\bar{4}$	2.45	0

the origin set (Table 3.8), further phases being determined in terms of symbols (Table 3.7). Eleven phases were generated in terms of the origin phases and symbol a, 20 after adding letter b, and 47 after adding the third symbolic phase c. Table 3.8 illustrates the initial stages of this process. The criteria for accepting a phase were as follows:

(i)   that $V(\phi_h)$, irrespective of the actual choice for c and a (b is a structure invariant with phase 0 or π), should be less than 0.5 rad^2, no matter how many contributors there were to the sum in (3.2);

(ii) that where there were two or more different indications for a phase, the phase would be accepted only when indications of one type predominated strongly.

During the phase analysis it soon became clear that in order to avoid a large number of inconsistencies, the structure invariant b (ϕ_{206}) should have a value of zero. This result is shown by Table 3.9, which illustrates also an example of phase determination for the case of $|E(63\bar{3})|$. The indications

$$\phi(63\bar{3}) = c - 2a - b \qquad (3.27)$$

and

$$b = 0 \qquad (3.28)$$

TABLE 3.7. Course of the Phase Determination Procedure for Tubercidin

hkl	ϕ_h	$|E_h|$	Number of numerical or symbolic phases
Origin set	See Tables 3.6 and 3.8		5
$13\bar{8}$	a	2.99	11
206	b	2.20	20
790	c	2.76	47

TABLE 3.8. Initial Development of Phases for Tubercidin

h [a]	$\mid E_h \mid$	ϕ_h	$\mid E_h \mid \mid E_k \mid \mid E_{h-k} \mid$
Origin set			
* $40\bar{1}$	2.09	0	
* $10\bar{6}$	1.95	0	
* $71\bar{4}$	2.45	0	
New phases			
$40\bar{1}$		0	
$40\bar{1}$		0	
* $80\bar{2}$	2.33		10.2
$\bar{1}06$ [b]		0 [b]	
$71\bar{4}$		0	
* 612	1.83	0	8.7
First letter			
$13\bar{8}$	2.99	a	
612		0	
* $74\bar{6}$	2.20	a	12.0
$7\bar{1}4$		π [c]	
$\bar{1}38$		$\pi + a$ [c]	
* 624	2.19	a	16.0

[a] An asterisk denotes a new phase in the list.
[b] Symmetry-related phase used from (3.25).
[c] Symmetry-related phase used from (3.26).

are strong, because they both come from multiple indications (6 and 3, respectively). By reiteration of the phase addition procedure described above, the results in Table 3.10 indicate relationships between a and c.

Bearing in mind that the objective is a self-consistent set of phases, it is well to consider how this might now be achieved. Refinement of phases could in principle be achieved by application of (3.5). However, this would be possible only if numerical values for a and c (taking b as zero) were available. Alternatively, if a working formula relating a and c could be found, (3.5) could be implemented by substitution of values for one symbol

TABLE 3.9. Phase Indications for $|E(6\bar{3}3)|$ for Tubercidin: $\mathbf{h} = 6\bar{3}3$, $|E_\mathbf{h}| = 2.37$ [a]

$\mathbf{k}$	$|E_\mathbf{k}|$	$\phi_\mathbf{k}$	$\mathbf{h} - \mathbf{k}$	$|E_{\mathbf{h}-\mathbf{k}}|$	$\phi_{\mathbf{h}-\mathbf{k}}$	$|E_\mathbf{h}||E_\mathbf{k}||E_{\mathbf{h}-\mathbf{k}}|$	$\phi_\mathbf{h} \approx \phi_\mathbf{k} + \phi_{\mathbf{h}-\mathbf{k}}$
790	2.76	c	$\bar{1}\bar{6}\bar{3}$	2.08	$-2a - b$	13.6	$c - 2a - b$
$\bar{2}0\bar{6}$	2.20	b	833	2.09	$c - 2a - b$	10.9	$c - 2a$
624	2.19	a	$01\bar{7}$	1.59			*
$\bar{2}\bar{2}\bar{5}$	2.13	a	812	1.49			*
840	1.67	$c - 2a$	$\bar{2}\bar{1}\bar{3}$	1.69	0	6.7	$c - 2a$
$\bar{4}22$	1.60	$a + b$	$101\bar{5}$	1.82			*
$7\bar{1}\bar{4}$	2.45	π	$\bar{1}41$	1.78	$\pi + c - 2a - b$	10.3	$c - 2a - b$
$7\bar{4}\bar{6}$	2.20	$-a$	$\bar{1}73$	1.96	$c - a - b$	10.2	$c - 2a - b$
$0\bar{2}\bar{1}$	2.12	$-a - b$	$65\bar{2}$	2.06	$c - a$	10.4	$c - 2a - b$
$6\bar{1}2$	1.83	π	$04\bar{5}$	1.61	$\pi + c - 2a - b$	7.0	$c - 2a - b$
$\bar{1}\bar{4}0$	1.64	$a - b$	$57\bar{3}$	2.49	$c - a$	9.7	$c - 2a - b$
$4\bar{2}\bar{2}$	1.60	$a - b$	$25\bar{1}$	1.66	$c - a - b$	6.3	$c - 2a$
$5\bar{1}3$	1.59		140	1.64	$a + b$		*
$5\bar{1}\bar{1}$	1.53		$14\bar{2}$	1.66	$c - a$		*

[a] There are six indications that $\phi(6\bar{3}3) = c - 2a - b$ and three indications that $\phi(6\bar{3}3) = c - 2a$. Accept $\phi(6\bar{3}3) = c - 2a - b$. Letter indication (3): $b = 0$. Five interactions marked with an asterisk do not contribute.

TABLE 3.10. Relationships between Letter Symbols

Form of relationship	Number of indications
$c = \pi + 2a$	7
$c = \pi + 3a$	15
$c = \pi + 4a$	19
$c = 3a$	5
$c = 4a$	2
$c = -3a$	4 or 5
$a = 0$	2
$a = \pi$	2
$b = 0$	Many
$b = \pi$	None

only. Table 3.10 shows that there were 41 indications that

$$c = \pi + pa \qquad (3.29)$$

where a numerical value for p has to be found. Stroud (1968) used the value $p = 3.29$ as a weighted average. Hence,

$$c = \pi + 3.29a \qquad (3.30)$$

The symbol a was then limited to the range

$$0 < a < \pi \qquad (3.31)$$

in order to fix the enantiomorph (Chapter 2). Values for a were chosen such that

$$a = n\pi/8 \qquad (n = 1, 2, \ldots, 8) \qquad (3.32)$$

and converted into phases by (3.30); each set was expanded and refined by (3.5) (taking $w_h = 1$) for up to 419 reflections with $|E_{min}| \geq 1.0$. Some phases were rejected because of inconsistencies in their phase indications. An interpretable E map was obtained using the refined phase set with $a = 6\pi/8$; a composite diagram is given in Fig. 3.4.

In conclusion, we given in Table 3.11 a list of phase indications obtained for 103 reflections with $|E_h| > 1.59$, using (3.1) and (3.2), and $b = 0$. The list contains about two or three times the number of terms recommended by the Karles' for an initial phase analysis; 72 of the 103 phases have symbolic or numerical phase assignments. The reader is invited to con-

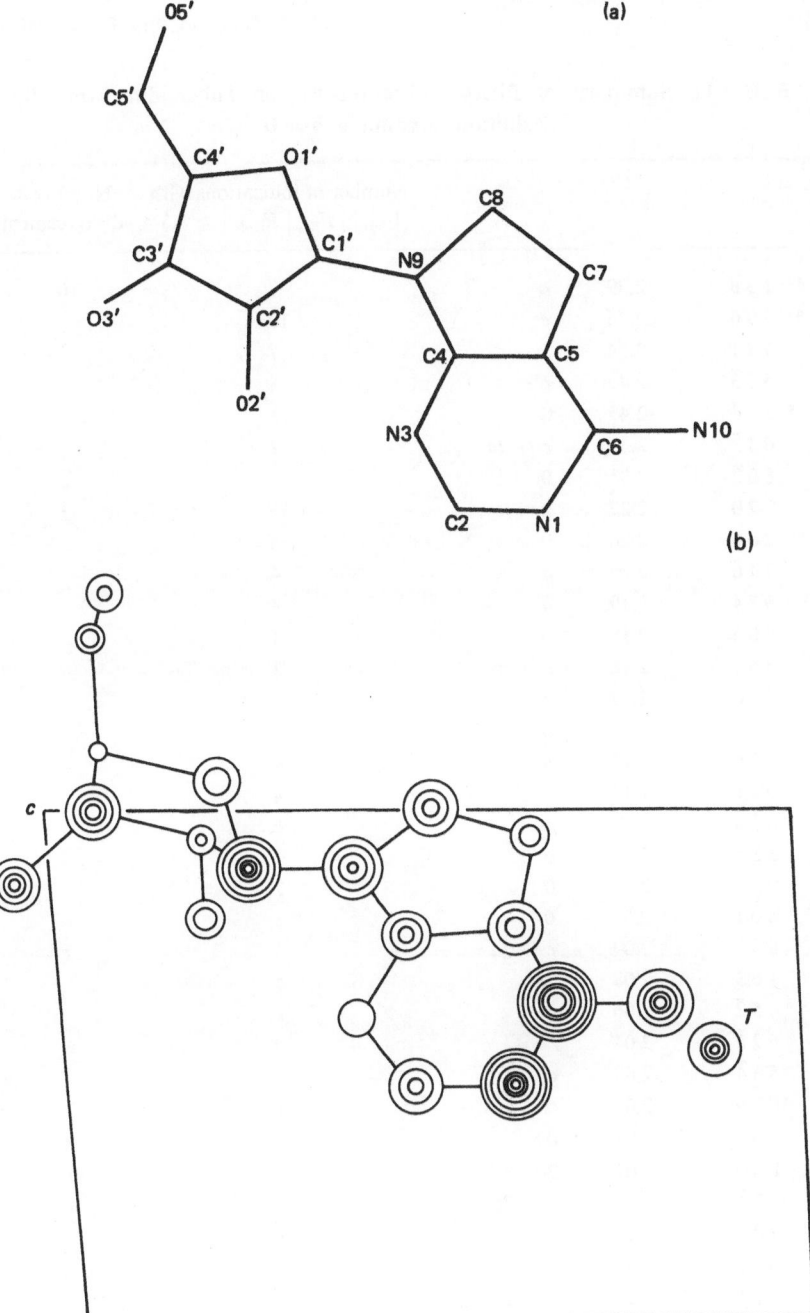

FIGURE 3.4. Tubercidin, $C_{11}H_{14}N_4O_4$: (a) structural formula in approximately the same orientation as in the E map, (b) composite E map; contours (idealized) are drawn at arbitrary equal intervals. Some peaks are heavier than others because of the limited data set used; peak T was the only significant spurious peak.

TABLE 3.11. Summary of Phase Indications [a] for Tubercidin from Phase Addition, Assuming $b = 0$

h [b]	$\lvert E_h \rvert$	ϕ_h	Number of indications with $\lvert E_h \rvert\,\lvert E_k \rvert\,\lvert E_{h-k} \rvert \geq 8.5$	Number in disagreement
** $1\,3\,\bar{8}$	2.99	a	7	0
** $7\,9\,0$	2.76	c	10	
$7\,7\,1$	2.54	$c - a$	7	
$5\,7\,\bar{3}$	2.49	$c - a$	5	
* $7\,1\,\bar{4}$	2.45	0	13	
$6\,3\,\bar{3}$	2.37	$c - 2a$	7	
$8\,0\,\bar{2}$	2.33	0	8	
$7\,7\,0$	2.22	$\pi + c - a$	7	1
** $2\,0\,6$	2.20	0	6	
$7\,4\,\bar{6}$	2.20	a	4	
$6\,2\,4$	2.19	a	8	
$8\,0\,\bar{3}$	2.18	π	4	
$8\,6\,1$	2.18	$c - a$	3	
$7\,0\,0$	2.17			
$7\,5\,5$	2.16	$\pi + 2a$	2	
$2\,2\,5$	2.13	a	6	
$0\,2\,1$	2.12	a	4	
$8\,2\,\bar{3}$	2.12	a	6	
$3\,3\,4$	2.12	a	1	
$9\,2\,\bar{1}$	2.11	0	2	
* $4\,0\,\bar{1}$	2.09	0	3	
$8\,3\,3$	2.09	$c - 2a$	4	1
$1\,6\,3$	2.08	$2a$	5	
$6\,5\,\bar{2}$	2.06	$c - a$	2	
$6\,2\,\bar{5}$	2.04		0	6
$5\,3\,\bar{4}$	2.03	a	1	
$10\,2\,0$	2.02	$\pi + a$	1	
$5\,0\,0$	2.01	0	1	2
$8\,6\,0$	2.01	$2a$		1
$1\,3\,\bar{4}$	1.98	$c - 2a$	2	1
$7\,0\,3$	1.97	π	1	
$6\,2\,\bar{7}$	1.97	a	2	
$0\,7\,3$	1.97	$\pi + c - a$	2	
$8\,6\,\bar{4}$	1.96	$\pi + 2a$	1	
$1\,7\,\bar{3}$	1.96	$\pi + c - a$	5	1
* $1\,0\,6$	1.95	0	1	
$8\,5\,\bar{5}$	1.95	$\pi + c - 2a$	1	

TABLE 3.11 (*continued*)

h [b]	$\lvert E_h \rvert$	ϕ_h	Number of indications with $\lvert E_h \rvert \lvert E_k \rvert \lvert E_{h-k} \rvert \geq 8.5$	Number in disagreement
0 2 6	1.94			
6 0 $\bar4$	1.88			
7 0 $\bar3$	1.87	0	0	2
6 4 $\bar4$	1.86			
8 5 $\bar1$	1.85	$\pi + 2a$	0	2
6 4 $\bar5$	1.84	a	2	1
9 5 $\bar2$	1.84	$\pi + 2a$	0	
9 6 $\bar3$	1.84			
6 2 $\bar6$	1.83	$\pi + a$	0	1
6 1 2	1.83	0	2	1
5 7 3	1.83			
2 0 $\bar7$	1.82	0	1	
10 1 $\bar5$	1.82			
5 3 $\bar5$	1.81			
12 0 $\bar2$	1.81	π	1	
1 5 $\bar4$	1.81	$\pi + c - 2a$	3	1
1 5 5	1.81	$c - a$	2	1
8 3 $\bar2$	1.80	$\pi + a$	0	2
6 0 7	1.79			
1 6 6	1.79	$\pi + c - a$	0	1
1 4 $\bar1$	1.78	$\pi + c - 2a$	3	1
6 6 0	1.78	$\pi + 2a$	0	1
2 4 2	1.77			
9 4 $\bar5$	1.77	$\pi + c - a$	0	1
7 0 4	1.76	0	1	0
7 2 2	1.75			
4 2 $\bar7$	1.75			
1 8 0	1.75			
7 2 5	1.74			
7 2 $\bar4$	1.71			
3 3 $\bar6$	1.71	$\pi + a$	0	
1 4 7	1.71	$\pi + c - 2a$	1	
5 5 4	1.71	$c - a$	0	1
11 0 $\bar1$	1.70			
1 2 $\bar6$	1.70	π or $\pi + a$	0	
2 1 3	1.69	0	1	1
0 4 7	1.69			

continued overleaf

TABLE 3.11 (*continued*)

h [b]	$\lvert E_{\mathbf{h}} \rvert$	$\phi_{\mathbf{h}}$	Number of indications with $\lvert E_{\mathbf{h}}\rvert\lvert E_{\mathbf{k}}\rvert\lvert E_{\mathbf{h-k}}\rvert \geq 8.5$	Number in disagreement
$5\,3\,\bar{2}$	1.68			
$6\,0\,5$	1.67	0	0	
$9\,0\,\bar{1}$	1.67	π	0	
$8\,4\,0$	1.67			
$8\,3\,1$	1.66			
$1\,4\,\bar{2}$	1.66	$c - 2a$	0	2
$2\,4\,\bar{3}$	1.66			
$2\,5\,\bar{1}$	1.66	$c - a$	0	
$7\,3\,2$	1.65	$\pi + c - 2a$	0	1
$1\,5\,\bar{3}$	1.65	$c - 3a$ $\bar{2}$	1	
$1\,3\,1$	1.64	$\pi + a$	3	
$1\,4\,0$	1.64	a	1	
$9\,3\,\bar{5}$	1.63			
$2\,4\,\bar{7}$	1.63			
$7\,8\,0$	1.63	$c - a$	0	1
$12\,0\,\bar{3}$	1.62	0	1	
$3\,2\,\bar{8}$	1.62			
$0\,6\,4$	1.62	$\pi + c - a$	2	
$3\,8\,0$	1.62			
$0\,4\,5$	1.61	$\pi + c - 2a$	2	
$4\,2\,\bar{2}$	1.60	a	0	2
$4\,2\,6$	1.60			
$6\,3\,3$	1.60			
$5\,7\,\bar{1}$	1.60			
$0\,1\,7$	1.59			
$5\,1\,\bar{3}$	1.59			
$9\,4\,4$	1.59	$c - 2a$	0	
$6\,5\,4$	1.59			
$4\,6\,\bar{3}$	1.59			

[a] Seventy-two of the 103 phases were assigned symbolic or numerical values.
[b] One asterisk indicates a symbolic phase assignment; two asterisks indicate the origin-determining phase.

struct a Σ_2 listing and work through this analysis. It is an excellent test of one's expertise in handling the basic formulas used in direct methods, including the proper use of symmetry relationships for the space group in question.

3.4. Advantages and Disadvantages of Symbolic Addition

It should be clear now that symbolic addition has several advantages and disadvantages. They may be summarized as follows:

(a) *Advantages*

 (i) The user is in control throughout the analysis. He has the responsibility of making sure that all formulas, including symmetry relationships, are correctly applied.

 (ii) The user can make decisions regarding criteria of acceptance of phase indications, the number of $|E|$ values to include, the number of symbolic phases, the choice of starting set, and so on.

(b) *Disadvantages*

 (i) The analysis can be carried out only by a specialist in crystallography.

 (ii) The procedure is slow to carry out and may require many hours of intense preparation before meaningful results emerge.

 (iii) If a large number of symbols is required, many phase sets will be produced, each of which requires refinement by the tangent formula.

Not surprisingly, alternative rapid and more automatic methods of applying direct methods formulas were sought in the late 1960s, leading to development of the multisolution methods (Woolfson and Germain, 1968; Germain, Main, and Woolfson, 1970a,b).

3.5. Multisolution Methods

3.5.1. Introduction: Multisolution Philosophy and Brief Description of the Program MULTAN

In symbolic addition, we saw that a new phase may be indicated several times by the same combination of symbols. Then, the individual indications reinforce one another to produce an improved joint probability, since α_h (3.6) is then given by

$$\alpha_h = \sum_{k_r} \kappa_{h,k} \tag{3.33}$$

taking $\kappa_{h,k}$ from (3.7).

The combination of indications involving entirely different symbols presents a problem. For instance, suppose two separate indications for ϕ_h are $a + b$ and $c + d$, where $a = \pi/4$, $b = 3\pi/4$, $c = -\pi/4$, and $d = -3\pi/4$, say. The individual indications $a + b$ and $c + d$ both predict $\phi_h = \pi$. Symbolic combination would yield an indication $\frac{1}{2}(a + b + c + d)$, which results in the false value of $\phi_h = 0$. Even if values of $c = -\pi/4 + 2\pi = 7\pi/4$, $d = -3\pi/4 + 2\pi = 5\pi/4$ are used the combination indication would again be false, since $\frac{1}{2}(a + b + c + d) \equiv 0$ modulo 2π. In such a situation, the usual practice with symbolic addition is to accept the strongest indication and put other indications aside for possible future use.

The multisolution philosophy gets around this problem by introducing numerical phases rather than symbols at an early stage. It is then always possible to combine individual phase indications by means of the tangent formula. This strategy was implemented in the program MULTAN (Main *et al.*, 1974), first described by Woolfson and Germain (1968). This phase-determining procedure is based on a starting set of phases, formed in a similar way to that used in symbolic addition, that may be categorized as follows:

(a)   origin and enantiomorph definition, requiring up to four phase assignments according to rules given in Chapter 2;

(b)   phases derived by a Σ_1 formula such as (3.15) or (3.16), or by any other valid method;

(c)   further phases required to initiate a continuous phase-determining process by tangent formula expansion; these phases are variables.

It is in category (c) where MULTAN differs fundamentally from the symbolic addition technique. Instead of introducing letter phases for these reflections, they are given numerical values. Specifically, according to space group symmetry, values such as 0, π, or $\pm\pi/2$ may be assigned. General noncentrosymmetric phases are assigned the values $\pm\pi/4$ and $\pm3\pi/4$, or for enantiomorph specification merely $\pm\pi/4$. In this way, a total of p variable phases, including enantiomorph specification, would therefore yield $2 \times 4^{p-1}$ possible phase sets for a noncentrosymmetric crystal. The method is justified numerically in that it gives a maximum error of 45° for any of the initial variable phases with a mean error of only 22.5°. The tangent formula is used to determine probable values of new phases, and the number of possible phase sets rises rapidly with p, as Table 3.12 shows.

The MULTAN program (Germain, Main, and Woolfson, 1971a,b) employs a modified tangent formula given by

$$\tan \phi_h = \frac{\sum_{k_r} Q_{h,k} \sin(\phi_k + \phi_{h-k})}{\sum_{k_r} Q_{h,k} \cos(\phi_k + \phi_{h-k})} = \frac{T_h}{B_h} \tag{3.34}$$

TABLE 3.12. Number of Phase Sets Generated by the Multisolution Method

Number of variable phases p	Number of phase sets generated	
	for a noncentrosymmetric crystal $2 \times 4^{p-1}$	for a centrosymmetric crystal 2^p
1	2	2
2	8	4
3	32	8
4	128	32
5	512	64

where

$$Q_{h,k} = w_k w_{h-k} \, | \, E_k \, | \, | \, E_{h-k} \, | / (1 - | \, U_h \, |^2) \qquad (3.35)$$

with

$$w_h = \tan[\sigma_3 \sigma_2^{-3/2} \, | \, E_h \, | \, (T_h^2 + B_h^2)^{1/2}] \qquad (3.36)$$

and

$$| \, U_h \, | = | \, F_h \, | \Big/ \sum_{j=1}^{N} f_j \qquad \text{(the unitary structure factor)} \qquad (3.37)$$

Thus, each phase assignment carries a weight so designed that poorly determined phases have little effect in the generation of new phases, while the fact that all phases are included leads to efficient propagation of phase information throughout the data set.

The choice of starting set reflections for MULTAN is made automatically by a subroutine called CONVERGE. This program attempts to apply principles given in Chapter 2 of this book. In the case of an unsatisfactory choice by CONVERGE, facilities are available for the user to make his own origin selection. CONVERGE forms an ordered list of reflections such that, from the starting set, each reflection may be determined in terms of all those preceding it. In the initial stages of phase determination, the first 60 phases from the convergence map are used and only those phase relationships determined with α_h (3.6) greater than 5 are accepted. Several passes are made through the tangent formula, lowering this limit each time. As phases are developed the α_h's increase, self-consistency being defined as a change of less than 2% in $\sum_h \alpha_h$ from one cycle to the next. Up to this stage the phases of the starting reflections have been kept constant. They are now allowed to vary and refine to produce a trial phase set.

A recent improvement in the weighting scheme has been introduced by Hull and Irwin (1978). It resulted from the observation that many wrong solutions were characterized by final values of α that were larger than those expected theoretically. The reason for this effect lies in the fact that even a random set of phases will produce a value of α (α_R, say) greater than 5, provided that there are at least 30–40 contributors:

$$\alpha_R = \left(\sum_r \kappa_r^2\right)^{1/2} \tag{3.38}$$

where κ is given by (3.7).

An $\alpha_R > 5$ is almost certain to produce unit weights which, once established, cannot be revised by the former weighting scheme. Hull and Irwin define the weight as

$$w_h = \min(0.2\alpha, 1.0) \tag{3.39}$$

and

$$w_h' = \psi \exp(-x^2) \int_0^x \exp(t^2)\, dt \tag{3.40}$$

where

$$x = \alpha/\alpha_E \tag{3.41}$$

α_E is given by (3.9) and ψ is chosen so that $w_h' = 1$ when $x = 1$. The graph of w' as a function of x is shown in Fig. 3.5. Evidently, there is a reduced weight when $\alpha > \alpha_E$ or $\alpha < \alpha_E$, thus increasing the probability of producing a reliable phase set. In practice, the minimum of w_h and w_h' is used as a weight in the tangent formula.

Phase sets are assessed, prior to calculation of E maps, by three figures of merit, ABS FOM, PSI ZERO, and RESID, computed for each phase set. In addition a fourth combined figure of merit COMBINED FOM is now produced in order to provide an overall picture of the three individual quantities.

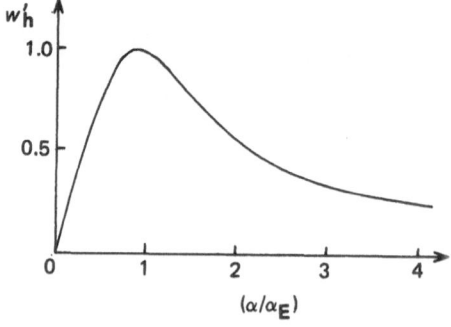

FIGURE 3.5. Hull–Irwin weighting scheme.

ABS FOM is a measure of the internal consistency amongst the Σ_2 relationships and is given by Z where

$$Z = \sum_{h} (\alpha_h - \alpha_{R_h}) \Big/ \sum_{h} (\alpha_{E_h} - \alpha_{R_h}) \qquad (3.42)$$

α_{R_h} is the value expected from random phases given by (3.38) and α_{E_h} is the estimated value of α_h calculated from (3.9) during the convergence procedure. Thus, ABS FOM is zero for random phases and unity if α_h is equal to its expectation value. For crystal structures containing translational symmetry elements, the correct set of phases should correspond to one of the higher values of ABS FOM, because the tangent formula tends to maximize phase relation consistencies. In practice, a correct set of phases has been found usually to correspond to values of ABS FOM in the range of 1.0–1.4, but phase sets with values as low as 0.7 have yielded interpretable E maps; there are also many instances of values of ABS FOM much larger than 1.5 leading to correct structures.

PSI ZERO is defined by Cochran and Douglas (1957) as

$$\psi_0 = \sum_{h} \sum_{k} | E_k | | E_{h-k} | \qquad (3.43)$$

where the $| E |$ values in this summation are either very small or zero. For small $| E_h |$, ψ_0 should have a small value for the correct phase set. It is independent of the tangent formula and therefore may be useful as a discriminator when ABS FOM yields similar values for different phase sets.[†] RESID corresponds to R_K (Karle and Karle, 1966), and is calculated in a similar way to the familiar crystallographic R factor:

$$\text{RESID} = R_K = \frac{\sum_h | | E_h | - | E_h |_{calc} |}{\sum_h | E_h |} \qquad (3.44)$$

where

$$| E_h |_{calc} = K \langle | E_k | | E_{h-k} | \rangle_h \qquad (3.45)$$

and K is a scale factor given by

$$K = \sum_{h} | E_h |^2 \Big/ \sum_{h} \langle | E_k | | E_{h-k} \rangle^2 \qquad (3.46)$$

The correct set of phases should correspond to that with the lowest RESID.

Experience has shown that discrimination between phase sets is often possible in terms of either Z or R_K. However, sometimes both may fail to

[†] Often with space groups containing no translational symmetry elements.

enable the correct phase set to be selected easily. A further useful indicator is the combined figure of merit C, given by

$$C = W_1 \frac{(Z - Z_{\min})}{(Z_{\max} - Z_{\min})} + W_2 \frac{[(\psi_0)_{\max} - \psi_0]}{[(\psi_0)_{\max} - (\psi_0)_{\min}]}$$
$$+ W_3 \frac{[(R_K)_{\max} - R_K]}{[(R_K)_{\max} - (R_K)_{\min}]} \tag{3.47}$$

where W_1, W_2, and W_3 are weights, often unity, which may be changed to give more emphasis to ψ_0 and less to Z for space groups without translational symmetry elements.

A more complete description of MULTAN is not possible here. Recent features include the automatic production of E maps and their interpretation in terms of molecular geometry, the use of known atomic positions, and subtraction of contributions from heavy atoms in special positions. The success of the program in numerous structure determinations speaks for itself. Those structures which have failed to yield provide the incentive for further developments, perhaps advancing to the solution of small protein structures in the foreseeable future.

3.5.2. Centrosymmetric Case: Papaverine Hydrochloride

The structure of papaverine hydrochloride (Fig. 3.6) was determined by MULTAN (Reynolds, Palmer and Gorinsky, 1974); crystal data are given in Table 3.13: 197 reflections with $|E| > 1.8$ were used, and the starting set (Table 3.14) was selected automatically.

Four phase sets were obtained from the permutations of the phases for reflections 513,1 and 254. One of them had the figure of merit, $Z = 1.14$ and $R_K = 0.14$; ψ_0 was not calculated. An E map computed with this set revealed the positions of all the nonhydrogen atoms in the molecule. The initial structure factor calculation based on these positions had a conventional R factor of 0.34. Subsequent least-squares refinement reduced it to 0.052; Fig. 3.7 shows a stereo view of the molecule.

3.5.3. Noncentrosymmetric Case: Methyl Warifteine and Dimethyl Warifteine

We describe now the analyses of two chemically similar molecules. Crystallographically the two problems appear similar but, in fact, the two structure determinations proved very different; possible reasons for these

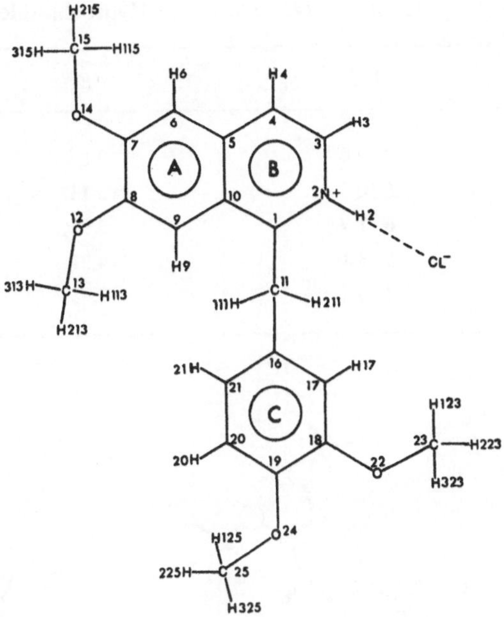

FIGURE 3.6. Structural formula of papaverine hydrochloride, showing the atomic numbering system.

differences will be discussed. Methyl warifteine (Borkakoti and Palmer, 1978a) provided a fairly straightforward application of MULTAN, whereas dimethyl warifteine (Borkakoti and Palmer, 1978b) required considerable manipulation before yielding the structure.

TABLE 3.13. Crystal Data for Papaverine Hydrochloride

Formula	$C_{20}H_{22}NO_4{}^+Cl^-$
M_r	375.85
Space group	$P2_1/c$
a	13.059(3) Å
b	15.620(3)
c	9.130(2)
β	92.14(1)°
V_c	1861 Å^3
D_m	1.33 g cm^{-3}
D_x	1.34
Z	4
$F(000)$	792
$\mu(C\mu\ K\alpha_1)$	15 cm^{-1}

TABLE 3.14. Papaverine Hydrochloride Starting-Set Reflections

hkl	$|E|$	ϕ/deg
3 12 4	3.86	0 ⎫
5 10 3	3.13	0 ⎬ origin
6 1 7	2.92	0 ⎭
5 13 1	4.21	0, 180
2 5 4	2.95	0, 180

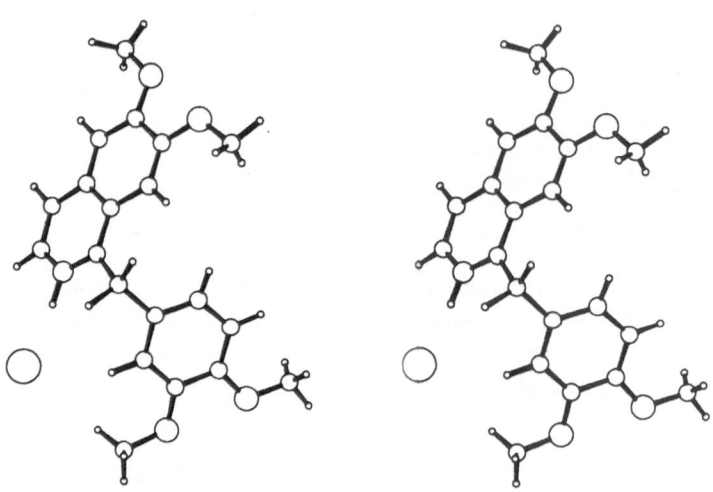

FIGURE 3.7. Stereoscopic diagram of the molecule of papaverine hydrochloride viewed along c; the circles, in increasing order of size, represent H, C, N, O, and Cl.

Methyl Warifteine (MEW)

Crystal data for methyl warifteine (Fig. 3.8, II) are listed in Table 3.15. The starting set reflections, generated automatically by CONVERGE, are given in Table 3.16. Using 228 data with $|E| > 1.70$, the number of Σ_2 interactions was limited (a facility available to the user) to 2000, and 16 phase sets were generated. The phase set having the second highest value of Z (0.96) and lowest R_K (0.23) (the combined figure of merit C was not calculated in the version of MULTAN used for this analysis) produced an E map from which 36 of the 45 nonhydrogen atoms were identified in

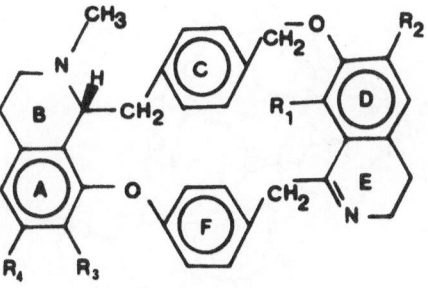

FIGURE 3.8. Structural formulas of the compounds (I) warifteine, $R_1=R_3=OH$; $R_2=R_4=OCH_3$; (II) methylwarifteine, $R_1=OH$; $R_2=R_3=R_4=OCH$; (III) dimethylwarifteine, $R_1=R_2=R_3=R_4=OCH_3$.

TABLE 3.15. Crystal Data for MEW

Formula	$C_{37}C_{38}N_2O_6$	V_c	3086 Å³
M_r	606.36	D_x	1.31 g cm⁻³
Space group	$P2_12_12_1$	Z	4
a	17.539(4) Å	$F(000)$	1288
b	12.224(3)	$\mu(C\mu\ K\alpha)$	5.6 cm⁻¹
c	14.393(3)	Crystal size	0.2, 0.3, 0.3 mm

TABLE 3.16. Starting Set Reflections for MEW

hkl	$|E|$	ϕ/deg
0 1 3	2.10	90 ⎫
1 0 3	2.30	90 ⎬ origin
14 3 0	4.73	360 ⎭
0 11 1	4.40	90, 270
1 1 7	2.36	45, 135, 225, 315
9 7 2	3.07	45, 315 (enantiomorph)

geometrically acceptable positions. These 36 sites were drawn from the highest 60 peaks in the map. The remaining 9 atoms of MEW were located by an electron density synthesis using all reflections, the initial R factor being 0.35. After least-squares refinement the final R factor was 0.057 for 2508 observed reflections. A stereoview of the molecule is shown in Fig. 3.9.

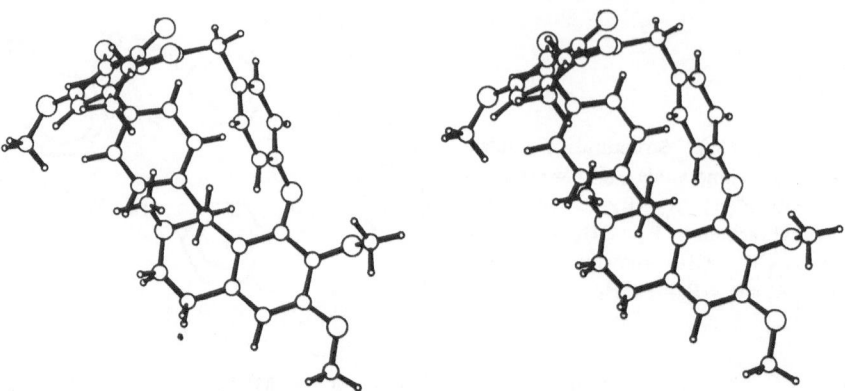

FIGURE 3.9. Stereoscopic view of MEW looking along c; the circles, in increasing order of size, represent H, C, N, and O.

Dimethyl Warifteine (DMW)

The crystal data for DMW are listed in Table 3.17. The solution of the phase problem proved unusually difficult, but was eventually achieved partly by the use of two special features available to users of MULTAN, namely, (i) the use of molecular scattering factors in the calculation of $|E|$ values and (ii) the use of Karle's (1968) recycling procedure. The final success of the analysis was also partly due to dogged determination, "green-fingered" crystallography, and, admittedly, a little bit of luck! (Borkakoti, 1978).

Initially several starting sets were tried, using different values for $|E_{min}|$. After many attempts the starting set A (Table 3.18) was expanded using the 145 $|E|$ values with $|E| > 1.87$ and 579 Σ_2 relationships to give 16 phase sets. Several phase sets with high Z (1.1–1.0) and low R_K (0.22–0.29) were used in the calculation of E maps, but no semblance of structure appeared in any map. Finally an E map was calculated [called E map (A)] with the phase set which was number 15 in the list of 16 arranged in descending order of the combined figure of merit C ($Z = 0.95$ and $R_K = 0.32$). For the first time an E map showed several peaks in a sensible chemical configuration, six atoms in a closed ring. Phases corresponding to selected $|F|$ data for the six-atom fragment were used in the Karle tangent formula recycling procedure described below (subroutine SFCALC), but it failed to produce any additional sensible structural features.

One variation in the method of analysis which had been tried earlier without success was the use of molecular scattering factors to produce a

TABLE 3.17. Crystal Data for DMW

Formula	$C_{38}H_{40}N_2O_6$
M_r	620.75
Space group	$P2_12_12_1$
a	14.714(4) Å
b	14.827(4)
c	15.365(4)
V_c	3351 Å^3
D_x	1.23 g cm^{-3}
Z	4
$F(000)$	1320
μ(Cu $K\alpha$)	5.5 cm^{-1}
Crystal size	0.2, 0.2, 0.1 mm

TABLE 3.18. Starting Set (A), Using $|E|$ Values from Wilson Plot for DMW

Reflection hkl	$\|E\|$	Phase/deg
0 3 1	2.62	90 ⎫
3 6 0	2.35	90 ⎬ origin
2 11 0	2.19	360 ⎭
2 8 13	2.88	45, 135, 225, 315
2 3 14	2.45	45, 135, 225, 315
3 0 2	3.44	0 (enantiomorph)

"Debye curve" for calculation of $|E|$ values, rather than the more usual Wilson plot (Wilson, 1942). This required the specification of molecular geometry, and was achieved with the use of standard bond lengths and angles. Production of $|E|$ values from $|F_o|$ data is carried out automatically in the first stage of MULTAN, if required. $|E|$ values may be calculated from the expression

$$|E_{\mathbf{h}}|^2 = K\,|F_{\mathbf{h}}|^2_{\text{obs}}/\varepsilon \sum_{i=1}^{N} f_i^2 \qquad (3.48)$$

where K is a scale factor and f_i is the atomic scattering factor for the ith atom type; the epsilon factor ε is defined in Chapter 2. Main *et al.* (1974) have suggested the use of molecular scattering factors g_i^M for groups of atoms of known stereochemistry in the structure, rather than f_i's. In (3.48),

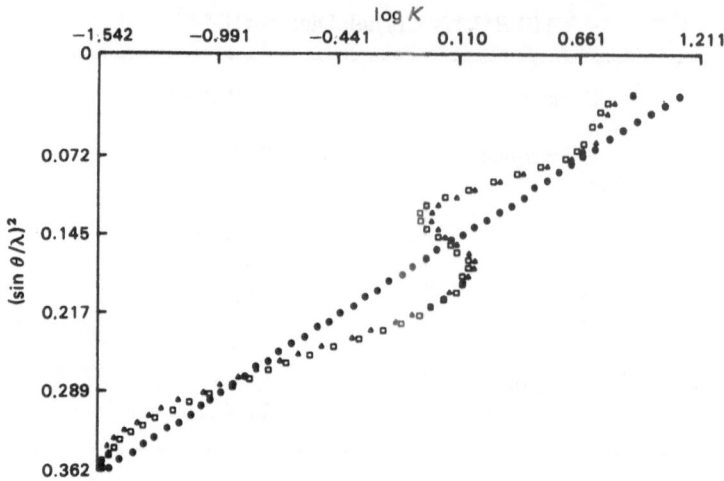

FIGURE 3.10. Plot of the Debye (six-atom fragment) (▲) and Wilson (□) curves and the least-squares straight line (●) through the Debye points. The effect on the starting set resulting from the use of (3.49) may be seen in Tables 3.18 and 3.19.

f_i is replaced by g_i^M given by the Debye scattering formula

$$(g_i^M)^2 = \sum_j \sum_k [f_j f_k (\sin sr_{jk})/sr_{jk}] \qquad (3.49)$$

where $s = 4\pi\lambda^{-1} \sin\theta$ and r_{jk} is the distance between atoms j and k in molecular group i. Lessinger (1976) has reported briefly on some direct phase determinations employing the Debye curve method.

In the structure determination of DMW, $|E|$ values were recalculated, using (3.49) with the six peaks located from E map (A); the Debye curve is shown in Fig. 3.10. Then using 145 $|E|$'s greater than 1.86 and 521 Σ_2 relationships, 64 phase sets were generated; the starting set (B) is shown in Table 3.19. The phase set with highest C (2.41), highest Z (1.28), and second lowest R_K (0.24) gave E map (B), in which 12 peaks appeared having reasonable molecular geometry; two of them were present also in E map (A). The Karle recycling procedure (Karle, 1968) was then employed. Phases from a proposed molecular fragment were subjected to tangent formula refinement and a new E map was calculated. The criteria for acceptance of a calculated phase were

$$|F_h|_{calc} \geq p\,|F_h|_{obs} \qquad (3.50)$$

and

$$|E_h| \geq |E_{min}| \qquad (3.51)$$

TABLE 3.19. Starting Set (B), Using |E| Values from the Debye Curve for DMW

Reflection hkl	|E|	Phase/deg
5 0 5	2.68	90 ⎤
0 11 2	2.65	90 ⎬ origin
0 3 1	2.25	90 ⎦
2 3 14	2.42	45, 135, 225, 315
7 2 12	2.62	45, 135, 225, 315
3 3 1	2.43	45, 135, 225, 315
3 0 2	3.11	0 (enantiomorph)

The value of p is less than or equal to unity, and is set by the MULTAN user. It depends on the fraction of the total structure represented by the trial molecular fragment; the more trial atoms, the larger the value of p. $|E_{min}|$ may be chosen so as to maintain the required phase probability, or variance.

In the present analysis, structure factors were calculated from the 12 initial atom positions, employing an overall temperature factor (B) of 3.6 Å²; p in (3.50) was taken to be 0.29 and $|E_{min}|$ was set at 1.95. In this way, 100 phases were accepted; they were expanded and refined through 1535 Σ_2 relationships by the tangent formula, using all reflections with $|E| > 1.70$. The resulting E map revealed 43 of the 46 nonhydrogen atoms in the molecule (Fig. 3.11). The remaining atoms were located from a weighted electron density synthesis (Sim, 1959, 1960). The R factor for the trial structure was 0.36, and least-squares refinement reduced it to 0.073. A stereoview of the molecule is shown in Fig. 3.12.

The procedure adopted in this analysis appears to be somewhat arbitrary but nevertheless possesses the virtue of having succeeded. Numerous earlier attempts to solve the DMW structure failed. Although both sets of data were measured by automatic four-circle diffractometry, the DMW crystal available for analysis was very small, being only one quarter the volume of the MEW crystal (Tables 3.15 and 3.17).

E map (A) was the first to show a closed-ring fragment, and it was felt to be a major breakthrough. However, it transpired that two of its six atoms were misplaced; the coordinates of the other four were at worst 1 Å from their final positions. The use of molecular scattering factors

(Lessinger, 1976) sometimes enables a correct tangent formula expansion to be obtained. However, in the present case it seems that the key to the success was that the small changes in $|E|$ values (Tables 3.18 and 3.19), induced by the use of partial molecular scattering factors, forced the expansion on to a correct pathway. More complete molecular scattering factors using the geometry of the whole molecule failed to reveal this pathway, whereas the six-atom fragment somewhat fortuitously did. We conclude that the CONVERGE method for assigning the starting set is very sensitive to the magnitudes of the $|E|$ values, and to the number of them

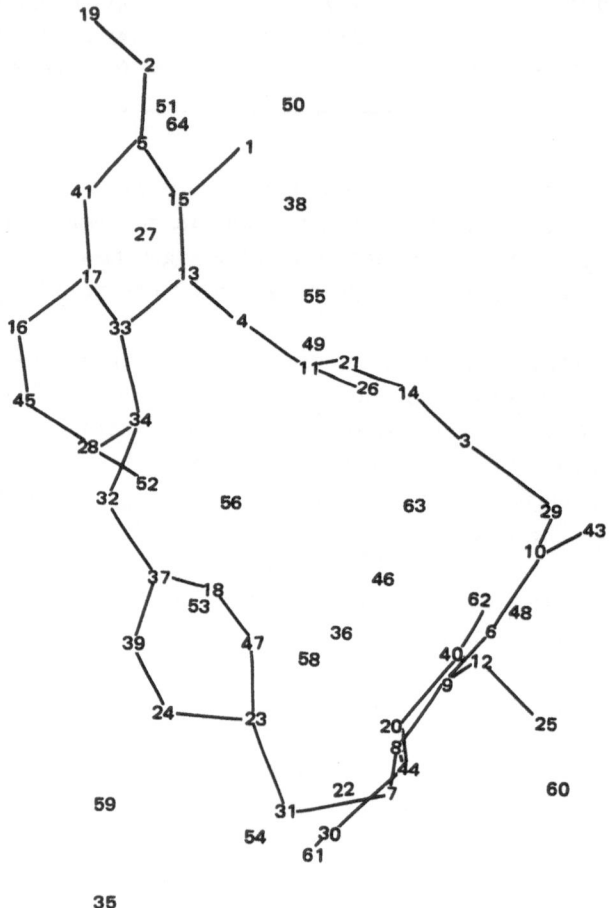

FIGURE 3.11. A representation of the computer output E map produced by the SFCALC routine of MULTAN. Peaks in atomic positions are shown connected; other peaks are spurious.

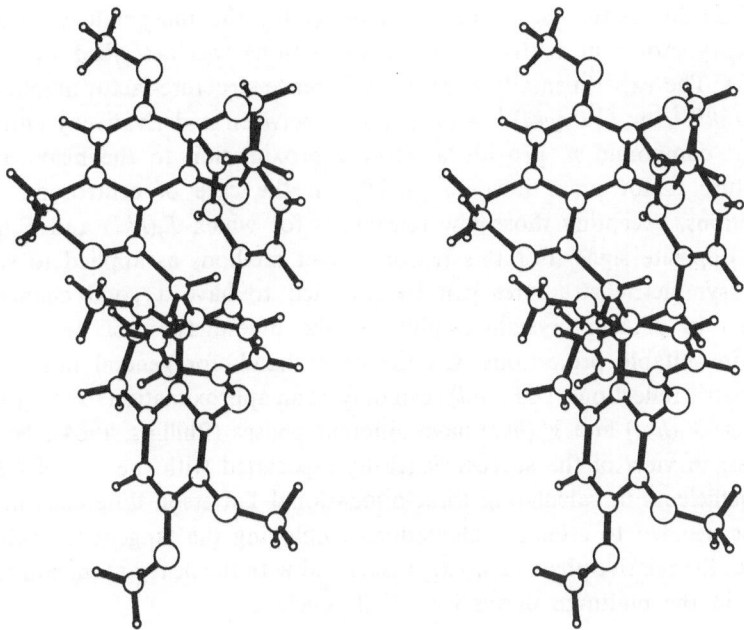

FIGURE 3.12. Stereoscopic view of DMW looking along a: the circles, in increasing order of size, represent H, C, N, and O.

employed in the phase run. This point will be expanded later on. There are several lessons to be learned from these two analyses, which typify situations often encountered in practice.

3.5.4. Experience with Large Structures: Ribonuclease-Potassium Hexachloroplatinate

This example (Palmer and Mazumdar, 1974) is included to demonstrate the use of direct methods in protein crystallography: it was one of the first attempts at using direct methods, in three dimensions, to determine or check the positions of heavy atoms prior to applying the method of iso-morphous replacement. The structure of the enzyme ribonuclease has been investigated at 2.5 Å resolution using the multiple isomorphous replace-ment technique (Carlisle *et al.*, 1974). The space group is $P2_1$ with $Z = 2$ $[M_r(\text{ribonuclease}) = 13700]$. The platinum derivative was prepared by soaking native crystals of ribonuclease in K_2PtCl_6 solution. Data from this derivative and the native crystal were used in the direct methods application described here.

The successful use of direct methods for the independent location of heavy atoms in centrosymmetric projections was described by Steitz (1968). The experimentally observed difference structure-factor amplitudes $|\Delta F_j(hkl)|$, or $||F_j(hkl)| - |F_p(hkl)||$, between a derivative j and the parent compound p, provide a good approximation to the heavy-atom structure factor amplitudes $|F_{Hj}(hkl)|$ in the case of centrosymmetric reflections, excepting those few reflections for which $F_j(hkl)$ and $F_p(hkl)$ have opposite signs. For this reason, direct methods as applied to small centrosymmetric structures can be expected to have a good chance of success in protein crystallography for the investigation of heavy-atom sites in suitable projections. On the other hand, for general noncentrosymmetric reflections, $|\Delta F_j(hkl)|$ can only be an approximation to $|F_{Hj}(hkl)|$ because $\mathbf{F}_j(hkl)$ and $\mathbf{F}_p(hkl)$ have different phases (Phillips, 1964). Nevertheless, in view of the success generally associated with the use of $(\Delta F)^2$ as coefficients in calculating three-dimensional Patterson difference maps, it was decided to attempt calculations employing the tangent formula to extract the relative phases $\phi_{Pt}(hkl)$ associated with the heavy-atom contributions in the platinum derivative of ribonuclease

$$\{\mathbf{F}_{Pt}(hkl) = |F_{Pt}(hkl)| \exp[i\phi_{Pt}(hkl)]\}$$

Such phases can be obtained from structure-factor calculations, once the heavy-atom parameters have been determined by the isomorphous replacement technique. The object of the present work was to compare the heavy-atom phases determined by the two methods.

Normalized Structure Factors

A set of $|\Delta E|$ values was calculated by Steitz's (1968) method:

$$|\Delta E_{Pt}(hkl)|^2 = |\Delta F_{Pt}|^2/f_{Pt} \exp[-B \sin^2(\theta/\lambda^2)]^2$$

where f_{Pt} is the atomic scattering factor for Pt computed from the coefficients of Cromer and Waber (1965); B was set at 20 Å^2, the approximate temperature factor associated with the native protein data $|F_p|$. The values of $|\Delta E|$ thus obtained were scaled in order to make the mean value of $|\Delta E|^2$ equal to unity. Scaling of $|\Delta F(h0l)|$, the centric reflections in space group $P2_1$, by a factor of 0.64 was considered as a method of improving the three-dimensional $|\Delta F(hkl)|$ distribution (Moews and Bunn, 1971), thus improving the $|\Delta E(hkl)|$ values, but never actually tried since it proved unnecessary. Statistical distributions are listed in Table 3.20. The

TABLE 3.20. $|\Delta E|$ Distribution for Ribonuclease-Potassium Hexachloroplatinate [a]

	Experimental	Theoretical	
		Centric	Acentric
$\%\,|\Delta E|>3$	1.1	0.30	0.012
$\%\,|\Delta E|>2$	5.5	5.00	1.83
$\%\,|\Delta E|>1$	26.4	32.0	36.8
$\langle|\Delta E|\rangle$	0.73	0.798	0.886
$\langle|\Delta E|^2\rangle$	1.00	1.000	1.000

[a] The theoretical values quoted are those for $|E|$-value distributions of equal-atom structures; the resolution limit is 2.5 Å.

$|\Delta E|$ distribution is clearly centric, in keeping with the pseudocentrosymmetric constellation of heavy-atom sites in this structure (Carlisle *et al.*, 1974).

Use of MULTAN

Using 230 reflections with $|\Delta E|>1.76$, phase calculations were carried out with MULTAN. The program also calculates signs of the structure-invariant centric reflections, of the type $2h, 0\ 2l$ in space group $P2_1$, using the Σ_1 formula (Hauptman and Karle, 1953), and attempts the automatic assignment of origin- and enantiomorph-fixing reflections by means of the convergence technique (Germain *et al.*, 1971a,b).

For the purpose of the Σ_1 calculations, the unit-cell contents were assumed to be 2 Pt atoms: $\phi_{Pt}(40\bar{1}\bar{4}) = 0$ (probability 0.9999) and $\phi_{Pt}(404) = \pi$ (probability 0.0000) were predicted. Although not included in the following tangent formula expansion, these Σ_1 results were used in the assessment of the validity of the various phase sets produced.

1073 Σ_2 interactions were found among the 230 reflections used, and the convergence routine made the assignments of the origin-fixing and variable-phase reflections given in Table 3.21. They were accepted and introduced by the program into the tangent formula routine.

Eight phase sets were generated, together with the corresponding values of Z and R_K. Table 3.22 summarizes the results obtained, together with an assessment[†] of the individual sets determined from the values of

[†] The combined figure of merit C was not calculated in the version of the program used.

TABLE 3.21. Origin-Fixing and Variable-Phase Reflections Selected by MULTAN

hkl	$|\Delta E|$	ϕ	Comment
11 4	3.23	0	
10 $\bar{6}$	2.43	0	Origin-fixing phases
40 $\bar{3}$	1.94	0	
8 9 $\overline{11}$	5.44	$\pm45, \pm135$	
27 2	3.91	$\pm45, \pm135$	Variable phases

TABLE 3.22. Summary of Phase Set Criteria Obtained from MULTAN with 230 Terms; $|\Delta E| > 1.76$

Phase set	Z	R_K	Σ_1 and tangent formula agree (Y) or not (N)
1	0.699	0.285	Y
2	0.733	0.289	Y
3	0.749	0.265	Y
4	0.745	0.266	Y
5	0.740	0.265	Y
6	0.747	0.277	N
7	0.571	0.299	N
8	0.601	0.302	N

R_K, and shows whether the phases of reflections 4014 and 404 from Σ_1 and tangent expansion were in agreement. The differences between the values of Z and R_K for the eight sets are only marginal, but the worst three sets, 6, 7, and 8, could be eliminated on the grounds of disagreement between the Σ_1 and tangent formula results. Of the other five phase sets, 3, 4, and 5 look the most promising.

Calculation of ΔE Maps

Because none of the phase sets obtained from the tangent expansion appeared to be outstanding, all eight sets were used in the calculation of a series of Fourier syntheses with coefficients $|\Delta E|_{Pt} \exp(i\phi_{DPt})$, where ϕ_{DPt} is the Pt phase contribution calculated by direct methods. The ΔE maps from phase sets, 1, 2, 3, 4, and 5 showed similarities among themselves, as did also those from sets 6, 7, and 8, but with different density distributions, reflecting the demarcation of the criteria in Table 3.22. Of the eight ΔE maps, the one calculated from phase set 5 was the cleanest in terms of peak/background ratio and was selected for detailed study. A further brief mention of ΔE maps 6, 7, and 8 may be made at this point. The interest in these syntheses is that they all show two regions of fairly high density, which, it turns out, simulates sites 1 and 2 in the Pt derivative. For example, in map 7 the two peaks have coordinates 1.145, 0.760, 0.507, and 0.410, 0.745, 0.223, and are related to the Pt sites by the arbitrary translation 0.374, 0.652, 0.142. Apart from this similarity, however, these two sites do not conform fully to the $(\Delta F)^2$ Patterson in that the cross peak 1–2 on the section $v = \frac{1}{2}$ is misplaced. The minimum function (Buerger, 1959), calculated using both of these sites with equal weight, retains only site 1 and eliminates site 2, unlike superposition on the correct sites, which confirm both positions (Palmer and Mazumdar, 1974). This suggests an alternative method for distinguishing between correct and incorrect phase sets determined by direct methods. Karle (1972) has described a method for positioning a geometrically correct molecule in the unit cell, having been misplaced by direct phasing.

Discussion of ΔE Map 5 and Origin Correlation

The main peaks present in ΔE map 5 occur near the section $y = 0.2$, Fig. 3.13; their coordinates are listed in Table 3.23.

This distribution of density was identified with the known Pt sites by its predicted vector distribution [comparison with the $(\Delta F)^2$ Patterson] and by determination of the shift required for correlation of the origins. This correlation is effected with respect to the four origins in the ac plane as $(\frac{1}{2} + x, z)$ and in the b direction as $y - 0.197$, the latter being calculated so as to effect coincidence of the y parameter of the major site; for this reason it is somewhat arbitrary. The coordinates of the site positions after transformation are given in Table 3.24, together with the refined isomorphous replacement parameters for comparison. The agreement in these parameters is very reasonable apart from the y coordinate of site 2, which nevertheless is comparable with the preliminary assignment from the Patterson analysis. What is somewhat surprising (and pleasing) is the identification of peak 3 in the ΔE map with site 3 from the isomorphous replacement analysis, which was included only after calculation of double difference electron density maps at 2.5 Å resolution. The fourth peak from the ΔE map was not identified in the previous analysis.

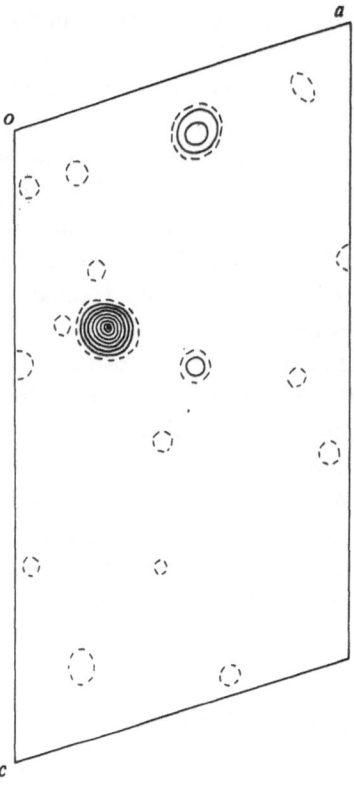

FIGURE 3.13. Section $y = 0.2$ through ΔE map 5 showing the main site 1 and site 2. The origin shift required to effect positioning with respect to isomorphous phase analysis is $\frac{1}{2} + x, y - 0.197, z$.

Phase Comparison

The correspondence of the heavy-atom density and site location from application of the isomorphous replacement and direct methods techniques suggests that the heavy-atom phases determined by both methods should

TABLE 3.23. Fractional Coordinates, Peak Weights, and Peak Heights from ΔE Map 5

x	y	z	Peak weight	Peak height
0.285	0.305	0.363	208	100
0.565	0.320	0.087	119	36
0.280	0.390	0.370	61	41
0.285	0.210	0.370	54	34
	Maximum background		46	27

TABLE 3.24. Correlated Heavy-Atom Parameters from Direct Phasing and Isomorphous Phasing [a]

| Direct methods: phase set 5 | | | | Isomorphous replacement: refined parameters | | | | |
x	y	z	Weight	x	y	z	$B/\text{Å}^2$	A
0.785	0.108	0.363	208	0.771	0.107	0.365	28.0	35.2
0.065	0.123	0.087	119	0.067	0.023	0.099	83.6	30.0
0.780	0.193	0.370	61	0.783	0.217	0.369	47.6	16.0

[a] B is the isotropic temperature factor and A the occupancy in electrons.

also be in good agreement after application of the above origin transformation. The isomorphous heavy-atom phases $\phi_I(hkl)$ were determined by structure-factor calculations. Transformed direct phases $\phi_D(hkl)$ were calculated from phase set 5 by adding the difference $\delta\phi = 360(h/2 - 0.197k)°$. The results of these two calculations are listed in Table 3.25. Figure 3.14 is a histogram of $|\Delta\phi|$, the difference in the phases obtained by the two methods. The pleasing aspect of these results is that the phase difference is quite small for most reflections, being less than 30° for 77% of the data, with an overall average difference $|\overline{\Delta\phi}|$ of 27°. For the centric reflections, the agreement is, as might be expected, even better, with only one change of sign out of 29, that being for the reflection 705 which has the smallest $|\Delta E|$ value (1.81) among the $h0l$ terms. This trend is not general, as can be seen by the plot of $|\Delta E|$ (Fig. 3.15). From this plot there is evidently no correlation between $|\Delta E|$ and $|\Delta\phi|$ and hence within the limits of $|\Delta E| > 1.76$ used in these calculations no loss of reliability associated with decreasing $|\Delta E|$ value. These results show clearly that the two phase sets apply to the same enantiomorph. Attempts to correlate the phase sets by reversal of enantiomorph (which resulted in a better placing of site 2) proved negative, and in view of the above discussion were not pursued further.

The analysis described above showed conclusively that heavy-atom phase information can be computed directly by application of the tangent formula, in a manner very much akin to that used in small molecule analysis. The phases obtained were reassuringly similar to those obtained by isomorphous replacement. Confirmation of heavy-atom sites in protein crystallography using similar methods is now undertaken on a fairly routine basis in cases where it is felt desirable.

TABLE 3.25. Results of the Two Phase-Determining Procedures for the Heavy-Atom Contributions in the K_2PtCl_6 Derivative of Ribonuclease [a]

h	k	l	$|\Delta E|$	$|F_{Pt}|$	$\phi_{D_{Pt}}$	$\phi_{I_{Pt}}$	$|\Delta\phi|$
8	9	−11	5.44	20.99	84.50	106.26	21.76
8	9	−4	4.00	27.70	245.84	237.35	8.49
9	2	−7	3.98	27.55	306.28	254.43	51.85
2	7	2	3.91	29.72	359.70	15.06	15.36
1	0	−20	3.90	17.76	180.00	180.00	0
1	6	−9	3.57	48.37	34.01	68.43	34.42
4	1	−13	3.54	39.37	140.02	126.09	13.93
4	2	−11	3.53	46.98	82.10	86.49	4.39
1	7	−8	3.51	61.41	155.26	158.38	3.12
6	2	−13	3.49	31.80	77.60	85.30	7.70
6	1	−8	3.44	71.64	310.20	310.92	0.72
9	9	−9	3.29	21.07	274.66	271.49	3.17
1	1	4	3.23	96.50	136.17	128.68	7.49
7	2	5	3.10	21.11	24.75	229.82	154.93
7	2	−13	3.09	16.83	282.30	265.49	16.81
6	0	−15	3.06	31.55	0	0	0
10	1	−3	3.04	18.94	336.76	310.87	25.89
10	1	−9	3.03	4.99	192.20	337.72	145.52
4	6	−8	2.99	44.18	223.22	164.18	59.04
1	9	8	2.97	45.27	275.42	276.33	0.91
9	2	3	2.96	29.73	74.34	69.24	5.10
9	2	−2	2.94	8.34	156.73	300.99	144.26
11	2	0	2.93	19.70	271.46	247.26	24.20
11	0	−10	2.91	14.97	0	0	0
3	1	0	2.90	118.21	124.06	132.87	8.81
0	1	3	2.90	98.62	118.96	121.73	2.77
1	2	−6	2.85	80.78	244.00	267.71	23.71
7	9	−4	2.83	10.24	229.53	161.57	67.96
5	2	−12	2.76	42.85	293.71	264.36	29.35
10	2	−12	2.75	14.61	252.85	267.73	14.88
4	10	−11	2.75	29.65	41.89	52.31	10.42
0	2	4	2.73	87.23	251.55	266.23	14.68
0	9	10	2.72	32.34	261.34	262.43	1.09
3	6	−10	2.68	23.37	39.80	91.95	52.15
3	2	−1	2.68	82.37	72.26	85.48	13.22
0	1	10	2.66	54.44	318.67	309.39	9.28
10	1	−11	2.65	30.28	312.67	311.75	0.92
6	9	−9	2.63	26.60	77.39	94.75	17.36
8	3	−10	2.62	0.28	46.85	65.03	18.18
10	0	−5	2.62	32.08	0	0	0
6	1	3	2.61	76.54	307.06	313.02	5.96

TABLE 3.25 (*continued*)

h	k	l	$|\Delta E|$	$|F_{Pt}|$	$\phi_{D_{Pt}}$	$\phi_{I_{Pt}}$	$|\Delta\phi|$
2	6	−7	2.61	69.07	216.16	218.14	1.98
8	0	−2	2.60	58.53	180.00	180.00	0
7	1	−14	2.60	39.77	130.94	131.01	0.07
6	7	−1	2.60	23.79	0.65	31.71	31.06
7	6	−1	2.59	41.38	218.81	230.45	11.64
5	9	−6	2.59	43.45	279.99	280.35	0.36
4	0	−15	2.59	28.53	180.00	180.00	0
10	2	−5	2.56	22.58	88.76	78.17	10.59
1	6	9	2.54	12.29	185.76	215.95	30.19
4	8	−8	2.54	33.64	274.85	259.95	14.90
3	6	−8	2.54	10.62	345.10	62.39	77.29
7	1	8	2.52	39.51	125.76	155.24	29.48
1	6	8	2.46	30.18	40.35	47.45	7.10
3	0	3	2.44	119.71	180.00	180.00	0
4	2	−3	2.44	83.28	68.12	81.30	13.18
4	1	0	2.43	81.14	120.60	130.56	9.96
1	0	−6	2.43	126.13	180.00	180.00	0
3	9	−11	2.42	37.54	92.30	104.45	12.15
5	10	−15	2.42	13.12	210.47	221.71	11.24
5	2	−5	2.42	63.86	78.55	90.77	12.22
2	1	−9	2.42	102.32	125.84	132.52	6.68
0	2	7	2.42	65.84	266.02	259.97	6.05
11	0	−8	2.41	15.07	180.00	180.00	0
4	0	−14	2.41	62.42	0	0	0
2	2	−14	2.41	37.01	277.29	267.66	9.63
8	2	−14	2.40	26.87	78.74	80.60	1.86
8	2	4	2.39	20.51	301.19	174.74	126.45
10	6	−8	2.39	4.02	102.76	251.74	148.98
0	9	7	2.39	26.43	300.64	324.30	23.66
5	0	2	2.07	92.30	180.00	180.00	0
4	11	−9	2.07	17.91	352.75	351.51	1.24
5	0	−12	2.06	40.67	180.00	180.00	0
7	6	−8	2.06	21.43	29.85	54.26	24.41
9	0	1	2.05	46.78	180.00	180.00	0
10	4	−9	2.04	14.75	351.28	332.48	18.80
5	2	−8	2.04	56.18	131.22	67.14	64.08
10	2	−4	2.04	2.97	310.58	273.59	36.99
7	10	−6	2.04	10.45	51.74	261.00	150.74
3	11	−4	2.04	38.24	353.14	356.58	3.44
8	1	−3	2.03	41.62	124.09	136.64	12.55
8	1	6	2.02	30.43	124.98	135.08	10.10

continued overleaf

TABLE 3.25 (*continued*)

h	k	l	$|\Delta E|$	$|F_{Pt}|$	$\phi_{D_{Pt}}$	$\phi_{I_{Pt}}$	$|\Delta\phi|$
4	9	9	2.02	24.41	98.66	156.67	58.01
7	2	−18	2.02	10.97	114.68	101.07	13.61
4	7	−18	2.02	2.74	141.47	340.25	161.22
5	9	0	2.01	32.19	235.60	252.37	16.77
4	6	−6	2.00	42.43	211.74	236.57	24.83
2	2	−6	2.00	63.49	237.53	265.07	27.54
9	6	−9	1.99	11.05	51.36	39.05	12.31
0	7	8	1.99	7.88	249.89	230.14	19.75
4	9	−9	1.98	36.87	244.92	262.78	17.86
9	2	−3	1.98	21.33	119.87	69.92	49.95
2	10	6	1.98	4.83	224.62	127.95	96.67
7	2	−9	1.97	30.74	123.49	61.30	62.19
0	10	11	1.97	37.73	46.04	47.31	1.27
3	9	14	1.97	8.97	62.27	327.40	94.87
9	7	−3	1.95	27.09	164.08	259.94	95.86
7	1	−5	1.95	31.15	320.59	306.31	14.28
9	1	−13	1.94	35.21	129.58	132.41	2.83
4	0	−3	1.94	129.82	0	0	0
10	1	−14	1.94	11.83	343.07	307.67	35.40
4	10	−14	1.94	27.38	44.08	50.56	6.48
1	6	−18	1.94	6.48	220.92	217.64	3.28
3	1	−13	1.94	32.59	308.06	298.23	9.83
2	11	−6	1.93	21.90	173.99	184.00	10.01
10	7	0	1.93	13.77	175.78	202.17	26.39
7	9	−14	1.93	21.59	87.61	93.04	5.43
4	11	−11	1.93	17.17	177.24	193.63	16.39
4	1	−15	1.93	40.87	305.62	288.57	17.05
5	6	1	1.93	31.58	229.88	56.34	173.54
4	0	1	1.93	109.89	180.00	180.00	0
1	3	−18	1.92	17.21	198.73	226.01	27.28
0	2	8	1.92	57.08	88.59	73.25	15.34
0	3	10	1.91	32.04	47.03	29.85	17.18
2	4	8	1.91	50.90	8.41	333.34	35.07
1	0	9	1.91	99.67	0	0	0
2	7	7	1.90	45.68	332.77	345.83	13.06
1	2	5	1.90	76.11	244.07	261.01	16.94
10	3	−14	1.89	13.14	76.40	25.79	50.61
7	6	−2	1.89	13.15	49.32	118.82	69.50
3	0	−16	1.88	50.67	180.00	180.00	0
7	1	9	1.88	24.05	331.32	311.30	20.02
2	6	−10	1.88	14.16	192.58	225.90	33.32
6	1	−1	1.88	86.81	126.34	135.54	9.20

TABLE 3.25 (*continued*)

h	k	l	$|\Delta E|$	$|F_{Pt}|$	$\phi_{D_{Pt}}$	$\phi_{I_{Pt}}$	$|\Delta\phi|$
10	5	−14	1.88	9.93	107.07	113.52	6.45
6	3	−16	1.88	20.54	17.33	34.97	17.64
4	10	−10	1.88	31.56	221.15	219.30	1.85
2	11	−5	1.88	43.35	355.09	361.74	6.65
3	9	−2	1.87	28.26	258.18	241.11	17.07
9	1	1	1.87	23.21	155.59	128.48	27.11
8	0	−17	1.86	29.87	0	0	0
9	1	−6	1.86	37.07	319.61	302.23	17.38
10	7	−1	1.85	17.88	349.50	327.66	21.84
3	11	−7	1.85	30.46	354.58	348.04	6.54
11	1	−2	1.85	25.66	313.39	324.01	10.62
0	10	15	1.85	24.86	224.97	223.85	1.12
6	10	−14	1.85	16.46	205.16	191.39	13.77
4	3	−12	1.85	39.42	64.54	31.23	33.31
11	6	−5	1.84	9.21	54.17	285.12	129.05
3	7	−4	1.84	18.33	197.93	197.94	0.01
9	0	−6	2.38	19.83	0	0	0
8	1	2	2.37	35.75	297.46	317.56	20.10
2	1	−16	2.37	40.54	314.83	319.64	4.81
2	2	−8	2.36	52.82	252.21	261.26	9.05
9	2	−8	2.35	32.88	82.75	84.54	1.79
4	7	−2	2.35	27.53	355.61	34.63	39.02
9	9	−6	2.35	23.19	258.13	269.45	11.32
11	2	−7	2.35	22.15	77.61	82.80	5.19
8	7	8	2.33	5.99	351.19	45.59	54.40
7	2	8	2.33	13.57	280.27	274.13	6.14
7	7	5	2.33	28.39	343.90	0.92	17.02
4	7	−11	2.33	4.12	38.76	63.20	24.44
3	1	−7	2.31	79.06	318.90	313.22	5.68
2	7	−12	2.31	35.64	340.06	351.01	10.95
10	2	−8	2.30	10.53	72.00	83.48	11.48
4	11	−10	2.29	17.52	169.23	173.62	4.39
4	9	−12	2.28	31.48	250.41	276.92	26.51
4	2	−14	2.27	42.72	78.07	95.62	17.55
2	1	2	2.26	124.92	126.28	134.75	8.47
2	2	−10	2.25	50.47	76.52	88.90	12.38
6	0	−7	2.25	48.16	0	0	0
5	1	11	2.24	20.80	326.13	228.33	97.80
8	3	−5	2.23	25.39	206.80	218.69	11.89
12	0	−4	2.20	3.59	0	0	0
0	0	4	2.20	136.41	180.00	180.00	0
5	6	−9	2.18	11.14	18.76	81.63	62.87

continued overleaf

TABLE 3.25 (*continued*)

h	k	l	$|\Delta E|$	$|F_{Pt}|$	$\phi_{D_{Pt}}$	$\phi_{I_{Pt}}$	$|\Delta\phi|$
8	2	9	2.18	23.25	258.00	256.19	1.81
3	0	−4	2.18	75.59	0	0	0
3	2	−17	2.18	22.76	121.86	71.61	50.25
8	9	−3	2.16	18.96	105.83	79.53	26.30
2	3	10	2.16	30.20	198.89	210.14	11.25
1	0	5	2.16	118.66	180.00	180.00	0
1	7	7	2.15	65.82	335.55	352.79	17.24
7	8	−1	2.14	39.15	290.47	286.76	3.71
11	2	−4	2.14	20.56	101.65	68.70	32.95
0	8	10	2.14	36.21	97.31	85.98	11.33
3	2	3	2.13	74.91	245.20	261.80	16.60
1	8	−6	2.13	41.20	151.36	153.51	2.15
6	1	5	2.13	6.33	213.22	315.42	102.20
4	2	−7	2.12	71.71	250.36	262.57	12.21
4	1	−9	2.12	47.38	337.97	312.29	25.68
5	1	−17	2.12	34.23	309.56	313.21	3.65
5	1	−13	2.11	37.42	134.70	132.85	1.85
3	7	0	2.11	23.64	8.73	20.53	11.80
7	1	−10	2.10	45.49	320.24	314.79	5.45
4	2	1	2.10	71.20	256.24	263.37	7.13
4	0	4	2.09	88.39	180.00	180.00	0
8	8	−10	2.09	23.46	139.74	156.03	16.29
7	10	−7	2.09	18.11	44.37	45.76	1.39
3	1	2	2.07	48.96	120.71	134.26	13.55
5	0	−16	1.84	37.58	0	0	0
8	1	4	1.84	33.24	315.08	304.60	10.48
8	2	−3	1.84	34.68	90.53	81.69	8.84
6	0	−5	1.83	54.18	0	0	0
2	0	13	1.83	27.23	180.00	180.00	0
6	8	5	1.82	37.60	108.50	127.82	19.32
10	3	−8	1.82	19.76	22.06	36.01	13.95
4	9	6	1.82	64.63	64.63	74.24	9.61
9	9	0	1.81	7.58	288.51	50.11	121.60
7	0	5	1.81	20.10	0	180.00	180.00
8	9	−8	1.81	26.20	63.15	66.39	3.24
4	1	−19	1.81	25.74	131.74	148.33	16.59
7	1	−8	1.81	17.10	279.48	297.84	18.36
7	10	1	1.80	3.42	282.64	40.47	117.83
3	1	−5	1.80	42.51	277.53	315.57	38.04
1	6	10	1.80	47.78	36.73	36.87	0.14
5	2	−18	1.79	9.96	235.15	288.10	52.95
7	7	−11	1.79	12.38	9.87	340.91	28.96
3	4	10	1.79	18.92	152.72	162.47	9.75
2	6	−8	1.79	13.52	18.82	68.91	50.09

TABLE 3.25 (*continued*)

h	k	l	$|\Delta E|$	$|F_{Pt}|$	$\phi_{D_{Pt}}$	$\phi_{I_{Pt}}$	$|\Delta\phi|$
8	8	−8	1.78	3.05	86.09	173.18	87.09
8	5	−11	1.78	11.29	262.60	286.05	23.45
3	7	−8	1.78	16.87	3.83	39.31	35.48
4	8	−4	1.78	44.36	103.83	101.98	1.85
0	6	6	1.78	42.73	219.55	123.50	96.05
5	10	−2	1.77	31.48	45.90	7.12	38.78
7	3	−13	1.77	28.85	73.40	30.75	42.65
3	9	−16	1.77	4.61	18.38	349.45	28.93
1	10	14	1.77	12.23	21.81	14.42	7.39
8	11	−4	1.77	13.70	353.19	346.16	7.03
9	2	−17	1.77	2.63	8.12	94.35	86.23
6	1	−20	1.77	19.85	131.59	130.89	0.70
2	2	17	1.77	2.57	15.59	70.03	54.44
10	1	6	1.77	6.34	336.81	138.48	161.67
7	7	−14	1.76	18.30	354.09	18.63	24.54
3	9	−10	1.76	34.77	237.29	245.86	8.57
2	11	−7	1.76	2.46	193.80	319.11	125.31
7	9	−6	1.76	31.74	67.13	74.55	7.42
8	4	−4	1.76	18.35	359.83	219.45	140.38
0	6	11	1.76	15.91	197.66	257.08	59.42

[a] $\phi_{D_{Pt}}$ is the phase determined by application of the tangent formula and $\phi_{I_{Pt}}$ is the phase calculated from the known heavy-atom constellation. The difference between the two phases is $|\Delta\phi|$. All phases are in degrees.

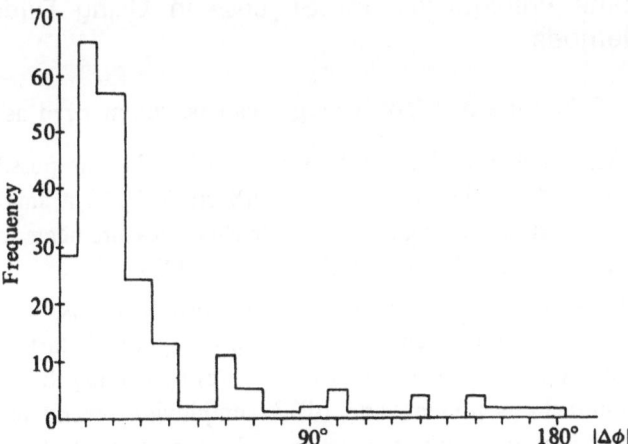

FIGURE 3.14. Histogram showing agreement between direct and isomorphous calculated structure-factor phases.

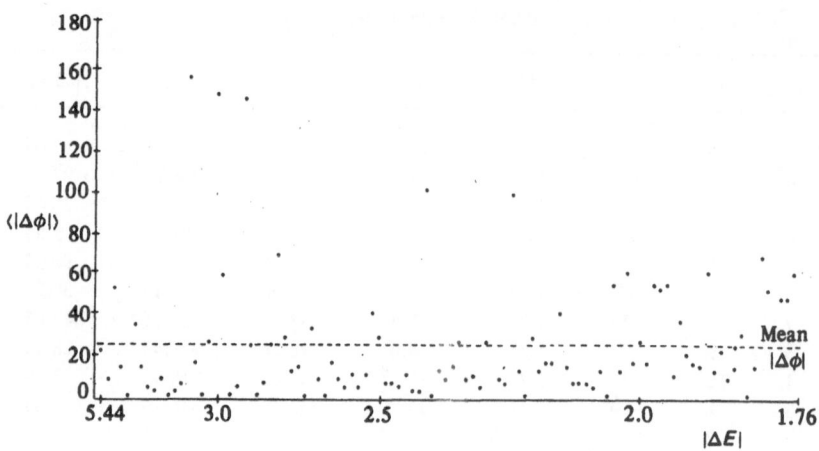

FIGURE 3.15. Variation of the local average $\langle |\Delta\phi| \rangle$ as a function of $|\Delta E|$.

3.6. Success Is Not Guaranteed

Direct methods have not yet been developed to the extent where all structures can be solved in a straightforward manner. The experience of many workers in this field does, however, permit the drawing up of a number of strategies which have often proved significant in structure determinations.

3.6.1. Some Prerequisites for Success in Using Direct Methods

Some of the rules that have emerged can be summarized as follows:

(i)   As complete and accurate a set of $|F_o|$ data as possible must be used. In the case of DMW (Section 3.5.3) a small crystal size resulted in many weak-intensity measurements and consequent difficulty in fulfilling this condition.

(ii) If automatic solution of the phase problem fails, we must be prepared to intervene (a) by selecting origin and starting phases, (b) by varying the value of $|E_{min}|$, (c) by varying the number of interactions present, and (d) by employing special techniques, possibly involving tangent-formula recycling, molecular scattering factors, and so on.

(iii)   Remember that a computer program may contain mistakes, even after considerable length of use. Always check the validity of results as far as possible, especially assignments of basic origin- and enantiomorph-fixing reflections and values of special phases.

(iv)   Do not give in!

3.6.2. Figures of Merit: a Practical Guide

We have mentioned the four criteria calculated in MULTAN for assessment and ordering of phase sets. This facility is necessary in order to overcome the basic disadvantage inherent in the multisolution method, namely, that many possible phase sets may have to be explored. Figures of merit allow one to give priority to the best phase sets for computation of E maps. In principle, the phase sets are explored in order until the structure is found; normally, one does not investigate them further.

Experience suggests that of the figures of merit currently calculated in MULTAN, Z and R_K are usually sufficient to enable the most likely E maps to be explored. C is a fairly recent innovation, and also appears to be very useful; ψ_0 can be used in difficult cases. As a guide, and not as an absolute measure, we may use the following values in selecting the order of calculation of E maps: $Z > 1.0$, $R_K < 0.20$, and $C > 2.5$. Other programs may calculate different figures of merit, and one should check with the author for the corresponding recommended values.

3.6.3. Signs of Trouble, and Past Remedies When the Structure Failed to Solve

This section summarizes some of our own experiences, and records factors involved when a structure solution failed to emerge from direct phase-determining methods. Although these comments apply primarily to the use of MULTAN, similar considerations would apply to other like program systems.

(a) *All values of Z are too low ($\ll 1.0$) or all values of R_K are too high ($\gg 0.20$)—both conditions can arise.* The criterion R_K applies strictly speaking only when all possible Σ_2 interactions have been used. The usual maximum number of Σ_2 interactions allowed in MULTAN is 2000, but a cutoff can be applied by the user; some versions allow up to 4000 Σ_2 interactions. If a cutoff is applied, the correct phase set could have a much higher value of R_K (up to 0.3 has been noted).

(b) *All phase sets and their figures of merit are similar.* This situation may occur if too few Σ_2 interactions are being used. Try more.

(c) *The origin-defining set is poor or incorrect.* Try another one, choosing it with the aid of the rules given in Chapter 2.

(d) *The E map contains one very large peak.* The phases are probably very inaccurate; the heavy peak may be located in the center of a closed ring. Start again; do not waste time trying to interpret the E map.

(e) *The E map is not interpretable or chemically sensible.* The phases are incorrect. Try again.

(f) *If heavy atoms are present in the structure, they alone may show up.* Proceed to Fourier methods using interpretable heavy-atom sites. A check against the Patterson function might prove useful here.

(g) *Only a small molecular fragment is discernible from the E map.* Try recycling, basing phases on the fragment found, or try to obtain more phases by increasing the initial data set.

(h) *The program selects an incorrect or poor starting set (too few Σ_2 interactions, for example).* Select your own starting set. If you suspect that the program may contain a fault, inform the author; do not attempt to correct it.

(i) *The solution still fails to emerge.* Review the calculation of the $|E|$ values. Try the Debye curve fitting, for example, or omit reflections that appear to have a bad influence on the phase-determining pathway.

(j) *All fails.* Go back to fundamentals. Check the space group, data collection, and processing and any other factor that might be at fault.

If you exhaust these possibilities without achieving success, try another program if available or another method for determining the structure. Or give it a rest and try again later.

3.6.4. Comments on Molecular Scattering Factors

It has been reported on several occasions (Main *et al.*, 1974, 1977; Lessinger, 1976) that the use of molecular scattering factors in calculating $|E|$ values has led to the solution of a structure by MULTAN where the use of atomic scattering factors failed. While there can be no doubt of the correctness of this statement, it is by no means clear that the success is not due rather to a different pathway through the CONVERGE routine. Recent calculations (Ladd, 1978) have shown that the distribution of $|E|$ values obtained with molecular scattering factors is very different from that calculated from the solved structure, but that $|E|$ values obtained with a "minimum profile" K curve followed the calculated distribution closely. It was found also that the "Debye" $|E|$ values may be obtained by a K curve drawn *through* the experimental points as well as by (3.49). The difference between the $|E|$ values obtained statistically and those calculated from the structure were significantly less for the large $|F_o|$ values: this fact seems to indicate a possible useful "rule" for the initial stage of a direct phase determination.

That the CONVERGE routine is very sensitive to the magnitudes of the $|E|$ values is illustrated in Table 3.26. The compound, $C_{26}H_{26}CuN_4O_8$, is triclinic with $a = 10.537$, $b = 10.521$, $c = 15.377\,\text{Å}$, $\alpha = 121.03$, $\beta = 111.96$, $\gamma = 86.11°$, $Z = 2$, $P\bar{1}$. $300\,|E|$'s > 1.80 were used and two runs of MULTAN were carried out with different E-map search distances. Inadvertently, one reflection (233, $|F_o| = 49.1$, $|E| = 0.38$) was omitted from the first run. The two sets of $|E|$ values obtained differed by only about 0.1%. These and other experiences indicate that absorption corrections might perhaps be applied to $|F_o|$ as a routine stage in data reduction, particularly if heavy atoms are present—another rule? Further comments on $|E|$ values are given by Ladd (1978). In Table 3.26, the large values for the combined FOMs arise from large values for PSI ZERO. Although this quantity should be small, especially as the space group lacks translational symmetry, it has been enhanced in this example because of the presence of heavy atoms (Cu) in special positions.

TABLE 3.26. Two Runs with MULTAN: $C_{26}H_{26}CuN_4O_8$

	I	II
Number of $|F_o|$ data	4572	4573
$2B/\text{Å}^2$	8.255	8.251
K	0.04269	0.04273
Σ_1 ($>99\%$)	$40\bar{4}$; 224	$40\bar{4}$; 224
Origin set	$81\bar{8}$; $3\bar{5}\bar{8}$; $30\bar{3}$	$10,11\bar{0}$; $30\bar{3}$; $3\bar{5}4$
Other reflections in the starting set	$20\bar{2}$; $11,2\bar{6}$; 055	$81\bar{8}$; $11,2\bar{6}$; $3\bar{2}\bar{4}$
	$3\bar{2}4$; $3\bar{4}4$	$25\bar{2}$; $3,10,\bar{1}\bar{0}$
Maximum bonding distance/Å	2.60	1.95
First three combined FOM's and result	2.58, correct	2.37, incorrect
	2.44, incorrect	2.00, correct
	2.34, correct	1.82, incorrect

3.7. More Recent Developments of the Multisolution Method: Magic Integers

So far we have described and commented upon both standard and fairly difficult crystal structure analyses carried out by the MULTAN program. Much of what we have discussed is relevant to current determinations of structures having less than about 100 atoms in the asymmetric unit. There is nothing particularly significant about the number 100, but it is at about this point where the probabilities associated with single tpsr relationships begin to become too small to be useful, and more and more contributors are needed in order to generate reliable phases in a tangent formula expansion. For complex structures, therefore, we shall need a large number of $|E|$ values for phase expansion (low $|E_{min}|$) and possibly a larger number of variable phases in the starting set, leading to correspondingly more phase sets. These factors increase computer requirements, both in store and execution time.

In 1975, White and Woolfson introduced a new technique, recently incorporated into MULTAN (Declerq, Germain, and Woolfson, 1975), which promises to reduce these difficulties and to increase the potential of MULTAN for solving difficult or complex structures. The device, an inspiration of Woolfson, is known as "magic integers." This technique depends upon the approximate representation of n phases (usually 3) by one nonintegral symbol and n integers. A general phase lies between 0 and 2π, or in cycles, between 0 and 1. Consider three phases ϕ_1, ϕ_2, and ϕ_3 expressed in cycles. Then suppose

$$\phi_1 = px \bmod 1$$

$$\phi_2 = qx \bmod 1 \qquad (3.52)$$

$$\phi_3 = rx \bmod 1$$

where p, q, r are integers and x has a value between 0 and 1. The nature of this arithmetic may be illustrated as follows.

(i) If $\phi = 2x \pmod 1$ where $\phi = 0.5$ then $x = 0.25$ or 0.75; or if $\phi = 3x \pmod 1$ where $\phi = 0.1$ then $x = 0.033, 0.367$ or 0.700.

We see that possible values of x satisfying (3.52) will depend on the integers p, q, r. This example was designed by selecting randomly a few numbers. Other examples, designed similarly, may turn out to be either better or worse!

(ii) Suppose that

$$\phi_1 = 3x$$
$$\phi_2 = 4x \qquad (3.53)$$
$$\phi_3 = 5x$$

with $\phi_1 = 0.1$, $\phi_2 = 0.7$, $\phi_3 = 0.4$. Following example (i), we see that possible values of x are

$$\phi_1: \quad x = 0.033, \ 0.367, \ 0.700$$
$$\phi_2: \quad x = 0.175, 0.425, 0.675, 0.925 \qquad (3.54)$$
$$\phi_3: \quad x = 0.080, 0.280, 0.480, 0.680, 0.880$$

Selecting close values and taking averages weighted on p, q, r, respectively, gives two possible values for x:

(a) $x = 0.433$, which returns values for ϕ of

$$\phi_1 = 0.30 \quad (\text{error } 72°)$$
$$\phi_2 = 0.73 \quad (\text{error } 11°) \qquad (3.55)$$
$$\phi_3 = 0.17 \quad (\text{error } 84°)$$

(b) $x = 0.683$, which returns values for ϕ of

$$\phi_1 = 0.05 \quad (\text{error } 18°)$$
$$\phi_2 = 0.73 \quad (\text{error } 11°) \qquad (3.56)$$
$$\phi_3 = 0.42 \quad (\text{error } 7°)$$

Hence (b) gives a better estimate of x than does (a). The reader may find it useful to try a few similar examples; other sets of integers in Table 3.27 may be used.

A single tpr (3.1) may be written as

$$\phi_h - \phi_k - \phi_{h-k} \equiv 0 \ (\text{mod } 2\pi) \qquad (3.57)$$

The cosine of a triplet such as (3.57) would tend to 1, although it must always be less than 1. We may write this result as

$$\cos(\phi_h - \phi_k - \phi_{h-k}) \to 1 \qquad (3.58)$$

As an application of magic integers we shall use part of the data for tuber-

TABLE 3.27. Comparison of Possible Sets of Magic Integers

Integers								R.m.s. error	Maximum error
1	2	3						47	90
1	3	4						42	79
1	3	6						35	65
1	4	16						23	43
2	3	4						38	72
2	3	7						33	65
3	4	5						33	65
3	4	10						34	65
3	4	13						25	50
3	5	6						33	65
3	5	7						30	68
4	5	6						31	65
4	5	7						29	54
4	5	21						23	43
5	6	7						30	61
5	6	31						23	54
2	3	7	13					39	75
3	4	5	6					46	83
3	4	5	7					45	83
3	4	5	13					40	72
3	4	13	25					33	65
4	5	7	10					38	75
4	5	7	13					37	69
4	5	21	41					28	56
3	4	5	8	13				49	102
3	4	5	9	12				51	93
3	4	5	9	20				46	88
3	4	5	21	28				42	83
3	5	7	14	21				46	88
4	5	6	11	17				48	93
5	7	9	11	13				45	92
5	7	9	11	20				45	90
6	7	8	13	21				48	89
13	15	17	19	21				45	85
3	4	5	11	16	27			50	98
3	4	5	12	15	19			53	98
3	4	5	15	20	25			50	100
3	4	5	30	40	50			44	87
4	5	6	7	8	9			56	106
7	7	9	11	13	15			53	99
7	9	11	13	17	30			49	91
7	10	13	16	19	22			52	98
7	9	11	13	17	23	43		52	100
17	23	29	33	37	63	100		44	87
9	11	13	15	17	19	21	23	61	111
13	17	19	23	25	27	34	61	54	106
21	23	24	25	26	27	33	60	53	100

TABLE 3.28. Partial Data Set for Tubercidin with Magic Integer Phases
Indicated

Code [a]	hkl	$\lvert E_h \rvert$	ϕ_h
2	790	2.76	px
6	63$\bar{3}$	2.37	qx
8	770	2.22	rx
12	80$\bar{3}$	2.18	π
17	021	2.12	py
23	163	2.08	qy
25	65$\bar{2}$	2.06	ry
31	703	1.97	π
33	073	1.97	pz
61	94$\bar{5}$	1.77	qz
80	14$\bar{2}$	1.66	rz

[a] This is the number in the list of Table 3.11.

cidin (Table 3.11). The data used in this example are given in Table 3.28.
A magic integer sequence p, q, r with variables x, y, z is used to represent
phases mod (1). The tpr's arising from this list, remembering the phase
symmetry for space group $P2_1$, are listed in Table 3.29; the magic integer
representation is also shown in the table.

The terms denoted by n in Table 3.29 are of the form

$$Hx + Ky + Lz + A \equiv 0 \ \mathrm{mod}(1) \tag{3.59}$$

where H, K, and L are each some combination of the magic integers p, q,

TABLE 3.29. Tpr's and Corresponding Magic Integer Representations Generated
from Linked Phases in Table 3.28

n	tpr	Magic integer representation mod 1		
1	$\phi_2 - \phi_6 - \phi_{23} \approx 0$	$(p - q)x -$	qy	$\equiv 0$
2	$\phi_6 + \phi_{17} - \phi_{25} \approx 0$	$qx + (p - r)y$		$\equiv 0$
3	$\phi_8 - \phi_{33} \approx 0$	rx	$-pz$	$\equiv 0$
4	$\phi_{61} - \phi_{80} + \pi \approx 0$		$(q - r)z + 0.5 \equiv 0$	

and r, and A is a constant, corresponding for instance to 0.5 cycles (π radians) in entry 4 of Table 3.29. In cosine form, (3.59) becomes, by analogy with (3.58),

$$\cos[2\pi(Hx + Ky + Lz) + \alpha] \to 1 \qquad (3.60)$$

where α is the constant in radians, corresponding to A. White and Woolfson (1965) proposed that a condition that relationships like (3.60) were satisfied as closely as possible corresponded to a maximum value of the function

$$\psi(x, y, z) = \sum_n [|E_1||E_2||E_3|]_n \cos[2\pi(H_n x + K_n y + L_n z) + \alpha_n] \qquad (3.61)$$

where $|E_1|$, $|E_2|$, $|E_3|$ are the $|E|$ values of the three reflections involved in a tpr, and $|E_1||E_2||E_3|$ takes into account the strength of a relationship, (3.7). $\psi(x, y, z)$ in (3.61) may be evaluated over a map of x, y, z values each in the range 0 to 1 by Fourier summation (see for example Ladd and Palmer, 1978). The number of subdivisions along each axial direction should be about 4 times the corresponding observed maximum Miller index value. A high peak in this map corresponds to a possible solution in terms of x, y, z, which can then be used to generate phases as laid out in Table 3.28. Such a device, if successful, should lead rapidly to an extensive set of numerical phases suitable for tangent formula expansion and refinement. Woolfson (1975) has described a successful application of the method outlined above, but points out that the technique is not generally applicable because of phasing errors introduced by the approximations involved in using the magic integers themselves. Accuracy may be improved by determining shifts for the ϕ's which maximize ϕ.

A further refinement of the method is the $P - S$ sets method (Woolfson, 1975). This involves first determining a primary (P) set of reflections, which contains the origin- and enantiomorph-fixing phases together with a number reflections, that gives rise to as many determinations of new phases as possible from the strongest single tpr's. The dependent reflections, whose probable phases are given in terms of pairs of P set reflections, are placed in a secondary S set. Finally, all the tpr's that link the phases in the combined P and S sets, other than those tpr's that were used directly to generate S reflections from P reflections, are collected together and used in the calculation of a ψ map. Any primary reflection that does not meet the above requirements is removed from the P set prior to calculation of $\psi(x, y, z)$. Selected peaks from the ψ map are then used to generate limited trial phase sets only for reflections in the P and S sets. For all phase sets, all relationships linking P and S reflections, except origin-fixing reflections, are next used in a maximization parameter-shift routine to produce shifted phases for each phase set. Each of the phase sets derived after the parameter shift process is used as a starting point for tangent refinement and extension in MULTAN, and the final phase sets so produced are used for calculation of E maps in the usual way.

Woolfson (1975) has described successful applications of the method outlined above, which is a facility of the MULTAN 77 system. The new rationale promises to be a more powerful means of solving difficult structures than is the straight multisolution approach. By the rapid generation of a more extensive and potentially better starting set of phases, the chances of successful phase generation appear to be greatly enhanced. Further enhancements of MULTAN may soon be implemented by use of both the maximum determinant principle (Chapter 6) to produce more extensive starting sets, and the least-squares refinement of phases. A recent innovation (YZARC) employs random values for the initial phases.

References

Barrett, A. N., and Palmer, R. A. (1969). *Acta Crystallogr. Sect. B* **25**, 688.

Buerger, M. J. (1959). *Vector Space*, John Wiley, New York.

Borkakoti, N. (1978). Ph.D. thesis, University of London.

Borkakoti, N., and Palmer, R. A. (1978a). *Acta Crystallogr. Sect. B* **34**, 482.

Borkakoti, N., and Palmer, R. A. (1978b). *Acta Crystallogr. Sect. B* **34**, 490.

Carlisle, C. H., Palmer, R. A., Mazumdar, S. K., Yeates, D. G. R., and Gorinsky, B. A. (1974). *J. Mol. Biol.* **35**, 1.

Cochran, W. (1952). *Acta Crystallogr.* **5**, 65.

Cochran, W. (1953). *Acta Crystallogr.* **6**, 260.

Cochran, W. (1955). *Acta Crystallogr.* **8**, 473.

Cochran, W., and Douglas, A. S. (1957). *Proc. R. Soc. London Ser. A* **227**, 486.

Cromer, D. T., and Waber, J. T. (1965). *Acta Crystallogr.* **18**, 104.

Declerq, J. P., Germain, G., and Woolfson, M. M. (1975). *Acta Crystallogr. Sect. A* **31**, 367.

Germain, G., Main, P., and Woolfson, M. M. (1970a). *Acta Crystallogr. Sect. B* **26**, 274.

Germain, G., Main, P., and Woolfson, M. M. (1970b). *Acta Crystallogr. Sect. B* **26**, 275.

Germain, G., Main, P., and Woolfson, M. M. (1971a). *Acta Crystallogr. Sect. A* **27**, 368.

Germain, G., Main, P., and Woolfson, M. M. (1971b). *Acta Crystallogr. Sect. A* **27**, 1040.

Hauptman, H., and Karle, J. (1953). *The Solution of the Phase Problem. I. The Centrosymmetric Crystal*: American Crystallographic Association Monograph No. 3, Polycrystal Book Service, Pittsburgh, Pennsylvania.

Hull, S. E., and Irwin, M. J. (1978). *Acta Crystallogr. Sect. A* **34**, 863.

Karle, I. L. (1970). In *Crystallographic Computing*, Ed. F. R. Ahmed, Munksgaard, Copenhagen.

Karle, J. (1968). *Acta Crystallogr. Sect. B* **24**, 182.

Karle, J. (1972). *Acta Crystallogr. Sect. B* **28**, 820.

Karle, J., and Hauptman, H. (1956). *Acta Crystallogr.* **9**, 635.

Karle, J., and Karle, I. L. (1963). *Acta Crystallogr.* **17**, 835.

Karle, J., and Karle, I. L. (1966). *Acta Crystallogr.* **21**, 849.

Ladd, M. F. C. (1978). *Zt. Kristallogr.* **147**, 279.

Ladd, M. F. C., and Palmer, R. A. (1978). *Structure Determination by X-ray Crystallography*, Plenum, New York.

Lessinger, L. (1976). *Acta Crystallogr. Sect. A* **32**, 538.

Main, P., Woolfson, M. M., Lessinger, L., Germain, S., and Declerq, J. P. (1974, 1977). *MULTAN 74, A System of Computer Programs for the Automatic Solution of Crystal Structures*, University of York (1974). The latest version is MULTAN 78 (1978).

Moews, P. C., and Bunn, C. W. (1971). *Acta Crystallogr. Sect. B* **27**, 1780.

Palmer, R. A., and Mazumdar, S. K. (1974). *J. Cryst. Mol. Struct.* **4**, 107.

Phillips, D. C. (1964). Advances in protein crystallography, in Vol. 2 of *Advances in Structure Research by Diffraction Methods*, Eds. R. Brill and R. Mason.

Reynolds, C. D., Palmer, R. A., and Gorinsky, B. A. (1974). *J. Cryst. Mol. Struct.* **4**, 213.

Sayre, D. (1952). *Acta Crystallogr.* **5**, 60.

Sim, G. A. (1959). *Acta Crystallogr.* **12**, 813.

Sim, G. A. (1960). *Acta Crystallogr.* **13**, 511.

Steitz, T. A. (1968). *Acta Crystallogr. Sect. B* **24**, 504.

Stroud, R. M. (1968). Ph.D. Thesis, University of London.

Stroud, R. M. (1973). *Acta Crystallogr. Sect. B* **29**, 690.

White, P. S., and Woolfson, M. M. (1975). *Acta Crystallogr. Sect. A* **31**, 53.

Wilson, A. J. C. (1942). *Nature* **150**, 151.

Woolfson, M. M. (1954). *Acta Crystallogr.* **7**, 61.

Woolfson, M. M. (1975). Notes of the Meeting of the Commission of Crystallographic Computing of the International Union of Crystallography, Prague. Personal communication.

Woolfson, M. M., and Cochran, W. (1955). *Acta Crystallogr.* **8**, 1.

Woolfson, M. M., and Germain, G. (1968). *Acta Crystallogr. Sect. B* **24**, 91.

Zachariasen, W. H. (1952). *Acta Crystallogr.* **5**, 68.

Probabilistic Theory of the Structure Seminvariants

HERBERT HAUPTMAN

4.1. Major Goal

The major goal of this chapter is to exhibit the structure seminvariant as the central concept, not only in the development of the theoretical basis of direct methods, but also as a practical tool that is useful in phase determination as well. To this end conditional probability distributions of selected seminvariants are described, given, in the first instance, an appropriate set of structure-factor magnitudes $|E|$ and, in the second instance, the values of one or more structure seminvariants as well as magnitudes $|E|$. Each of these distributions leads to an estimate for the structure seminvariant. This estimate is particularly good when the variance of the distribution is small. Since the structure seminvariants lead in turn directly and unambiguously to the values of the individual phases ϕ, the structure seminvariants serve to link the known magnitudes $|E|$ with the desired phases ϕ. Thus the derivation of appropriate conditional probability distributions of the structure seminvariants constitutes in effect a program for phase determination.

The most important new idea, the neighborhood concept described in Section 4.9, serves to identify the small sets of magnitudes $|E|$ on which the value of a structure seminvariant primarily depends. In view of numerous illustrations presented in Sections 4.7, 4.8, 4.10, and 4.11, the evidence in support of the neighborhood principle, Section 4.9, appears now to be overwhelming.

HERBERT HAUPTMAN ● Medical Foundation of Buffalo, Inc., Buffalo, New York, 14203.

4.2. Introduction

If one denotes the phase of the structure factor F_h by ϕ_h, then

$$F_h = |F_h| \exp(i\phi_h) \qquad (4.1)$$

F_h and the electron density function $\varrho(r)$ are related by

$$F_h = \int_V \varrho(r) \exp(2\pi i h \cdot r) \, dV \qquad (4.2)$$

and

$$\varrho(r) = \frac{1}{V} \sum_h F_h \exp(-2\pi i h \cdot r) = \frac{1}{V} \sum_h |F_h| \exp(i\phi_h - 2\pi i h \cdot r) \qquad (4.3)$$

in which V represents the unit cell volume. Only the magnitudes $|F_h|$ of a finite number of structure factors are obtainable from experiment; the values of the phases ϕ_h, which are also needed if one is to determine $\varrho(r)$ from (4.3), cannot be determined experimentally. Thus the density function $\varrho(r)$ is not uniquely determined by the observed magnitudes $|F_h|$ alone. Even if one imposes the known condition that $\varrho(r) \geq 0$ for all r, the $|F_h|$ are still not sufficient to determine $\varrho(r)$ uniquely. In view of (4.1) and (4.2) it follows that the phase problem, to determine the values of the phases ϕ_h of the structure factors F_h when only the magnitudes $|F_h|$ are given, is, in principle, unsolvable when formulated in these terms.

Next, suppose that the real crystal, with continuous electron density $\varrho(r)$, is replaced by an idealized one, the unit cell of which consists of N discrete, nonvibrating, point atoms. Then the structure factor F_h is replaced by the normalized structure factor E_h and (4.1)–(4.3) are replaced by

$$E_h = |E_h| \exp(i\phi_h) \qquad (4.4)$$

$$E_h = \frac{1}{\sigma_2^{1/2}} \sum_{j=1}^N f_j^0 \exp(2\pi i h \cdot r_j) \qquad (4.5)$$

$$\langle E_h \exp(-2\pi i h \cdot r) \rangle_h = \frac{1}{\sigma_2^{1/2}} \left\langle \sum_{j=1}^N f_j^0 \exp(2\pi i h \cdot (r_j - r)) \right\rangle_h$$

$$\left.
\begin{aligned}
&= \frac{f_j^0}{\sigma_2^{1/2}} \quad \text{if } r = r_j \\
&= 0 \quad\quad\ \text{if } r \neq r_j
\end{aligned}
\right\} \qquad (4.6)$$

where f_j^0 is the zero-angle atomic scattering factor, r_j is the position vector of the atom labeled j, and

$$\sigma_2 = \sum_{j=1}^{N} (f_j^0)^2 \tag{4.7}$$

In X-ray diffraction the f_j^0 are equal to the atomic numbers Z_j and are therefore all positive; in neutron diffraction some of the f_j^0 may be negative. It is noteworthy that the methods to be described here are equally valid whether or not some of the f_j^0 are negative, so that the application to neutron diffraction is automatic.

In practice the magnitudes $|E_\mathbf{h}|$ of the normalized structure factors $E_\mathbf{h}$ are obtainable, at least approximately, from the observed magnitudes $|F_\mathbf{h}|$, while the phases $\phi_\mathbf{h}$, as defined by (4.4) and (4.5), cannot be determined experimentally. Since one now requires only the $3N$ components of the N position vectors $\mathbf{r}_j$, rather than the much more complicated electron density function $\varrho(\mathbf{r})$, it turns out that, in general, the known magnitudes are more than sufficient. This is most readily seen from (4.5) which, if real and imaginary parts are equated, is in reality a system of $2n$ equations, where n is the number of known magnitudes $|E_\mathbf{h}|$. The unknowns consist of the n phases $\phi_\mathbf{h}$ and the $3N$ components of the vectors $\mathbf{r}_j$, or $n + 3N$ unknowns in all. Since the number of equations (4.5), $2n$, usually exceeds by far the number of unknowns, $n + 3N$, the problem to determine the phases $\phi_\mathbf{h}$ when only the magnitudes $|E_\mathbf{h}|$ are given (the phase problem) is now greatly overdetermined in general. Thus the phase problem is, in principle, solvable when reformulated in terms of fixed, point atoms. In what follows it will be assumed that the phase problem is uniquely solvable, that is, no homometric structures, other than enantiomorphs, exist.

4.3. Structure Invariants

Equation (4.6) implies that the normalized structure factors $E_\mathbf{h}$ determine the crystal structure. However, (4.5) does not imply that, conversely, the crystal structure determines the values of the normalized structure factors $E_\mathbf{h}$ since the position vectors $\mathbf{r}_j$ depend not only on the structure but on the choice of origin as well. It turns out nevertheless that the magnitudes $|E_\mathbf{h}|$ of the normalized structure factors are in fact uniquely determined by the crystal structure and are independent of the choice of origin, but that the values of the phases $\phi_\mathbf{h}$ depend also on the choice of origin. Although the values of the individual phases depend on the structure and the choice of origin, there exist certain linear combinations of the phases, the so-called structure invariants, whose values are determined by the structure alone and are independent of the choice of origin. Our next task is to determine those linear combinations of the phases that are structure invariants for all the space groups.

If the origin of coordinates is shifted to a point whose position vector with respect to the initial origin is r_0, then r_j, the position vector of the jth atom with respect to the first origin, is replaced by

$$r_j' = r_j - r_0 \tag{4.8}$$

and $E_\mathbf{h}$ of (4.5) is replaced by

$$E_\mathbf{h}' = \frac{1}{\sigma_2^{1/2}} \sum_{j=1}^{N} f_j^0 \exp(2\pi i\mathbf{h} \cdot r_j') \tag{4.9}$$

Hence,

$$E_\mathbf{h}' = \frac{1}{\sigma_2^{1/2}} \sum_{j=1}^{N} f_j^0 \exp[2\pi i\mathbf{h} \cdot (r_j - r_0)] \tag{4.10}$$

$$E_\mathbf{h}' = \frac{\exp(-2\pi i\mathbf{h} \cdot r_0)}{\sigma_2^{1/2}} \sum_{j=1}^{N} f_j^0 \exp(2\pi i\mathbf{h} \cdot r_j) \tag{4.11}$$

In view of (4.5),

$$E_\mathbf{h}' = E_\mathbf{h} \exp(-2\pi i\mathbf{h} \cdot r_0) \tag{4.12}$$

whence

$$|E_\mathbf{h}'| = |E_\mathbf{h}| \quad \text{and} \quad \phi_\mathbf{h}' = \phi_\mathbf{h} - 2\pi\mathbf{h} \cdot r_0 \tag{4.13}$$

and $\phi_\mathbf{h}'$ is the phase of the structure factor $E_\mathbf{h}$ with respect to the new origin. Consider any finite linear combination of both sides of (4.13) having integer coefficients $A_\mathbf{h}$ which depend upon $\mathbf{h}$:

$$\sum_\mathbf{h} A_\mathbf{h}\phi_\mathbf{h}' = \sum_\mathbf{h} A_\mathbf{h}\phi_\mathbf{h} - 2\pi\left(\sum_\mathbf{h} A_\mathbf{h}\mathbf{h}\right) \cdot r_0 \tag{4.14}$$

Clearly, if

$$\sum_\mathbf{h} A_\mathbf{h}\mathbf{h} = 0 \tag{4.15}$$

then

$$\sum_\mathbf{h} A_\mathbf{h}\phi_\mathbf{h}' = \sum_\mathbf{h} A_\mathbf{h}\phi_\mathbf{h} \tag{4.16}$$

no matter what the vector r_0 may be, and the linear combination of the phases (4.16) is a structure invariant since it has the same value for every choice of origin. The proof of the fundamental theorems is now completed:

Theorem 1. If the origin of coordinates is shifted to a new point having position vector $\mathbf{r}_0$ with respect to the old origin, then the phase $\phi_\mathbf{h}$ of the normalized structure factor $E_\mathbf{h}$ with respect to the old origin is replaced by the new phase $\phi_\mathbf{h}'$ with respect to the new origin given by

$$\phi_\mathbf{h}' = \phi_\mathbf{h} - 2\pi\mathbf{h} \cdot \mathbf{r}_0 \tag{4.13}$$

Theorem 2. The linear combination of the phases

$$\sum_\mathbf{h} A_\mathbf{h}\phi_\mathbf{h} \tag{4.17}$$

where the $A_\mathbf{h}$ are integers satisfying

$$\sum_\mathbf{h} A_\mathbf{h}\mathbf{h} = 0 \tag{4.18}$$

is a structure invariant for every space group.

Corollary 2.1. If

$$\mathbf{h}_1 + \mathbf{h}_2 + \mathbf{h}_3 = 0 \tag{4.19}$$

then

$$\phi_{\mathbf{h}_1} + \phi_{\mathbf{h}_2} + \phi_{\mathbf{h}_3} \tag{4.20}$$

is a structure invariant for every space group.

Corollary 2.2. If

$$\mathbf{h}_1 + \mathbf{h}_2 + \mathbf{h}_3 + \mathbf{h}_4 = 0 \tag{4.21}$$

then

$$\phi_{\mathbf{h}_1} + \phi_{\mathbf{h}_2} + \phi_{\mathbf{h}_3} + \phi_{\mathbf{h}_4} \tag{4.22}$$

is a structure invariant for every space group.

4.4. Structure Seminvariants

For all space groups other than $P1$ the origin may not be chosen arbitrarily if the simplification permitted by the space-group symmetries is to be realized. For example, if a crystal has a center of symmetry it is natural to place the origin at such a center, while if a twofold screw axis, but no other symmetry element, is present, the origin would normally be situated on this axis. In such cases the permissible origins are greatly restricted and it is therefore plausible to assume that many linear combinations of the phases, in addition to those permitted by Theorem 2, will remain unchanged in value when the origin is shifted only in the restricted ways allowed by the space group symmetries. One is thus led to the notion of the structure seminvariant, those linear combinations of the phases whose values are independent of the choice of permissible origin.

(a) *Equivalence Concept.* For any space group the coordinates of equivalent positions depend upon the choice of origin. Hence the functional form for the geometric structure factor also depends on the choice of origin. Two origins will be said to be equivalent if they give rise to the same functional form for the geometric structure factor. Alternatively, two points are equivalent if they are geometrically situated in the same way with respect to the symmetry elements. Thus, in the space group $P1$ all points are equivalent; in $P\bar{1}$ all eight centers of symmetry are equivalent, but no other point is equivalent to any of these eight; in $P2_1$ all corresponding points on any of the four twofold screw axes are equivalent, but no other point is equivalent to a point on a twofold screw axis; in $P4$ all points on either of the two fourfold axes are equivalent, all points on either of the two twofold axes are equivalent, but no point on a twofold axis is equivalent to any point on a fourfold axis. All points equivalent to a given point are equivalent to each other and are said to form an equivalence class.

(b) *Primary Origin.* The reader is referred to the *International Tables for X-Ray Crystallography*, Vol. I, for the definition of the primary origin for each space group.

(c) *Permissible Origins.* All points equivalent to the primary origin constitute the permissible origins for each space group.

(d) *Structure Seminvariants.* The structure seminvariants are those linear combinations of the phases whose values are uniquely determined by the crystal structure alone, no matter what the choice of permissible origin. Alternatively, for a given functional form for the geometric structure factor, the values of the structure seminvariants are determined by the structure alone.

The structure invariants and seminvariants have been tabulated for all the space groups (Hauptman and Karle, 1953, 1956, 1959; Karle and Hauptman, 1961; Lessinger and Wondratschek, 1975).

(e) *Space Group $P\bar{1}$.* In view of Theorem 2 the linear combination

$$2\phi_{-\mathbf{h}} + \phi_{2\mathbf{h}} \qquad (4.23)$$

is a structure invariant. In $P\bar{1}$, $\phi = 0$ or π if the origin is permissible. Hence $2\phi_{-\mathbf{h}} = 0$ and, in view of (4.23), the single phase $\phi_{2\mathbf{h}}$ is a structure seminvariant for every reciprocal vector $\mathbf{h}$. In other words the value of $\phi_{2\mathbf{h}}$ is determined uniquely by the crystal structure no matter which of the eight permissible origins (centers of symmetry) is selected.

Next, suppose that $\mathbf{h}_1$ and $\mathbf{h}_2$ are two reciprocal vectors the corresponding components of which have the same parity. Then $\mathbf{h}_1 + \mathbf{h}_2 = 2\mathbf{h}$, the linear combination

$$\phi_{\mathbf{h}_1} + \phi_{\mathbf{h}_2} + \phi_{-2\mathbf{h}} \qquad (4.24)$$

is (in view of Theorem 2) a structure invariant and, since $\phi_{-2\mathbf{h}}$ is itself a structure seminvariant, the linear combination

$$\phi_{\mathbf{h}_1} + \phi_{\mathbf{h}_2} \qquad (4.25)$$

is also a structure seminvariant. In other words, if corresponding components of $\mathbf{h}_1$ and $\mathbf{h}_2$ have the same parity, then the value of the linear combination (4.25) depends only on the structure and is independent of the choice of permissible origin.

In a similar way it may be shown that if $\mathbf{h}_1$, $\mathbf{h}_2$, and $\mathbf{h}_3$ are reciprocal vectors the components of whose sum are even integers, then the linear combination

$$\phi_{\mathbf{h}_1} + \phi_{\mathbf{h}_2} + \phi_{\mathbf{h}_3} \qquad (4.26)$$

is a structure seminvariant.

(f) *Space Groups P2 and $P2_1$.* If b is the unique axis, the single phase ϕ_{h0l} is a structure seminvariant if h and l are both even. If $h_1 + h_2$ is even, $k_1 + k_2 = 0$, $l_1 + l_2$ is even, then $\phi_{\mathbf{h}_1} + \phi_{\mathbf{h}_2}$ is a structure seminvariant, where $\mathbf{h}_1 = (h_1 k_1 l_1)$, $\mathbf{h}_2 = (h_2 k_2 l_2)$. Finally, if $h_1 + h_2 + h_3$ is even, $k_1 + k_2 + k_3 = 0$, $l_1 + l_2 + l_3$ is even, then $\phi_{\mathbf{h}_1} + \phi_{\mathbf{h}_2} + \phi_{\mathbf{h}_3}$ is a structure seminvariant, where $\mathbf{h}_1 = (h_1 k_1 l_1)$, $\mathbf{h}_2 = (h_2 k_2 l_2)$, $\mathbf{h}_3 = (h_3 k_3 l_3)$.

(g) *Exercise.* What are the permissible origins for each of $P222$, $P2_1 2_1 2_1$, and $Pmm2$? Determine the structure seminvariants for these space groups.

(h) *Space Group P4.* In order to determine those single phases $\phi_{\mathbf{h}}$ that are structure seminvariants, one refers to Theorem 1. The permissible origins consist of all points lying on either of the two fourfold axes so that $\mathbf{r}_0$ must be of the form

$$0,\ 0,\ z \qquad \text{or} \qquad \tfrac{1}{2},\ \tfrac{1}{2},\ z \qquad (4.27)$$

where z is arbitrary. Then, in view of Theorem 1,

$$2\pi\mathbf{h} \cdot \mathbf{r}_0 = 2\pi n \tag{4.28}$$

where n is an integer and $\mathbf{r}_0$ may take any of the values (4.27). Writing $\mathbf{h} = (hkl)$, (4.28) becomes

$$h \cdot 0 + k \cdot 0 + lz = \text{an integer}$$
$$h \cdot \tfrac{1}{2} + k \cdot \tfrac{1}{2} + lz = \text{an integer} \tag{4.29}$$

Clearly (4.29) is satisfied for all z if and only if $h + k$ is an even integer and $l = 0$. Hence the single phase $\phi_\mathbf{h} = \phi_{hkl}$ is a structure seminvariant if $h + k$ is even and $l = 0$.

Similarly, $\phi_{\mathbf{h}_1} + \phi_{\mathbf{h}_2}$ is a structure seminvariant if $(h_1 + h_2) + (k_1 + k_2)$ is an even integer and $l_1 + l_2 = 0$, where $\mathbf{h}_1 = (h_1 k_1 l_1)$, $\mathbf{h}_2 = (h_2 k_2 l_2)$.

(*i*) *Exercise.* Determine the permissible origins and the structure seminvariants, in particular the single phases that are structure seminvariants, for each of $P4_1$, $P\bar{4}$, $P422$, $P3$, $P3_1$, $P312$, $P3_112$, $P321$, $P3_121$, $R3$, and $R32$.

In any space group any structure invariant is also a structure seminvariant, but, in general, the converse is not true. Hence the term structure seminvariant encompasses also the structure invariants: in space group $P1$ the two concepts coincide.

4.5. The Structure Seminvariants Link the Observed Magnitudes $|E|$ with the Desired Phases ϕ

It is known that the values of a sufficiently extensive set of cosine seminvariants, the cosines of the structure seminvariants, lead unambiguously to the values of the individual phases (Hauptman, 1972). The magnitudes $|E|$ are capable of yielding estimates of only the cosine seminvariants, or, equivalently, the magnitudes of the structure seminvariants: the signs of the structure seminvariants are ambiguous because the two enantiomorphous structures permitted by the observed magnitudes $|E|$ correspond to two values of each structure seminvariant differing only in sign. However, once the enantiomorph has been selected by specifying arbitrarily the sign of a particular structure seminvariant different from 0 or π, then the existence of conditional distributions which assume the values of one or more structure seminvariants to be known, as well as the magnitudes $|E|$, permits the estimation of both signs and magnitudes of structure seminvariants consistent with the chosen enantiomorph. In this way both signs and magnitudes of an extensive class of structure seminvariants, rather than the magnitudes alone, are available for phase determination. In the remainder of this chapter, only the problem of estimating the values of the structure seminvariants through suitable conditional probability distributions is studied; the problem of determining the values of the individual phases, assuming as known the values of the structure seminvariants, is relatively straightforward (Hauptman, 1972) and is not treated here.

4.6. Probabilistic Background

It is assumed throughout that a crystal structure is fixed. Reciprocal space is denoted by W. Then the n-fold Cartesian product $W \times W \times \cdots$ consists of the collection of all ordered n-tuples of reciprocal vectors $(\mathbf{h}, \mathbf{k}, \ldots)$. The primitive random variable is the ordered n-tuple $(\mathbf{h}, \mathbf{k}, \ldots)$ which is assumed to be uniformly distributed over a well-defined subset of the n-fold Cartesian product $W \times W \times \cdots$. Then the structure seminvariant

$$\phi = \phi_{\mathbf{h}} + \phi_{\mathbf{k}} + \cdots \tag{4.30}$$

as a function of the primitive random variables $\mathbf{h}, \mathbf{k}, \ldots$, is itself a random variable, and it is on this basis that the conditional probability distribution of ϕ is derived. The distribution leads to an estimate for ϕ which is particularly good in the favorable case that the variance of the distribution happens to be small.

4.7. Three-Phase Structure Invariant

4.7.1. Space Group $P1$

Suppose that a crystal structure consisting of N atoms, not necessarily identical, per unit cell in $P1$ is fixed. Denote by $\phi_{\mathbf{h}}$ the phase of the normalized structure factor $E_{\mathbf{h}}$. Assume also that R_1, R_2, and R_3 are fixed nonnegative numbers. Finally, let the primitive random variable (vector) be the ordered triple $(\mathbf{h}, \mathbf{k}, \mathbf{l})$ of reciprocal vectors $\mathbf{h}, \mathbf{k}, \mathbf{l}$ which is assumed to be uniformly distributed over the subset of the threefold Cartesian product $W \times W \times W$ defined by

$$|E_{\mathbf{h}}| = R_1, \qquad |E_{\mathbf{k}}| = R_2, \qquad |E_{\mathbf{l}}| = R_3 \qquad (4.31)$$

and

$$\mathbf{h} + \mathbf{k} + \mathbf{l} = 0 \qquad (4.32)$$

Note that, in order to ensure that the domain of the primitive random variable be nonvacuous, it is necessary to interpret the exact equality $|E_{\mathbf{h}}| = R_1$ of (4.31), for example, as the inequality $R_1 \le |E_{\mathbf{h}}| \le R_1 + dR_1$, where dR_1 is a small quantity, etc. Then, the structure invariant

$$\phi_3 = \phi_{\mathbf{h}} + \phi_{\mathbf{k}} + \phi_{\mathbf{l}} \qquad (4.33)$$

is a function of the primitive random variables $\mathbf{h}, \mathbf{k}, \mathbf{l}$ and therefore is itself a random variable. Denote by

$$P_{1|3} = P(\Phi \mid R_1, R_2, R_3) \qquad (4.34)$$

the conditional probability distribution of ϕ_3, given the three magnitudes (4.31). Then (Cochran, 1955; Hauptman, 1976a,b),

$$P_{1|3} \approx \frac{1}{2\pi I_0(A)} \exp(A \cos \Phi) \qquad (4.35)$$

where I_0 is the modified Bessel function, A is defined by

$$A = \frac{\sigma_3}{\sigma_2^{3/2}} R_1 R_2 R_3 \qquad (4.36)$$

and

$$\sigma_n = \sum_{j=1}^{N} f_j^{0n} \qquad (4.37)$$

where f_j^0 is the zero-angle atomic scattering factor of the atom labeled j.

Graphs of the distribution (4.35) for $A = 2.316$ and $A = 0.731$ are shown in Figs. 4.1 and 4.2. Clearly this distribution has a unique maximum at $\Phi = 0$ in the interval $-\pi$ to π, so that the most probable value of ϕ_3 is zero. The larger the value of A the smaller is the variance of the distribution and the more reliable is the estimate of ϕ_3—zero in this case.

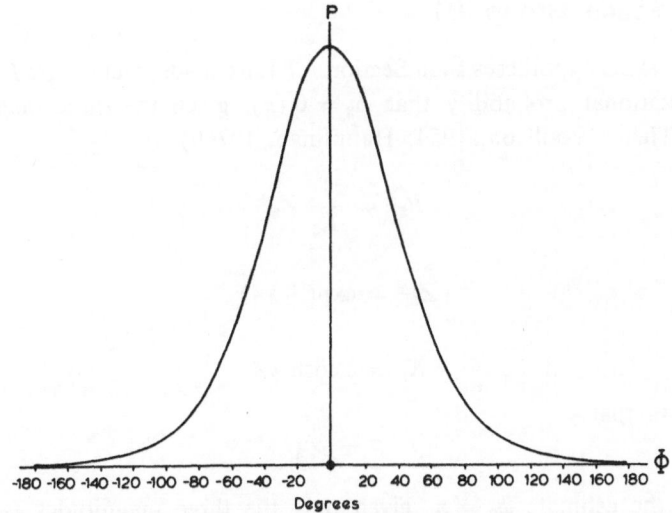

FIGURE 4.1. Distribution $P_{1|3}$, (4.35), for $A = 2.316$.

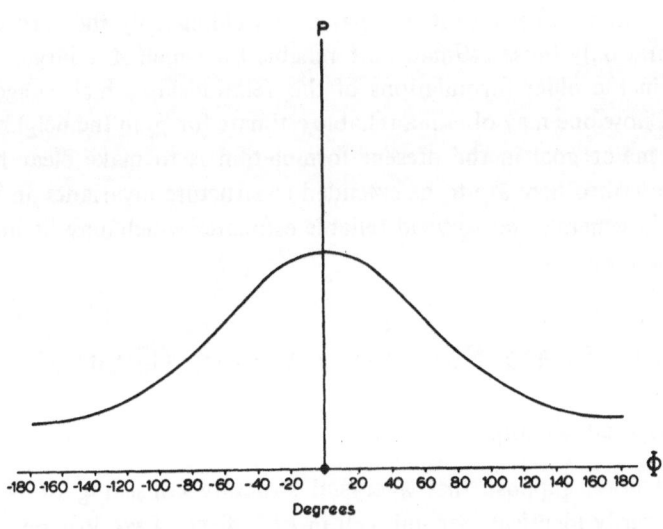

FIGURE 4.2. Distribution $P_{1|3}$, (4.35), for $A = 0.731$.

4.7.2. Space Group $P\bar{1}$

The same hypotheses as in Section 4.7.1 are made. Denote by P_3^+ (P_3^-) the conditional probability that $\phi_3 = 0$ (π), given the three magnitudes (4.31). Then (Woolfson, 1954; Hauptman, 1976b),

$$P_3^\pm = \frac{1}{K_3} Z_3^\pm \qquad (4.38)$$

where

$$Z_3^\pm = \exp(\pm \tfrac{1}{2}A) \qquad (4.39)$$

and

$$K_3 = 2 \cosh \tfrac{1}{2}A \qquad (4.40)$$

It follows that

$$P_3^+ > \tfrac{1}{2} \qquad (4.41)$$

so that the estimate $\phi_3 = \pi$, given only the three magnitudes (4.31), is impossible.

The results described in this section are often written

$$\phi_3 = \phi_\mathbf{h} + \phi_\mathbf{k} + \phi_\mathbf{l} \approx 0 \qquad (4.42)$$

which, for $P1$, means that ϕ_3 is probably close to zero, and, for $P\bar{1}$, that ϕ_3 is probably equal to zero. Furthermore, the larger the value of A, (4.36), the more likely is the probabilistic statement (4.42). It is remarkable how useful this relationship has proved to be in applications; and yet (4.42) is severely limited because it is capable of yielding only the zero estimate for ϕ_3, and only those estimates are reliable for which A is large. There is nothing in the older formulations of this relationship which suggests, for example, how one may obtain a reliable estimate for ϕ_3 in the neighborhood of π. A major goal in the present formulation is to make clear how the results described here are to be extended to structure invariants and seminvariants in general, and to yield reliable estimates which may lie anywhere in the interval $-\pi$ to π.

4.8. Four-Phase Structure Invariant (Quartet)

4.8.1. Space Group $P1$

As before, suppose that a crystal structure consisting of N atoms, not necessarily identical, per unit cell in $P1$ is fixed. Two distributions will be obtained. The first, in strict analogy with Section 4.7.1, is the conditional probability distribution of the quartet, given four magnitudes; the second, marking a radical departure from all earlier ones, is the conditional probability distribution of the quartet, assuming that an appropriate set of seven magnitudes is known.

4.8.1.1. Four-Magnitude Distribution

Assume that R_1, R_2, R_3, and R_4 are fixed nonnegative numbers. Next, suppose that the primitive random variable (vector) is the ordered quadruple $(\mathbf{h}, \mathbf{k}, \mathbf{l}, \mathbf{m})$ of reciprocal vectors $\mathbf{h}, \mathbf{k}, \mathbf{l}, \mathbf{m}$, which is assumed to be uniformly distributed over the subset of the fourfold Cartesian product $W \times W \times W \times W$ defined by

$$|E_\mathbf{h}| = R_1, \qquad |E_\mathbf{k}| = R_2, \qquad |E_\mathbf{l}| = R_3, \qquad |E_\mathbf{m}| = R_4 \qquad (4.43)$$

and

$$\mathbf{h} + \mathbf{k} + \mathbf{l} + \mathbf{m} = 0 \qquad (4.44)$$

In view of (4.44), the linear function of four phases

$$\phi_4 = \phi_\mathbf{h} + \phi_\mathbf{k} + \phi_\mathbf{l} + \phi_\mathbf{m} \qquad (4.45)$$

is a structure invariant which, as a function of the primitive random variables $\mathbf{h}, \mathbf{k}, \mathbf{l}, \mathbf{m}$, is itself a random variable. Denote by

$$P_{1|4} = P(\Phi \mid R_1, R_2, R_3, R_4) \qquad (4.46)$$

the conditional probability distribution of ϕ_4, given the four magnitudes (4.43). Then (Hauptman, 1975a,b, 1976b),

$$P_{1|4} \approx \frac{1}{2\pi I_0(B)} \exp(B \cos \Phi) \qquad (4.47)$$

where B is defined by

$$B = \frac{\sigma_4}{\sigma_2{}^2} R_1 R_2 R_3 R_4 \qquad (4.48)$$

and σ_n by (4.37). Thus, $P_{1|4}$ is identical with $P_{1|3}$, but B replaces A. Hence, similar remarks apply to $P_{1|4}$. In particular, (4.47) always has a unique maximum at $\Phi = 0$, so that the most probable value of the structure invariant (4.45) is zero, and the larger the value of B, the more likely it is that $\phi_4 \approx 0$. Since B values, of order $1/N$, tend to be less than A values, of order $1/N^{1/2}$, the estimate of zero for the quartet (4.45) is in general less reliable than the estimate of zero for the triple (4.33). Hence (4.47) is certainly no improvement over (4.35) and the goal of obtaining a reliable nonzero estimate for a structure invariant is not realized by (4.47). The decisive step in this direction is made in Section 4.8.1.2.

4.8.1.2. Seven-Magnitude Distribution

If one assumes as known not only the four magnitudes (4.43), but the additional three magnitudes $|E_{h+k}|$, $|E_{k+l}|$, and $|E_{l+h}|$, then, in favorable cases, one obtains a more reliable estimate for the quartet (4.45) and, furthermore, the estimate may lie anywhere in the interval 0 to π. The idea of using these three additional magnitudes is suggested by the following heuristic argument (Schenk and de Jong, 1973; Schenk, 1973, 1974).

Assume first that the four magnitudes (4.43)

$$|E_h|, \qquad |E_k|, \qquad |E_l|, \qquad |E_m| \qquad \text{are all large} \qquad (4.49)$$

where h, k, l, m satisfy (4.44). If it should happen that $|E_{h+k}|$ is also large then Section 4.7 implies

$$\phi_h + \phi_k + \phi_{-h-k} \approx 0 \qquad (4.50)$$

and, in view of (4.44),

$$\phi_l + \phi_m + \phi_{h+k} \approx 0 \qquad (4.51)$$

Then, by addition of (4.50) and (4.51),

$$\phi_h + \phi_k + \phi_l + \phi_m \approx 0 \qquad (4.52)$$

If $|E_{k+l}|$ is large of if $|E_{l+h}|$ is large then, by symmetry, the relationship (4.52) would again hold. If it should happen that

$$|E_{h+k}|, \qquad |E_{k+l}|, \qquad |E_{l+h}| \qquad \text{are all large} \qquad (4.53)$$

then one would again expect (4.52) to hold, but now with a very high probability. However, the distribution (4.47) implies that the value of the quartet (4.45) must occasionally be in the neighborhood of π, even if B is quite large. How is one to identify the small fraction of quartets which are in fact near π? In view of the foregoing argument, it is plausible to suppose that

$$\phi_h + \phi_k + \phi_l + \phi_m \approx \pi \qquad (4.54)$$

precisely in the case that

$$|E_{h+k}|, \qquad |E_{k+l}|, \qquad |E_{l+h}| \qquad \text{are all small} \qquad (4.55)$$

Although the argument presented here is only heuristic and proves nothing, it does suggest the question: what is the conditional probability distribution of the quartet (4.45), assuming as known the seven magnitudes $|E_h|$, $|E_k|$, $|E_l|$, $|E_m|$, $|E_{h+k}|$, $|E_{k+l}|$, and $|E_{l+h}|$?

In order to answer this question assume that the seven nonnegative numbers R_1, R_2, R_3, R_4, R_{12}, R_{23}, and R_{31} are fixed. Suppose next that the ordered quadruple of reciprocal vectors $(\mathbf{h}, \mathbf{k}, \mathbf{l}, \mathbf{m})$ is a random variable which is uniformly distributed over the subset of the fourfold Cartesian product $W \times W \times W \times W$ defined by

$$|E_h| = R_1, \qquad |E_k| = R_2, \qquad |E_l| = R_3, \qquad |E_m| = R_4 \qquad (4.56)$$

and

$$|E_{h+k}| = R_{12}, \qquad |E_{k+l}| = R_{23}, \qquad |E_{l+h}| = R_{31} \qquad (4.57)$$

with

$$\mathbf{h} + \mathbf{k} + \mathbf{l} + \mathbf{m} = 0 \qquad (4.58)$$

Then the quartet (4.45) is a structure invariant which, as a function of the primitive random variable $(\mathbf{h}, \mathbf{k}, \mathbf{l}, \mathbf{m})$, is itself a random variable. Denote by

$$P_{1|7} = P(\Phi \mid R_1, R_2, R_3, R_4, R_{12}, R_{23}, R_{31}) \qquad (4.59)$$

the conditional probability distribution of the quartet (4.45), given the seven magnitudes (4.56) and (4.57). Then (Hauptman, 1975a,b, 1976b; Giacovazzo, 1975, 1976b),

$$P_{1|7} \approx \frac{1}{L} \exp(-2B' \cos \Phi) I_0\left(\frac{2\sigma_3}{\sigma_2^{3/2}} R_{12} X_{12}\right) I_0\left(\frac{2\sigma_3}{\sigma_2^{3/2}} R_{23} X_{23}\right)$$
$$\times I_0\left(\frac{2\sigma_3}{\sigma_2^{3/2}} R_{31} X_{31}\right) \qquad (4.60)$$

where

$$B' = \frac{1}{\sigma_2^3} (3\sigma_3^2 - \sigma_2\sigma_4) R_1 R_2 R_3 R_4 \qquad (4.61)$$

$$X_{12} = [R_1^2 R_2^2 + R_3^2 R_4^2 + 2R_1 R_2 R_3 R_4 \cos \Phi]^{1/2} \qquad (4.62)$$

$$X_{23} = [R_2^2 R_3^2 + R_1^2 R_4^2 + 2R_1 R_2 R_3 R_4 \cos \Phi]^{1/2} \qquad (4.63)$$

$$X_{31} = [R_3^2 R_1^2 + R_2^2 R_4^2 + 2R_1 R_2 R_3 R_4 \cos \Phi]^{1/2} \qquad (4.64)$$

σ_n is defined by (4.37), and L is a normalizing parameter, independent of Φ, which is not needed for the present purpose.

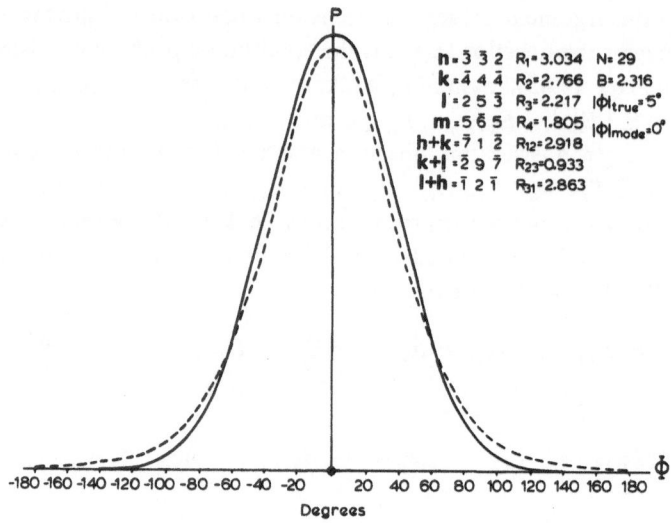

FIGURE 4.3. Distributions (4.60) (——) and (4.47) (- - -) for the values of the seven parameters (4.56) and (4.57) shown. The mode of (4.60) is 0, of (4.47) always 0.

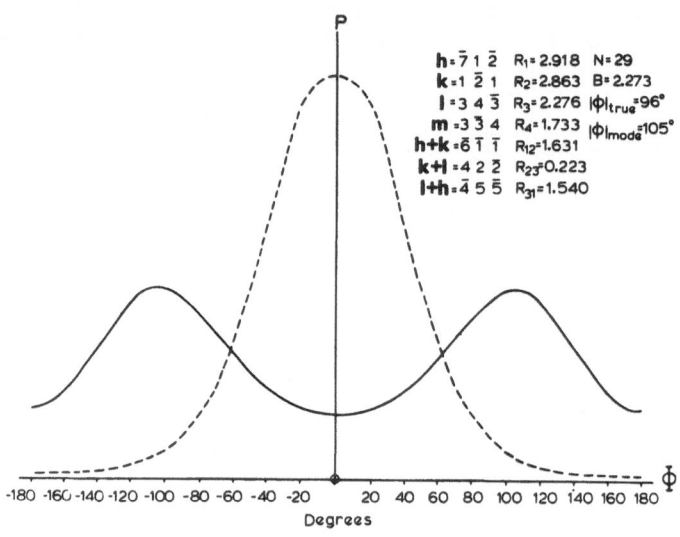

FIGURE 4.4. Distributions (4.60) (——) and (4.47) (- - -) for the values of the seven parameters (4.56) and (4.57) shown. The mode of (4.60) is 105°, of (4.47) always 0.

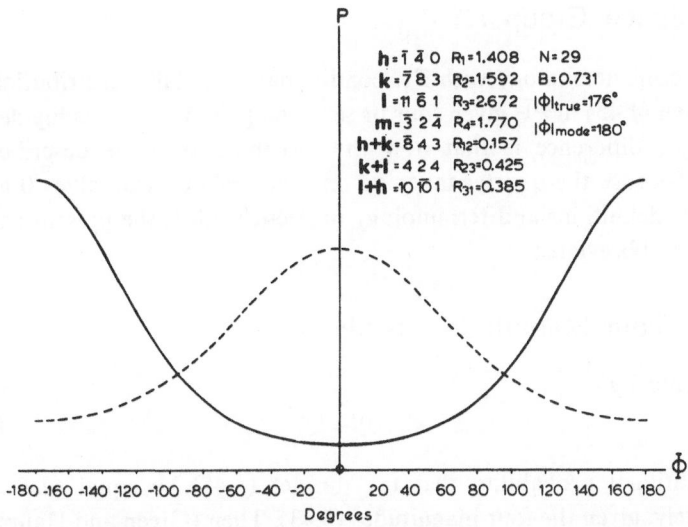

FIGURE 4.5. Distributions (4.60) (——) and (4.47) (– – –) for the values of the seven parameters (4.56) and (4.57) shown. The mode of (4.60) is 180°, of (4.47) always 0.

Figures 4.3–4.5 show the distribution (4.60) for typical values of the seven parameters (4.56) and (4.57). For comparison the distribution (4.47) is shown. Since the magnitudes $|E|$ have been obtained from a real structure, comparison with the true value of the quartet is also possible. As already emphasized, the distribution (4.47) always has a unique maximum at $\Phi = 0$. The distribution (4.60), on the other hand, may have a maximum at $\Phi = 0$, or π, or any value between these extremes, as shown by Figs. 4.3 to 4.5. Roughly speaking, the maximum of (4.60) occurs at 0 or π according as the three parameters R_{12}, R_{23}, and R_{31} are all large or all small, respectively, thus confirming the heuristic argument given earlier. The figures clearly show also the improvement which may result when, in addition to the four magnitudes (4.56), the three magnitudes (4.57) are also assumed to be known. Finally, in the special case that

$$R_{12} \approx R_{23} \approx R_{31} \approx 0 \tag{4.65}$$

the distribution (4.60) reduces to

$$P_{1|7} \approx \frac{1}{L} \exp(-2B' \cos \Phi) \tag{4.66}$$

which has a unique maximum at $\Phi = \pi$ (Fig. 4.5).

4.8.2. Space Group $P\bar{1}$

In complete analogy with $P1$, conditional probability distributions for the quartet (4.45) in $P\bar{1}$, given four or seven magnitudes, are readily derived. The major difference is that now the distributions to be described are discrete because the quartet takes on only one of the two values 0 and π. Using the definitions and terminology of Section 4.8.1, the present account is greatly abbreviated.

4.8.2.1. Four-Magnitude Distribution

Denote by

$$P_4^+ \text{ or } P_4^- \tag{4.67}$$

the conditional probability that the quartet (4.45) be equal to 0 or π, respectively, given the four magnitudes (4.43). Then (Green and Hauptman, 1976a; Hauptman and Green, 1976; Hauptman, 1976b),

$$P_4^{\pm} = \frac{1}{K_4} Z_4^{\pm} \tag{4.68}$$

where

$$Z_4^{\pm} = \exp(\pm \tfrac{1}{2}B) \tag{4.69}$$

and

$$K_4 = Z_4^+ + Z_4^- \tag{4.70}$$

The distribution (4.68)–(4.70) should be compared with the $P1$ analog (4.47). The variance of the distribution (4.68) is given by

$$\text{Var} = 4P_4^+ P_4^- = \frac{4Z_4^+ Z_4^-}{(Z_4^+ + Z_4^-)^2} \tag{4.71}$$

Clearly

$$P_4^+ > \tfrac{1}{2} \tag{4.72}$$

so that (4.68) can give only the zero estimate for the quartet. Finally, the smaller the value of the variance, the more reliable is the estimate, zero in this case.

4.8.2.2. Seven-Magnitude Distribution

Denote by

$$P_7^+ \text{ or } P_7^- \tag{4.73}$$

the conditional probability that the quartet (4.45) be equal to 0 or π, respectively, given the seven magnitudes (4.56) and (4.57). Then (Green and Hauptman, 1976a; Hauptman and Green, 1976; Hauptman, 1976b; Giacovazzo, 1976a),

$$P_7^{\pm} = \frac{1}{K_7} Z_7^{\pm} \tag{4.74}$$

where

$$Z_7^{\pm} = \exp(\mp B') \cosh\left(\frac{\sigma_3}{\sigma_2^{3/2}} R_{12} X_{12}^{\pm}\right) \cosh\left(\frac{\sigma_3}{\sigma_2^{3/2}} R_{23} X_{23}^{\pm}\right)$$
$$\times \cosh\left(\frac{\sigma_3}{\sigma_2^{3/2}} R_{31} X_{31}^{\pm}\right) \tag{4.75}$$

$$X_{12}^{\pm} = R_1 R_2 \pm R_3 R_4 \tag{4.76}$$

$$X_{23}^{\pm} = R_2 R_3 \pm R_1 R_4 \tag{4.77}$$

$$X_{31}^{\pm} = R_3 R_1 \pm R_2 R_4 \tag{4.78}$$

$$K_7 = Z_7^+ + Z_7^- \tag{4.79}$$

and the variance is given by

$$\text{Var} = 4 P_7^+ P_7^- = \frac{4 Z_7^+ Z_7^-}{(Z_7^+ + Z_7^-)^2} \tag{4.80}$$

In contrast to (4.68), P_7^+ may lie anywhere in the interval 0 to 1 so that (4.74)–(4.79) may give either 0 or π for the quartet. In complete analogy with $P1$, the estimate is 0 or π according as R_{12}, R_{23}, and R_{31} are all large or all small, respectively. In the extreme case that

$$R_{12} \approx R_{23} \approx R_{31} \approx 0 \tag{4.81}$$

(4.75) reduces to

$$Z_7^+ = \exp(\mp B') \tag{4.82}$$

Again, the smaller the value of the variance, the more reliable is the estimate.

4.8.2.3. An Application

A crystal structure in $P\bar{1}$ consisting of $N = 90$ identical atoms in the unit cell was constructed and 6701 normalized structure factors E_h calculated. Using the 299 $|E|$'s > 2 ($|E_h| > 2$, $|E_k| > 2$, $|E_l| > 2$, $|E_m| > 2$), the 1000 quartets

$$\mathbf{h} + \mathbf{k} + \mathbf{l} + \mathbf{m} = 0 \tag{4.83}$$

TABLE 4.1. Fifty Representative Values of P_7^+, (4.74)–(4.79), in Ascending Order of $(Var)^{1/2}$, (4.80), (SIG), for a Structure in $P\bar{1}$ Consisting of $90(N)$ Identical Atoms in the Unit Cell [a]

No.	h	k	l	m	Observed magnitudes, $|E|$							B	$\cos(T)$	$P(+)$	SIG
					h	k	l	m	h+k	k+l	l+h				
6	$8\ \bar{10}\ 0$	$\bar{3}\ \bar{6}\ 3$	$\bar{4}\ 6\ \bar{8}$	$\bar{1}\ 10\ 5$	3.68	2.95	2.81	2.39	3.00	4.25	3.56	1.62	0.999	0.999	0.002
7	$4\ \bar{4}\ \bar{8}$	$\bar{2}\ 1\ 10$	$1\ \bar{1}\ 5$	$\bar{3}\ 6\ 3$	3.56	3.46	3.39	2.95	4.17	2.39	2.41	2.74	0.999	0.999	0.003
8	$2\ \bar{11}\ 0$	$6\ 1\ 0$	$1\ \bar{1}\ 5$	$\bar{9}\ 11\ 5$	3.46	3.44	3.39	2.64	3.68	4.25	1.56	2.36	0.999	0.999	0.003
9	$1\ 12\ 0$	$\bar{2}\ \bar{6}\ \bar{10}$	$\bar{3}\ 0\ 1$	$8\ \bar{11}\ \bar{1}$	3.66	3.44	2.99	2.35	2.77	3.72	3.70	1.96	0.999	0.999	0.003
10	$2\ 7\ \bar{8}$	$2\ \bar{11}\ 0$	$2\ 6\ 7$	$\bar{6}\ \bar{2}\ 1$	4.17	3.46	3.33	2.23	3.56	4.18	1.50	2.38	0.999	0.999	0.007
186	$2\ 7\ 8$	$7\ \bar{6}\ 7$	$\bar{6}\ 2\ 3$	$3\ 3\ \bar{2}$	4.17	3.36	3.00	2.44	3.72	1.19	2.14	2.27	0.999	0.998	0.074
187	$2\ 7\ 8$	$\bar{9}\ \bar{1}\ 1$	$8\ \bar{11}\ \bar{1}$	$\bar{1}\ 5\ 8$	4.17	3.72	2.35	2.34	3.36	3.66	0.26	1.89	0.999	0.998	0.074
188	$4\ \bar{5}\ 7$	$\bar{6}\ 2\ 3$	$7\ \bar{7}\ \bar{10}$	$\bar{5}\ 10\ 0$	4.18	3.00	2.68	2.28	1.49	3.96	1.35	1.71	0.999	0.998	0.074
189	$1\ 12\ 0$	$6\ \bar{2}\ \bar{3}$	$\bar{3}\ 0\ 1$	$\bar{4}\ \bar{10}\ 2$	3.66	3.00	2.99	2.22	2.19	1.33	3.70	1.63	0.999	0.998	0.074
190	$4\ \bar{5}\ 7$	$3\ 0\ \bar{1}$	$1\ 11\ 0$	$\bar{8}\ 6\ \bar{6}$	4.18	2.99	2.76	2.03	1.89	3.02	2.30	1.55	0.999	0.998	0.075
401	$1\ \bar{5}\ 7$	$7\ 1\ 4$	$2\ 1\ 0$	$\bar{10}\ 3\ 3$	3.96	3.12	2.58	2.38	1.96	1.32	2.67	1.69	1.000	0.991	0.179
402	$4\ \bar{5}\ 7$	$\bar{7}\ 9\ 0$	$\bar{3}\ \bar{1}\ 0$	$6\ 3\ 7$	4.18	2.63	2.61	2.57	2.67	1.33	2.03	1.63	0.999	0.991	0.181
403	$8\ \bar{10}\ 0$	$1\ \bar{1}\ 5$	$\bar{5}\ 11\ 0$	$4\ 0\ 5$	3.68	3.39	2.77	2.09	2.64	0.97	2.61	1.60	0.999	0.991	0.181
404	$1\ \bar{1}\ 5$	$1\ 7\ 7$	$6\ 3\ \bar{9}$	$\bar{8}\ 11\ 3$	3.39	3.04	2.77	2.46	1.97	1.50	2.31	1.56	0.999	0.991	0.182
405	$7\ 0\ 5$	$\bar{1}\ \bar{11}\ 0$	$2\ 0\ \bar{2}$	$\bar{8}\ 11\ \bar{3}$	4.25	2.76	2.55	2.46	3.32	0.68	2.23	1.63	0.999	0.991	0.183
566	$4\ \bar{5}\ 7$	$5\ \bar{6}\ 7$	$3\ 9\ \bar{1}$	$2\ \bar{10}\ 1$	4.18	3.02	2.56	2.51	0.24	0.10	0.06	1.81	−1.000	0.029	0.339
567	$7\ 0\ 5$	$\bar{8}\ 10\ 0$	$3\ 0\ \bar{10}$	$2\ \bar{10}\ 5$	4.25	3.68	2.37	2.21	2.39	1.38	1.72	1.82	0.999	0.970	0.340
568	$4\ \bar{5}\ 7$	$\bar{8}\ 9\ 1$	$3\ 1\ 0$	$\bar{1}\ 5\ \bar{8}$	4.18	2.84	2.61	2.34	3.56	1.59	0.09	1.61	0.999	0.970	0.340
569	$1\ \bar{1}\ 5$	$3\ 10\ \bar{6}$	$7\ \bar{9}\ 0$	$\bar{11}\ 0\ 1$	3.39	3.16	2.63	2.51	1.35	1.53	2.09	1.56	0.999	0.970	0.340
570	$4\ \bar{4}\ 8$	$\bar{6}\ \bar{1}\ 0$	$5\ 2\ 8$	$7\ 3\ 0$	3.56	3.44	3.17	2.31	0.04	0.49	0.21	1.99	−0.999	0.030	0.342
636	$1\ \bar{5}\ 7$	$\bar{6}\ \bar{1}\ 0$	$1\ 7\ 7$	$4\ 13\ 0$	3.96	3.44	3.04	2.40	2.30	0.07	2.33	2.21	0.999	0.952	0.423
637	$7\ 0\ 5$	$\bar{3}\ 4\ 7$	$3\ 6\ 3$	$7\ 2\ 5$	4.25	3.00	2.95	2.26	0.01	0.38	0.58	1.89	−1.000	0.047	0.423

638	9 1 1̄	2 7 8̄	3 0 1̄	4 8̄ 8	8 7 8	2̄ 6̄ 4	3.72	2.99	2.80	2.36	1.25	1.04	2.51	1.63	0.999	0.952	0.423
639	10 5 1̄	7 0 5̄	3 0 1̄	10̄ 11̄ 2	3 6̄ 9	7 6̄ 1	3.26	2.99	2.77	2.64	2.68	1.34	0.66	1.59	0.999	0.953	0.423
640	4 5 7̄	6 6̄ 7̄	7 6̄ 7̄	2 1̄0̄ 3	5 9 1̄	5 9̄ 1̄	4.18	3.36	2.51	2.05	0.29	0.19	0.10	1.60	−0.999	0.047	0.424
641	8 1̄0̄ 0̄	9 1 1̄	1 1̄ 5	9 2 1̄	0 13̄ 3̄	0 13̄ 6̄	3.68	3.39	2.77	2.15	2.64	0.49	1.90	1.65	0.999	0.951	0.427
642	9 1 1̄	2 7 8̄	2̄ 11 0	4 1̄1̄ 1̄	3 1̄ 1̄	3 1̄ 1̄0̄	3.72	3.46	3.02	2.61	1.14	1.17	2.28	2.25	0.999	0.951	0.430
643	2 1̄2̄ 1̄	7 0 5̄	8 1̄0̄ 0̄	3 1 1̄	3 3 1̄	3 1̄ 3̄	3.70	3.68	2.67	2.44	0.14	0.62	0.43	1.96	−0.999	0.049	0.434
644	9 1 1̄	2 1̄1̄ 0̄	1̄ 12̄ 0̄	0 4 7̄	8 7 8̄	8̄ 7̄ 6̄	3.72	3.66	2.45	2.36	2.35	1.04	1.61	1.75	0.999	0.950	0.435
645	1 9 7̄	4 5 7̄	2 1̄1̄ 0̄	6 1̄ 0̄	3 3 7̄	3 3̄ 7̄	3.54	3.46	3.44	2.26	0.28	0.47	0.68	2.12	−0.999	0.050	0.439
781	2 7 8̄	3 0 1̄	3 0 1̄	2 1 0̄	7 8̄ 9̄	9 8̄ 5̄	4.17	2.99	2.58	2.22	0.74	0.68	2.80	1.59	1.000	0.871	0.670
782	7 0 5̄	9 1̄ 1̄	9 1̄ 1̄	2̄ 12̄ 1̄	0 13̄ 5	5 13̄ 7̄	4.25	3.72	3.70	2.05	1.49	1.54	1.46	2.66	0.999	0.869	0.673
783	6 6̄ 7̄	6 6̄ 7̄	2 6̄ 10̄	3 0 3̄	3 7 1̄	7̄ 3̄ 6̄	3.57	3.44	2.99	2.08	1.42	1.25	1.51	1.69	0.999	0.867	0.677
784	9 1 1̄	1 7̄ 7̄	1 7̄ 7̄	5 3̄ 8̄	5 3̄ 2̄	2 3̄ 2̄	3.72	3.04	2.99	2.07	2.08	0.83	1.35	1.55	0.999	0.867	0.678
785	2 7 8̄	7 1 4̄	7 1 4̄	3 6̄ 4̄	6 2̄ 0̄	2̄ 2̄ 0̄	4.17	3.12	2.64	2.06	0.86	0.58	0.14	1.57	−1.000	0.133	0.680
786	7 0 5̄	3̄ 10 6	3̄ 10 6	7 9 0̄	3 1̄ 11̄	3 1̄1̄ 1̄	4.25	3.16	2.63	2.18	1.18	1.53	1.45	1.71	−1.000	0.865	0.681
787	2 7 8̄	6 1 0̄	6 1 0̄	9 2 1̄	1 6̄ 7̄	7 6̄ 8̄	4.17	3.44	2.77	2.03	0.20	1.79	1.93	1.79	0.999	0.865	0.683
788	2 1̄1̄ 0̄	1̄ 1̄ 5̄	1̄ 1̄ 5̄	6 2 3̄	3 10̄ 8̄	3̄ 10̄ 0̄	3.46	3.39	3.00	2.43	1.56	1.23	1.28	1.90	0.999	0.863	0.686
789	6 1 0̄	0 9 8̄	0 9 8̄	8 9̄ 8̄	2 1̄ 0̄	2 1̄ 3̄	3.44	3.05	2.70	2.55	0.81	0.76	2.38	1.60	0.999	0.860	0.693
790	4 5 7̄	2 6̄ 10̄	2 6̄ 10̄	8 5̄ 1̄0̄	6 1̄ 3̄	6̄ 1̄ 3̄	4.18	3.44	2.54	2.24	1.39	1.26	1.52	1.81	0.999	0.859	0.694
881	2 7 8̄	4 4 8̄	4 4 8̄	3 3 2̄	1̄ 14̄ 2̄	2 14̄ 2̄	4.17	3.56	2.44	2.12	3.46	0.54	0.50	1.70	0.999	0.763	0.850
882	4 5 7̄	6 3̄ 9̄	6 3̄ 9̄	8 7̄ 2̄	2 1̄ 0̄	2 1̄ 0̄	4.18	2.77	2.59	2.55	1.52	0.25	0.36	1.70	−0.999	0.238	0.852
883	1 5 7̄	3 0 1̄	3 0 1̄	0 11̄ 2̄	2 6̄ 8̄	2 6̄ 8̄	3.96	2.99	2.65	2.30	0.97	0.50	2.18	1.60	1.000	0.760	0.853
884	1 8 7̄	5 2̄ 8̄	5 2̄ 8̄	8 9̄ 1̄	2 1̄ 0̄	2 1̄ 0̄	3.29	3.17	2.84	2.58	0.94	0.75	0.56	1.70	−1.000	0.243	0.858
885	8 1̄0̄ 0̄	1̄ 1 5	1̄ 1 5	6 5̄ 6̄	1̄ 14̄ 1̄	2 14̄ 1̄	3.68	3.39	2.50	2.28	0.69	1.00	0.50	1.58	−1.000	0.246	0.861
886	7 0 5̄	8 9̄ 4̄	8 9̄ 4̄	3 1 0̄	2 8̄ 9̄	8 8̄ 9̄	4.25	2.97	2.61	2.13	0.80	1.60	1.23	1.56	0.999	0.753	0.862
887	2 1̄2̄ 1̄	4̄ 9̄ 6̄	4̄ 9̄ 6̄	2 1̄ 0̄	0 4̄ 7̄	4 4̄ 7̄	3.70	3.06	2.55	2.45	2.41	0.23	0.99	1.57	−0.999	0.752	0.863
888	7 0 5̄	6 2 3̄	6 2 3̄	0 1̄ 1̄0̄	1̄ 3̄ 2̄	1̄ 3̄ 2̄	4.25	3.00	2.77	2.11	0.96	0.48	0.88	1.66	−1.000	0.251	0.867
889	10 8 1̄	1 1̄ 5	1 1̄ 5	6 2̄ 3̄	5̄ 7̄ 7̄	7̄ 7̄ 1̄	3.80	3.39	3.00	2.51	0.85	1.23	0.41	2.15	0.999	0.253	0.870
890	2 7 8̄	3 2 7̄	3 2 7̄	4̄ 13̄ 0̄	1̄ 4̄ 1̄	1̄ 4̄ 1̄	4.17	3.10	2.40	2.32	2.05	0.32	1.39	1.61	0.999	0.745	0.871

(a) One thousand values, arranged in increasing order of $(\mathrm{Var})^{1/2}$, were calculated and compared with the true values of the cosine, $\cos(T)$. The first error occurred at quartet number 786.

were constructed having the largest values of B (4.48). The 1000 values of P_7^+ [(4.74)–(4.79)] were calculated and arranged in increasing order of the standard deviation, $(Var)^{1/2}$, (4.80); (see column headed SIG in Table 4.1). A representative sample of 50 of the P_7^+ are shown in Table 4.1 [see column headed $P(+)$] together with the true values of $\cos(\phi_h + \phi_k + \phi_l + \phi_m)$, the column headed $\cos(T)$. It is noteworthy that the first incorrect estimate occurs at invariant 786. Thus, if one ignores the three special quartets with $h = k$ not shown in Table 4.1, the first 782 invariants are calculated with perfect accuracy and, of these, 745 are equal to zero and 37 equal to π. Towards the end of Table 4.1, as P_7^+ values tend toward $\frac{1}{2}$, one observes increasing numbers of incorrect indications, as is to be expected. Naturally, in applications, one would not use the latter invariants since, in view of the large standard deviations, they are known not to be reliably determined.

It is particularly noteworthy that, employing some 400 of the most reliably estimated values of the quartets (and a single phase obtained by means of Σ_1), unique values were obtained, with perfect accuracy, for 230 of the phases having $|E|$ values greater than 2. Thus, this structure is solvable solely through the estimated values of the quartets.

4.9. The Neighborhood Principle

It has been known for many years that, for a chosen enantiomorph, the values of the structure factor magnitudes $|E|$, where no homometric solutions exist, determine uniquely the values of the structure invariants. Alternatively, for a fixed functional form for the geometric structure factor and for fixed choice of enantiomorph, the magnitudes $|E|$ determine uniquely the values of the structure seminvariants. A number of formulas relating explicitly the cosine invariants to the magnitudes of all, or large numbers of, structure factors E have been known for some time, and they have played an important role in the applications of direct methods. However, their value has been somewhat limited by the requirement that certain kinds of rational dependence among atomic coordinates not be present, and for very complex structures their usefulness is greatly reduced. The results described in the earlier sections point strongly to a deeper insight. Instead of seeking to express the structure seminvariants in terms of all observed structure-factor magnitudes $|E|$, or even in terms of large numbers of $|E|$'s, as the earlier formulas did, it is now suggested that the value of each structure seminvariant ϕ is primarily determined, in favorable cases, by the values of one or more small sets of appropriately chosen magnitudes $|E|$, the so-called neighborhoods of ϕ, and is relatively insensitive to the vast bulk of the remaining magnitudes (the neighborhood principle).

Thus, in the favorable case that A, (4.36), is large, the value of the three-phase structure invariant (4.33) is determined primarily by the three magnitudes (4.31) and the larger the value of A the more reliable is the estimate for ϕ_3, zero in this case, as reference to (4.35) shows. Hence, the first neighborhood of the three-phase structure invariant (4.33) is defined to consist of the three magnitudes (4.31).

Next, if one is given only the four magnitudes (4.43), then the most probable value of the quartet ϕ_4 is zero in view of (4.47), and in the favorable case that B, (4.48), is large, the zero estimate is good. Thus, the first neighborhood of the quartet ϕ_4 is defined to consist of the four magnitudes (4.43).

Suppose next that the seven magnitudes (4.56) and (4.57) are given. In this case, the most probable value of the quartet ϕ_4 is given by the maximum of the distribution (4.60) and this estimate may lie anywhere in the range 0 to π, or, because of symmetry, in the range $-\pi$ to 0, depending on the values of the seven parameters (4.56) and (4.57). In the favorable case that the variance of the distribution (4.60) is very small, then the estimate for ϕ_4 is particularly good. The second neighborhood of the quartet ϕ_4 is therefore defined to consist of the seven magnitudes (4.56) and (4.57).

The variance of the distribution (4.60) may be greater than or less than that of (4.47). If B, (4.48), is fixed, then in the favorable cases that the three magnitudes (4.57) are all large or all small the variance of (4.60) will be smaller than that of (4.47) and the estimate for ϕ_4 given by (4.60) will then be more reliable than that given by (4.47). Thus, the gain in going from the first to the second neighborhood is that the potential for obtaining a distribution with a small variance is increased; in short it becomes possible to find a more reliable estimate for ϕ_4, and the estimate is no longer restricted to be zero but may instead lie anywhere in the interval 0 to π.

4.10. More on Quartets: Higher Neighborhoods

In Section 4.8 conditional probability distributions associated with the first two neighborhoods of the quartet ϕ_4, (4.45), were described. In the present section an account is given of the probability distributions derived from the third (13-magnitude) neighborhoods. Of major importance is the existence of joint conditional distributions of two or more structure invariants given the magnitudes $|E|$ alone, as well as of conditional distributions of a single structure invariant, given not only the magnitudes $|E|$ but also the values of one or more structure invariants. These distributions permit the estimation of the values, that is, both magnitudes and signs, of a large number of structure invariants ϕ_4, consistent with a specified enantiomorph, and not merely the estimation of the magnitudes of ϕ_4. In this way enantiomorph specification is made prior to the process leading from the values of the structure invariants to the values of the individual phases rather than being made part of this process. In effect then, both the values of $\cos \phi_4$ and $\sin \phi_4$ are available for phase determination rather than only the values of $\cos \phi_4$, thus making the process of phase determination a better conditioned one. Next, the higher neighborhoods of ϕ_4 will be defined.

4.10.1. Third Neighborhoods[†] of the Following Structure Invariant:

$$\phi_4 = \phi_h + \phi_k + \phi_l + \phi_m \tag{4.84}$$

If p and q are arbitrary reciprocal vectors that satisfy

$$\mathbf{h} + \mathbf{k} + \mathbf{p} + \mathbf{q} = 0 \tag{4.85}$$

then

$$\phi_{pq} = \phi_h + \phi_k + \phi_p + \phi_q \tag{4.86}$$

is a structure invariant and, in view of (4.44),

$$\mathbf{l} + \mathbf{m} - \mathbf{p} - \mathbf{q} = 0 \tag{4.87}$$

so that

$$\Psi_{pq} = \phi_l + \phi_m - \phi_p - \phi_q \tag{4.88}$$

is also a structure invariant. In view of Section 4.8, ϕ_4 is estimated by means of the seven magnitudes in its second neighborhood,

$$|E_h|, \quad |E_k|, \quad |E_l|, \quad |E_m|, \quad |E_{h+k}|, \quad |E_{k+l}|, \quad |E_{l+h}| \tag{4.89}$$

ϕ_{pq} by means of the seven magnitudes in its second neighborhood,

$$|E_h|, \quad |E_k|, \quad |E_p|, \quad |E_q|, \quad |E_{h+k}|, \quad |E_{k+p}|, \quad |E_{p+h}| \tag{4.90}$$

and Ψ_{pq} by means of the seven magnitudes in its second neighborhood,

$$|E_l|, \quad |E_m|, \quad |E_p|, \quad |E_q|, \quad |E_{l+m}|, \quad |E_{m-p}|, \quad |E_{p-l}| \tag{4.91}$$

[†] Hauptman (1977a).

However, from (4.45), (4.86), and (4.88) it is clear that

$$\phi - \phi_{pq} - \Psi_{pq} \equiv 0 \tag{4.92}$$

It is therefore to be expected that, in the favorable case that the seven-magnitude estimates yield values for ϕ_4, ϕ_{pq}, and Ψ_{pq} in accord with the identity (4.92), ϕ_4 will be well estimated in terms of the 21 magnitudes (4.89), (4.90), and (4.91), of which only the following 13 are distinct:

$$|E_h|, \quad |E_k|, \quad |E_l|, \quad |E_m|, \quad |E_p|, \quad |E_q| \tag{4.93}$$

$$|E_{h+k}|, \quad |E_{k+l}|, \quad |E_{l+h}|, \quad |E_{h+p}|, \quad |E_{k+p}|, \quad |E_{l-p}|, \quad |E_{m-p}| \tag{4.94}$$

Hence the third (13-magnitude) neighborhood of ϕ is obtained by adjoining to the second (seven-magnitude) neighborhood (4.56) and (4.57) the additional six magnitudes

$$|E_p|, \quad |E_q|, \quad |E_{h+p}|, \quad |E_{k+p}|, \quad |E_{l-p}|, \quad |E_{m-p}| \tag{4.95}$$

where $\mathbf{p}$ is an arbitrary reciprocal vector; thus there are many third neighborhoods.

One naturally anticipates that the conditional variance of the structure invariant ϕ_4, given the 13 magnitudes in its third neighborhood, will be small if the three seven-magnitude subsets of the third neighborhood which are the respective second neighborhoods of the structure invariants ϕ_4, ϕ_{pq}, Ψ_{pq} yield reliable estimates for the latter in accord with the identity (4.92). Thus only those third neighborhoods are useful for which $|E_p|$ and $|E_q|$ are both large, where p and q satisfy (4.85).

4.10.2. Higher Neighborhoods[†]

Using the same kind of argument as that described in Section 4.10.1, the fourth and higher neighborhoods of the quartet ϕ_4 are readily found. These neighborhoods are conveniently exhibited in Fig. 4.6. The first neighborhood consists of the four magnitudes shown in the first shell; the second shell contains the three magnitudes to be added to the first neighborhood in order to obtain the second, seven-magnitude, neighborhood; the third shell shows the six magnitudes that are adjoined to the second neighborhood to obtain the third, 13-magnitude, neighborhood, and so on.

† Hauptman (1977a).

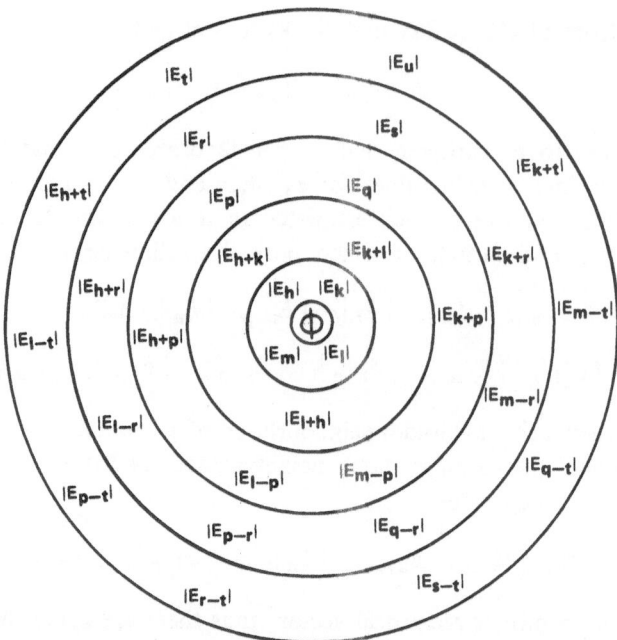

FIGURE 4.6. Sequence of nested neighborhoods of the structure invariant $\phi_4 = \phi_h + \phi_k + \phi_l + \phi_m$. The reciprocal vectors $h, k, l, m, p, q, r, s, t,$ and u satisfy $h + k + l + m = h + k + p + q = h + k + r + s = h + k + t + u = 0$, but are otherwise arbitrary.

4.10.3. Probability Distributions in $P\bar{1}$ Derived from the Third (Thirteen-Magnitude) Neighborhoods[†]

(a) *Joint Conditional Probability Distribution of the Pair of Structure Invariants* $\phi_4 = \phi_h + \phi_k + \phi_l + \phi_m$, $\phi_{pq} = \phi_h + \phi_k + \phi_p + \phi_q$, *Given the 13 Magnitudes* $|E_h|$, $|E_k|$, $|E_l|$, $|E_m|$, $|E_p|$, $|E_q|$; $|E_{h+k}|$, $|E_{k+l}|$, $|E_{l+h}|$; $|E_{h+p}|$, $|E_{k+p}|$, $|E_{l-p}|$, $|E_{m-p}|$. Suppose that the 13 nonnegative numbers $R_1, R_2, R_3, R_4, R_5, R_6$; R_{12}, R_{23}, R_{31}; and $R_{15}, R_{25}, R_{35}, R_{45}$ are specified. Assume that the ordered sextuple (h, k, l, m, p, q) of reciprocal vectors is a random variable (vector) which is uniformly distributed over the subset of the sixfold Cartesian product $W \times W \times W \times W \times W \times W$ defined by

$$|E_h| = R_1, \qquad |E_k| = R_2, \qquad |E_l| = R_3, \qquad |E_m| = R_4,$$
$$|E_p| = R_5, \qquad |E_q| = R_6 \qquad\qquad (4.96)$$
$$|E_{h+k}| = R_{12}, \qquad |E_{k+l}| = R_{23}, \qquad |E_{l+h}| = R_{31}; \qquad |E_{h+p}| = R_{15},$$
$$|E_{k+p}| = R_{25}, \qquad |E_{l-p}| = R_{35}, \qquad |E_{m-p}| = R_{45} \qquad (4.97)$$
$$h + k + l + m = 0 \qquad\qquad (4.98)$$

[†] Hauptman (1977b).

and

$$h + k + p + q = 0 \qquad (4.99)$$

In order that the domain of the primitive random variable (h, k, l, m, p, q) be nonvacuous, it is necessary to interpret the equation $|E_h| = R_1$ of (4.96), for example, as the inequalities $R_1 \le |E_h| \le R_1 + dR_1$, etc. In view of (4.98) and (4.99),

$$\phi_4 = \phi_h + \phi_k + \phi_l + \phi_m \qquad (4.100)$$

and

$$\phi_{pq} = \phi_h + \phi_k + \phi_p + \phi_q \qquad (4.101)$$

are structure invariants, which, as functions of the primitive random variables h, k, l, m, p, q, are themselves random variables. Denote by

$$P_{2|13} = P(\Phi_{34}, \Phi_{56} | R_1, R_2, R_3, R_4, R_5, R_6; R_{12}, R_{23}, R_{31}; R_{15}, R_{25}, R_{35}, R_{45})$$
$$(4.102)$$

the joint conditional probability distribution of the pair of structure invariants ϕ_4, ϕ_{pq}, given the 13 magnitudes (4.96) and (4.97). Then $P_{2|13}$ turns out to be, correct up to and including terms of order $1/N$,

$$P_{2|13} \approx \frac{1}{K} \exp[-2B_{1234} \cos \Phi_{34} - 2B_{1256} \cos \Phi_{56} - 2B_{3456} \cos(\Phi_{34} - \Phi_{56})]$$

$$\times I_0\left(\frac{2\sigma_3 R_{12} X'_{12}}{\sigma_2^{3/2}}\right) I_0\left(\frac{2\sigma_3 R_{23} X_{23}}{\sigma_2^{3/2}}\right) I_0\left(\frac{2\sigma_3 R_{31} X_{31}}{\sigma_2^{3/2}}\right) I_0\left(\frac{2\sigma_3 R_{15} X_{15}}{\sigma_2^{3/2}}\right)$$

$$\times I_0\left(\frac{2\sigma_3 R_{25} X_{25}}{\sigma_2^{3/2}}\right) I_0\left(\frac{2\sigma_3 R_{35} X_{35}}{\sigma_2^{3/2}}\right) I_0\left(\frac{2\sigma_3 R_{45} X_{45}}{\sigma_2^{3/2}}\right) \qquad (4.103)$$

where

$$B_{1234} = \frac{3\sigma_3^2 - \sigma_2\sigma_4}{\sigma_2^3} R_1 R_2 R_3 R_4 \qquad (4.104)$$

$$B_{1256} = \frac{3\sigma_3^2 - \sigma_2\sigma_4}{\sigma_2^3} R_1 R_2 R_5 R_6 \qquad (4.105)$$

$$B_{3456} = \frac{3\sigma_3^2 - \sigma_2\sigma_4}{\sigma_2^3} R_3 R_4 R_5 R_6 \qquad (4.106)$$

σ_n is defined by (4.37), and

$$X'_{12} = [R_1^2 R_2^2 + R_3^2 R_4^2 + R_5^2 R_6^2 + 2R_1 R_2 R_3 R_4 \cos \Phi_{34}$$
$$+ 2R_1 R_2 R_5 R_6 \cos \Phi_{56} + 2R_3 R_4 R_5 R_6 \cos(\Phi_{34} - \Phi_{56})]^{1/2} \qquad (4.107)$$

$$X_{23} = [R_2{}^2 R_3{}^2 + R_1{}^2 R_4{}^2 + 2R_1 R_2 R_3 R_4 \cos \Phi_{34}]^{1/2} \qquad (4.108)$$

$$X_{31} = [R_3{}^2 R_1{}^2 + R_2{}^2 R_4{}^2 + 2R_1{}^2 R_2{}^2 R_3{}^2 R_4{}^2 \cos \Phi_{34}]^{1/2} \qquad (4.109)$$

$$X_{15} = [R_1{}^2 R_5{}^2 + R_2{}^2 R_6{}^2 + 2R_1 R_2 R_5 R_6 \cos \Phi_{56}]^{1/2} \qquad (4.110)$$

$$X_{25} = [R_2{}^2 R_5{}^2 + R_1{}^2 R_6{}^2 + 2R_1 R_2 R_5 R_6 \cos \Phi_{56}]^{1/2} \qquad (4.111)$$

$$X_{35} = [R_3{}^2 R_5{}^2 + R_4{}^2 R_6{}^2 + 2R_3 R_4 R_5 R_6 \cos(\Phi_{34} - \Phi_{56})]^{1/2} \qquad (4.112)$$

$$X_{45} = [R_4{}^2 R_5{}^2 + R_3{}^2 R_6{}^2 + 2R_3 R_4 R_5 R_6 \cos(\Phi_{34} - \Phi_{56})]^{1/2} \qquad (4.113)$$

and K is a suitable normalizing parameter independent of Φ_{34} and Φ_{56} and not relevant for the present purpose. Clearly the 13 magnitudes (4.96) and (4.97) are parameters of the distribution.

In general, (4.103) has two maxima in the domain

$$-\pi < \Phi_{34} \leq \pi \qquad (4.114)$$

$$-\pi < \Phi_{56} \leq \pi \qquad (4.115)$$

related to each other by inversion through the origin, because (4.103) is unchanged when Φ_{34} and Φ_{56} are replaced by their negatives. One maximum yields the most probable values of the pair of invariants ϕ_4, ϕ_{pq}, (4.100) and (4.101), respectively, given the 13 magnitudes (4.96) and (4.97), for one enantiomorph, the other maximum the most probable values for the other enantiomorph. By choosing one or the other maximum, one selects the enantiomorph. In the case that the maximum occurs at $\Phi_{34} = \Phi_{56} = 0$ or π, or at $\Phi_{34} = 0$, $\Phi_{56} = \pi$, or at $\Phi_{34} = \pi$, $\Phi_{56} = 0$, the most probable values of the pair (ϕ_4, ϕ_{pq}) are the same for both enantiomorphs and (4.103) is not suitable for enantiomorph discrimination. It should be emphasized that when (4.103) is suitable for enantiomorph discrimination then, in general, the values, both signs and magnitudes, of two structure invariants consistent with the chosen enantiomorph are available, in contrast to the usual case when the value of only one structure invariant is used for enantiomorph selection.

(b) *Conditional Probability Distribution of the Structure Invariant* $\phi_4 = \phi_h + \phi_k + \phi_l + \phi_m$, *Given the Value of the Structure Invariant* $\phi_{pq} = \phi_h + \phi_k + \phi_l + \phi_m$ *and the 11 Magnitudes* $|E_h|$, $|E_k|$, $|E_l|$, $|E_m|$, $|E_p|$, $|E_q|$; $|E_{h+k}|$, $|E_{k+l}|$, $|E_{l+h}|$; $|E_{l-p}|$, $|E_{m-p}|$. Suppose that Φ_{56} ($-\pi < \Phi_{56} \leq \pi$) and the eleven nonnegative numbers, R_1, R_2, R_3, R_4, R_5, R_6; R_{12}, R_{23}, R_{31}; and R_{35}, R_{45}, are specified and that the ordered sextuple $(\mathbf{h}, \mathbf{k}, \mathbf{l}, \mathbf{m}, \mathbf{p}, \mathbf{q})$ of reciprocal vectors is the primitive random variable (vector) which is assumed to be uniformly distributed over the subset of $W \times W \times W \times W \times W \times W$ defined by

$$\phi_{pq} = \Phi_{56} \tag{4.116}$$

$$|E_h| = R_1, \qquad |E_k| = R_2, \qquad |E_l| = R_3, \qquad |E_m| = R_4,$$
$$|E_p| = R_5, \qquad |E_q| = R_6 \tag{4.117}$$

$$|E_{h+k}| = R_{12}, \qquad |E_{k+l}| = R_{23}, \qquad |E_{l+h}| = R_{31}; \qquad |E_{l-p}| = R_{35},$$
$$|E_{m-p}| = R_{45} \tag{4.118}$$

$$\mathbf{h + k + l + m = 0} \tag{4.119}$$

$$\mathbf{h + k + p + q = 0} \tag{4.120}$$

In view of (4.119) and (4.120),

$$\phi_4 = \phi_h + \phi_k + \phi_l + \phi_m \tag{4.121}$$

and

$$\phi_{pq} = \phi_h + \phi_k + \phi_p + \phi_q \tag{4.122}$$

are structure invariants. The structure invariant ϕ_4, as a function of the primitive random variables $\mathbf{h, k, l, m}$, is itself a random variable, and its conditional probability distribution, given the value of the structure invariant (4.116) and the 11 magnitudes (4.117) and (4.118), obtained from $P_{2|13}$ (4.103) by fixing Φ_{56} and multiplying by a suitable normalizing constant, is given by

$$P_{1|1,11} = P(\Phi_{34} \mid \Phi_{56}; R_1, R_2, R_3, R_4, R_5, R_6; R_{12}, R_{23}, R_{31}; R_{35}, R_{45})$$
$$\approx \frac{1}{L} \exp\{-2B_{1234} \cos \Phi_{34} - 2B_{3456} \cos(\Phi_{34} - \Phi_{56})\} I_0\left(\frac{2\sigma_3 R_{12} X'_{12}}{\sigma_2^{3/2}}\right)$$
$$\times I_0\left(\frac{2\sigma_3 R_{23} X_{23}}{\sigma_2^{3/2}}\right) I_0\left(\frac{2\sigma_3 R_{31} X_{31}}{\sigma_2^{3/2}}\right) I_0\left(\frac{2\sigma_3 R_{35} X_{35}}{\sigma_2^{3/2}}\right) I_0\left(\frac{2\sigma_3 R_{45} X_{45}}{\sigma_2^{3/2}}\right) \tag{4.123}$$

where $B_{1234}, B_{3456}, X'_{12}, X_{23}, X_{31}, X_{35}, X_{45}$ are given by (4.104), (4.106), (4.107), (4.108), (4.109), (4.112), and (4.113) respectively, L is a suitable normalizing parameter independent of Φ_{34}, and $\Phi_{56}, R_1, R_2, R_3, R_4, R_5, R_6; R_{12}, R_{23}, R_{31};$ and R_{35}, R_{45} are parameters of the distribution.

It should be noted that if $\Phi_{56} \neq 0$ or π, (4.123) is not an even function of Φ_{34} and has a unique maximum in the whole interval

$$-\pi < \Phi_{34} \leq \pi \tag{4.124}$$

of length 2π. In other words, once the enantiomorph has been fixed by proper choice of the value for Φ_{56}, then the most probable value, in both sign and magnitude, for Φ_{34}, given Φ_{56} and the 11 magnitudes (4.117) and (4.118), is given by the position of the unique maximum of (4.123). The initial estimate for Φ_{56}, to be used in (4.123), in terms of the magnitudes $|E|$ alone may be found, for example, from either (4.60) or the distribution (4.132) to be described.

If $\Phi_{56} = 0$ or π, then ϕ_{pq} has the same value for both enantiomorphs: (4.123) is an even function of Φ_{34} and, unless (4.123) has its maximum at $\Phi_{34} = 0$ or π, (4.123) is bimodal, one maximum corresponding to one enantiomorph and the second to the other enantiomorph. In this case enantiomorph selection may be made by specifying arbitrarily the sign of Φ_{34} in the interval $-\pi$ to π.

(c) *Conditional Probability Distribution of the Structure Invariant* $\phi_4 = \phi_h + \phi_k + \phi_l + \phi_m$, *Given the 13 Magnitudes* $|E_h|$, $|E_k|$, $|E_l|$, $|E_m|$, $|E_p|$, $|E_q|$; $|E_{h+k}|$, $|E_{k+l}|$, $|E_{l+h}|$; $|E_{h+p}|$, $|E_{k+p}|$, $|E_{l-p}|$, $|E_{m-p}|$. Suppose that the 13 nonnegative number $R_1, R_2, R_3, R_4, R_5, R_6$; R_{12}, R_{23}, R_{31}; and $R_{15}, R_{25}, R_{35}, R_{45}$ are specified and that the ordered sextuple (h, k, l, m, p, q) is the primitive random variable vector which is assumed to be uniformly distributed over the subset of the sixfold Cartesian product $W \times W \times W \times W \times W \times W$ defined by

$$|E_h| = R_1, \qquad |E_k| = R_2, \qquad |E_l| = R_3, \qquad |E_m| = R_4,$$
$$|E_p| = R_5, \qquad |E_q| = R_6 \qquad\qquad\qquad (4.125)$$

$$|E_{h+k}| = R_{12}, \qquad |E_{k+l}| = R_{23}, \qquad |E_{l+h}| = R_{31} \qquad (4.126)$$

$$|E_{h+p}| = R_{15}, \quad |E_{k+p}| = R_{25}, \quad |E_{l-p}| = R_{35}, \quad |E_{m-p}| = R_{45} \quad (4.127)$$

$$h + k + l + m = 0 \qquad\qquad\qquad (4.128)$$

$$h + k + p + q = 0 \qquad\qquad\qquad (4.129)$$

In view of (4.128),

$$\phi_4 = \phi_h + \phi_k + \phi_l + \phi_m \qquad\qquad (4.130)$$

is a structure invariant which is a function of the primitive random variables h, k, l, m. Hence ϕ_4 is itself a random variable and its conditional probability distribution

$$P_{1|13} = P(\Phi_{34} \mid R_1, R_2, R_3, R_4, R_5, R_6; R_{12}, R_{23}, R_{31}; R_{15}, R_{25}, R_{35}, R_{45})$$
$$(4.131)$$

given the 13 magnitudes (4.125) to (4.127), is obtained from $P_{2|13}$ (4.103), by integrating with respect to Φ_{56} from 0 to 2π:

$$P_{1|13} = \int_0^{2\pi} P_{2|13}\, d\Phi_{56} \tag{4.132}$$

Although this integration can be carried out exactly, the resulting expression is a complicated infinite multiple series which does not appear to be suitable for numerical calculation. For this reason it is suggested that the indicated integration (4.132) be carried out numerically in any given case.

4.10.4. Probability Distributions in $P\bar{1}$ Derived from the Third (Thirteen-Magnitude) Neighborhoods[†]

(a) *Joint Conditional Probability Distribution of the Pair of Structure Invariants* $\phi_4 = \phi_h + \phi_k + \phi_l + \phi_m$, $\phi_{pq} = \phi_h + \phi_k + \phi_p + \phi_q$, *Given the Thirteen Magnitudes* $|E_h|$, $|E_k|$, $|E_l|$, $|E_m|$, $|E_p|$, $|E_q|$; $|E_{h+k}|$, $|E_{k+l}|$, $|E_{l+h}|$; $|E_{h+p}|$, $|E_{k+p}|$, $|E_{l-p}|$, $|E_{m-p}|$. Suppose that the 13 non-negative numbers R_1, R_2, R_3, R_4, R_5, R_6; R_{12}, R_{23}, R_{31}; and R_{15}, R_{25}, R_{35}, R_{45} are specified. Assume that the ordered sextuple $(\mathbf{h, k, l, m, p, q})$ of reciprocal vectors is a random variable which is uniformly distributed over the subset of the sixfold Cartesian product $W \times W \times W \times W \times W \times W$ of reciprocal space W defined by

$$|E_h| = R_1, \quad |E_k| = R_2, \quad |E_l| = R_3, \quad |E_m| = R_4,$$
$$|E_p| = R_5, \quad |E_q| = R_6 \tag{4.133}$$

$$|E_{h+k}| = R_{12}, \quad |E_{k+l}| = R_{23}, \quad |E_{l+h}| = R_{31} \tag{4.134}$$

$$|E_{h+p}| = R_{15}, \ |E_{k+p}| = R_{25}, \ |E_{l-p}| = R_{35}, \ |E_{m-p}| = R_{45} \tag{4.135}$$

$$\mathbf{h + k + l + m} = 0 \tag{4.136}$$

and

$$\mathbf{h + k + p + q} = 0 \tag{4.137}$$

In order that the domain of the primitive random variable $(\mathbf{h, k, l, m, p, q})$ be nonvacuous, it is necessary to interpret the equality $|E_h| = R_1$ of (4.133), for example, as the inequalities $R_1 \leq |E_h| \leq R_1 + dR_1$, etc. In view of (4.136) and (4.137),

$$\phi_4 = \phi_h + \phi_k + \phi_l + \phi_m \tag{4.138}$$

[†] Hauptman (1977c).

and

$$\phi_{pq} = \phi_h + \phi_k + \phi_p + \phi_q \qquad (4.139)$$

are structure invariants which, as functions of the primitive random variables h, k, l, m, p, q, are themselves random variables. Denote by

$$P_{2|13} = P(\Phi_{34}, \Phi_{56} | R_1, R_2, R_3, R_4, R_5, R_6; R_{12}, R_{23}, R_{31}; R_{15}, R_{25}, R_{35}, R_{45}) \qquad (4.140)$$

the joint conditional probability distribution of the pair of structure invariants ϕ_4, ϕ_{pq}, given the 13 magnitudes (4.133)–(4.135). Then $P_{2|13}$ is found to be, correct up to and including terms of order $1/N$,

$$P_{2|13} \approx \frac{1}{K} \exp[-B_{1234} \cos \Phi_{34} - B_{1256} \cos \Phi_{56} - B_{3456} \cos(\Phi_{34} - \Phi_{56})]$$

$$\times \cosh\left(\frac{\sigma_3}{\sigma_2{}^{3/2}} R_{12} X_{12}'\right) \cosh\left(\frac{\sigma_3}{\sigma_2{}^{3/2}} R_{23} X_{23}\right) \cosh\left(\frac{\sigma_3}{\sigma_2{}^{3/2}} R_{31} X_{31}\right)$$

$$\times \cosh\left(\frac{\sigma_3}{\sigma_2{}^{3/2}} R_{15} X_{15}\right) \cosh\left(\frac{\sigma_3}{\sigma_2{}^{3/2}} R_{25} X_{25}\right) \cosh\left(\frac{\sigma_3}{\sigma_2{}^{3/2}} R_{35} X_{35}\right)$$

$$\times \cosh\left(\frac{\sigma_3}{\sigma_2{}^{3/2}} R_{45} X_{45}\right) \qquad (4.141)$$

where

$$B_{\mu\nu\varrho\sigma} = \frac{3\sigma_3{}^2 - \sigma_2\sigma_4}{\sigma_2{}^3} R_\mu R_\nu R_\varrho R_\sigma \qquad (4.142)$$

$$X_{12}' = [R_1 R_2 + R_3 R_4 \cos \Phi_{34} + R_5 R_6 \cos \Phi_{56}] \qquad (4.143)$$

$$X_{23} = [R_2 R_3 + R_1 R_4 \cos \Phi_{34}] \qquad (4.144)$$

$$X_{31} = [R_3 R_1 + R_2 R_4 \cos \Phi_{34}] \qquad (4.145)$$

$$X_{15} = [R_1 R_5 + R_2 R_6 \cos \Phi_{56}] \qquad (4.146)$$

$$X_{25} = [R_2 R_5 + R_1 R_6 \cos \Phi_{56}] \qquad (4.147)$$

$$X_{35} = [R_3 R_5 + R_4 R_6 \cos(\Phi_{34} - \Phi_{56})] \qquad (4.148)$$

$$X_{45} = [R_4 R_5 + R_3 R_6 \cos(\Phi_{34} - \Phi_{56})] \qquad (4.149)$$

and σ_n is given by (4.37); the normalizing parameter K is obtained by summing the right side of (4.141) over the four possible values of the pair (Φ_{34}, Φ_{56}), that is, Φ_{34} and Φ_{56} each take on the two values 0 or π independently, and then setting the result equal to unity. Equations (4.141)–(4.149) should be compared with (4.103)–(4.113) of the previous section, but the present distribution (4.141) is discrete since Φ_{34} and Φ_{56} each take on only the two values 0 and π. The 13 R's, (4.133)–(4.135) are seen to be parameters of the distribution.

(b) *Conditional Probability Distribution of the Structure Invariant*
$\phi_4 = \phi_h + \phi_k + \phi_l + \phi_m$, *Given the Value of the Structure Invariant*
$\phi_{pq} = \phi_h + \phi_k + \phi_p + \phi_q$ *and the 11 Magnitudes* $|E_h|$, $|E_k|$, $|E_l|$, $|E_m|$,
$|E_p|$, $|E_q|$; $|E_{h+k}|$, $|E_{k+l}|$, $|E_{l+h}|$; $|E_{l-p}|$, $|E_{m-p}|$. Suppose that
Φ_{56} ($= 0$ or π) and the 11 nonnegative numbers, $R_1, R_2, R_3, R_4, R_5, R_6$;
R_{12}, R_{23}, R_{31}; and R_{35}, R_{45}, are specified and that the ordered sextuple
(h, k, l, m, p, q) of reciprocal vectors is the primitive random variable
which is assumed to be uniformly distributed over the subset of $W \times W \times W$
$\times W \times W \times W$ defined by

$$\phi_{pq} = \Phi_{56} \tag{4.150}$$

$$|E_h| = R_1, \qquad |E_k| = R_2, \qquad |E_l| = R_3, \qquad |E_m| = R_4,$$
$$|E_p| = R_5, \qquad |E_q| = R_6 \tag{4.151}$$

$$|E_{h+k}| = R_{12}, \qquad |E_{k+l}| = R_{23}, \qquad |E_{l+h}| = R_{31} \tag{4.152}$$

$$|E_{l-p}| = R_{35}, \qquad |E_{m-p}| = R_{45} \tag{4.153}$$

$$\mathbf{h + k + l + m} = 0 \tag{4.154}$$

$$\mathbf{h + k + p + q} = 0 \tag{4.155}$$

It follows from (4.154) and (4.155) that

$$\phi_4 = \phi_h + \phi_k + \phi_l + \phi_m \tag{4.156}$$

and

$$\phi_{pq} = \phi_h + \phi_k + \phi_p + \phi_q \tag{4.157}$$

are structure invariants. The structure invariant ϕ_4, as a function of the
primitive random variables **h, k, l, m**, is itself a random variable, and its
conditional probability distribution $P_{1|1,11}$, given the value of the structure
invariant (4.150) and the 11 magnitudes (4.151)–(4.153), is obtained from
$P_{2|13}$, (4.141), by fixing Φ_{56} and multiplying by a suitable normalizing
constant:

$$P_{1|1,11} = P(\Phi_{34} \mid \Phi_{56}; R_1, R_2, R_3, R_4, R_5, R_6; R_{12}, R_{23}, R_{31}; R_{35}, R_{45})$$

$$\approx \frac{1}{L_{11}} \exp[-B_{1234} \cos \Phi_{34} - B_{3456} \cos(\Phi_{34} - \Phi_{56})]$$

$$\times \cosh\left(\frac{\sigma_3}{\sigma_2^{3/2}} R_{12} X_{12}'\right) \cosh\left(\frac{\sigma_3}{\sigma_2^{3/2}} R_{23} X_{23}\right) \cosh\left(\frac{\sigma_3}{\sigma_2^{3/2}} R_{31} X_{31}\right)$$

$$\times \cosh\left(\frac{\sigma_3}{\sigma_2^{3/2}} R_{35} X_{35}\right) \cosh\left(\frac{\sigma_3}{\sigma_2^{3/2}} R_{45} X_{45}\right) \tag{4.158}$$

where B_{1234}, B_{3456}, X'_{12}, X_{23}, X_{31}, X_{35}, and X_{45} are given by (4.142)–(4.145), (4.148), and (4.149), and the normalizing parameter L_{11} is obtained by summing the right side of (4.158) over the two possible values of Φ_{34} (0 and π) and setting the result equal to unity. Clearly Φ_{56} and the 11 magnitudes (4.151)–(4.153) are parameters of the distribution. The value of the quartet Φ_{56} may be obtained, for example, from the seven-magnitude distribution (4.74) or the 13-magnitude distribution (4.171), to be described.

(c) *Conditional Probability Distribution of the Structure Invariant* $\phi_4 = \phi_h + \phi_k + \phi_l + \phi_m$, *Given the* 13 *Magnitudes* $|E_h|$, $|E_k|$, $|E_l|$, $|E_m|$, $|E_p|$, $|E_q|$; $|E_{h+k}|$, $|E_{k+l}|$, $|E_{l+h}|$; $|E_{h+p}|$, $|E_{k+p}|$, $|E_{l-p}|$, $|E_{m-p}|$. Suppose that the 13 nonnegative numbers R_1, R_2, R_3, R_4, R_5, R_6; R_{12}, R_{23}, R_{31}; and R_{15}, R_{25}, R_{35}, R_{45} are specified, and that the ordered sextuple $(\mathbf{h}, \mathbf{k}, \mathbf{l}, \mathbf{m}, \mathbf{p}, \mathbf{q})$ is the primitive random variable which is assumed to be uniformly distributed over the subset of the sixfold Cartesian product $W \times W \times W \times W \times W \times W$ defined by

$$|E_h| = R_1, \qquad |E_k| = R_2, \qquad |E_l| = R_3, \qquad |E_m| = R_4,$$
$$|E_p| = R_5, \qquad |E_q| = R_6 \tag{4.159}$$

$$|E_{h+k}| = R_{12}, \qquad |E_{k+l}| = R_{23}, \qquad |E_{l+h}| = R_{31} \tag{4.160}$$

$$|E_{h+p}| = R_{15}, \quad |E_{k+p}| = R_{25}, \quad |E_{l-p}| = R_{35}, \quad |E_{m-p}| = R_{45} \tag{4.161}$$

$$\mathbf{h} + \mathbf{k} + \mathbf{l} + \mathbf{m} = 0 \tag{4.162}$$

$$\mathbf{h} + \mathbf{k} + \mathbf{p} + \mathbf{q} = 0 \tag{4.163}$$

In view of (4.162),

$$\phi_4 = \phi_h + \phi_k + \phi_l + \phi_m \tag{4.164}$$

is a structure invariant which is a function of the primitive random variables $\mathbf{h}$, $\mathbf{k}$, $\mathbf{l}$, $\mathbf{m}$. Hence ϕ_4 is itself a random variable and its conditional probability distribution

$$P_{1|13} = P(\Phi_{34} \mid R_1, R_2, R_3, R_4, R_5, R_6; R_{12}, R_{23}, R_{31}; R_{15}, R_{25}, R_{35}, R_{45}) \tag{4.165}$$

given the 13 magnitudes (4.159)–(4.161), is obtained from $P_{2|13}$, (4.141), by summing over the two values 0 and π for Φ_{56}:

$$P_{1|13} \approx \frac{1}{K_{13}} Z_{1|13} \tag{4.166}$$

where

$$Z_{1|13} = M(A_1 + A_2) \tag{4.167}$$

$$M = \exp(-B_{1234} \cos \Phi_{34}) \cosh\left[\frac{\sigma_3}{\sigma_2^{3/2}} R_{23}(R_2 R_3 + R_1 R_4 \cos \Phi_{34})\right]$$
$$\times \cosh\left[\frac{\sigma_3}{\sigma_2^{3/2}} R_{31}(R_3 R_1 + R_2 R_4 \cos \Phi_{34})\right] \tag{4.168}$$

$$A_1 = \exp(-B_{1256} - B_{3456} \cos \Phi_{34})$$
$$\times \cosh\left[\frac{\sigma_3}{\sigma_2^{3/2}} R_{12}(R_1 R_2 + R_3 R_4 \cos \Phi_{34} + R_5 R_6)\right]$$
$$\times \cosh\left[\frac{\sigma_3}{\sigma_2^{3/2}} R_{15}(R_1 R_5 + R_2 R_6)\right] \cosh\left[\frac{\sigma_3}{\sigma_2^{3/2}} R_{25}(R_2 R_5 + R_1 R_6)\right]$$
$$\times \cosh\left[\frac{\sigma_3}{\sigma_2^{3/2}} R_{35}(R_3 R_5 + R_4 R_6 \cos \Phi_{34})\right]$$
$$\times \cosh\left[\frac{\sigma_3}{\sigma_2^{3/2}} R_{45}(R_4 R_5 + R_3 R_6 \cos \Phi_{34})\right] \tag{4.169}$$

$$A_2 = \exp(+B_{1256} + B_{3456} \cos \Phi_{34})$$
$$\times \cosh\left[\frac{\sigma_3}{\sigma_2^{3/2}} R_{12}(R_1 R_2 + R_3 R_4 \cos \Phi_{34} - R_5 R_6)\right]$$
$$\times \cosh\left[\frac{\sigma_3}{\sigma_2^{3/2}} R_{15}(R_1 R_5 - R_2 R_6)\right] \cosh\left[\frac{\sigma_3}{\sigma_2^{3/2}} R_{25}(R_2 R_5 - R_1 R_6)\right]$$
$$\times \cosh\left[\frac{\sigma_3}{\sigma_2^{3/2}} R_{35}(R_3 R_5 - R_4 R_6 \cos \Phi_{34})\right]$$
$$\times \cosh\left[\frac{\sigma_3}{\sigma_2^{3/2}} R_{45}(R_4 R_5 - R_3 R_6 \cos \Phi_{34})\right] \tag{4.170}$$

$B_{\mu\nu\varrho\sigma}$ is defined by (4.142) and the normalizing parameter K_{13} by (4.176) below.

A more suggestive form of the distribution is obtained by defining P_{13}^+ (P_{13}^-) to be the conditional probability that ϕ_4 be 0 (π), given the 13 magnitudes (4.159)–(4.161). Then, from (4.166)–(4.170),

$$P_{13}^{\pm} = \frac{1}{K_{13}} Z_{13}^{\pm} \tag{4.171}$$

where

$$Z_{13}^{\pm} = M^{\pm}(A_1^{\pm} + A_2^{\pm}) \tag{4.172}$$

$$M^{\pm} = \exp(\mp B_{1234}) \cosh\left[\frac{\sigma_3}{\sigma_2^{3/2}} R_{23}(R_2 R_3 \pm R_1 R_4)\right]$$
$$\times \cosh\left[\frac{\sigma_3}{\sigma_2^{3/2}} R_{31}(R_3 R_1 \pm R_2 R_4)\right] \tag{4.173}$$

$$A_1^{\pm} = \exp(-B_{1234} \mp B_{3456}) \cosh\left[\frac{\sigma_3}{\sigma_2^{3/2}} R_{12}(R_1 R_2 \pm R_3 R_4 + R_5 R_6)\right]$$
$$\times \cosh\left[\frac{\sigma_3}{\sigma_2^{3/2}} R_{15}(R_1 R_5 + R_2 R_6)\right] \cosh\left[\frac{\sigma_3}{\sigma_2^{3/2}} R_{25}(R_2 R_5 + R_1 R_6)\right]$$
$$\times \cosh\left[\frac{\sigma_3}{\sigma_2^{3/2}} R_{35}(R_3 R_5 \pm R_4 R_6)\right] \cosh\left[\frac{\sigma_3}{\sigma_2^{3/2}} R_{45}(R_4 R_5 \pm R_3 R_6)\right] \tag{4.174}$$

$$A_2^{\pm} = \exp(+B_{1256} \pm B_{3456}) \cosh\left[\frac{\sigma_3}{\sigma_2^{3/2}} R_{12}(R_1 R_2 \pm R_3 R_4 - R_5 R_6)\right]$$
$$\times \cosh\left[\frac{\sigma_3}{\sigma_2^{3/2}} R_{15}(R_1 R_5 - R_2 R_6)\right] \cosh\left[\frac{\sigma_3}{\sigma_2^{3/2}} R_{25}(R_2 R_5 - R_1 R_6)\right]$$
$$\times \cosh\left[\frac{\sigma_3}{\sigma_2^{3/2}} R_{3\bar{5}}(R_3 R_5 \mp R_4 R_6)\right] \cosh\left[\frac{\sigma_3}{\sigma_2^{3/2}} R_{45}(R_4 R_5 \mp R_3 R_6)\right] \tag{4.175}$$

$$K_{13} = Z_{13}^{+} + Z_{13}^{-} \tag{4.176}$$

$B_{\mu\nu\varrho\sigma}$ is given by (4.142) and upper and lower signs go together.

4.11. Two-Phase Structure Seminvariants (Pairs)

In Sections 4.6–4.10 the probabilistic theory of the three- and four-phase structure invariants has been described in some detail for space groups $P1$ and $P\bar{1}$. In this account the neighborhood concept was seen to play the central role, in that it identified the small sets of magnitudes $|E|$ on which the value of the structure invariant, in favorable cases, primarily depends. In a similar way the neighborhood concept plays the fundamental role in extending the probabilistic theory to structure invariants and seminvariants in general, and to other space groups. Although the theory was initiated only recently (1975), a large number of distributions have already been obtained including, for example, distributions for quintets, sextets, septets, and selected one-, two-, and three-phase seminvariants. However, because of limitations of space, only the elementary theory of the two-phase structure seminvariants in space groups $P\bar{1}$, $P2_1$, and $P2_12_12_1$ will be briefly described here.

4.11.1. Space Group $P\bar{1}$[†]

(a) *First Neighborhood of the Two-Phase Structure Seminvariant in $P\bar{1}$*. In the space group $P\bar{1}$ the linear combination of two phases

$$\phi_2 = \phi_{\mathbf{h}} + \phi_{\mathbf{k}} \qquad (4.177)$$

is a structure seminvariant if and only if

$$\mathbf{h} + \mathbf{k} \equiv 0 \pmod{\boldsymbol{\omega}_s} \qquad (4.178)$$

where $\boldsymbol{\omega}_s$, the seminvariant modulus in $P\bar{1}$, is defined by

$$\boldsymbol{\omega}_s = (2, 2, 2) \qquad (4.179)$$

In other words, $\phi_{\mathbf{h}} + \phi_{\mathbf{k}}$ is a structure seminvariant if and only if the three components of the reciprocal vector $\mathbf{h} + \mathbf{k}$ are even.

Assume next that (4.178) holds, so that the components of each of $\frac{1}{2}(\mathbf{h} + \mathbf{k})$ are integers. Construct the two structure invariants

$$\phi_{\mathbf{h}} + \phi_{-\frac{1}{2}(\mathbf{h}+\mathbf{k})} + \phi_{-\frac{1}{2}(\mathbf{h}-\mathbf{k})} \qquad (4.180)$$

and

$$\phi_{\mathbf{k}} + \phi_{-\frac{1}{2}(\mathbf{h}+\mathbf{k})} + \phi_{\frac{1}{2}(\mathbf{h}-\mathbf{k})} \qquad (4.181)$$

Suppose further that the four magnitudes

$$|E_{\mathbf{h}}|, \qquad |E_{\mathbf{k}}|, \qquad |E_{\frac{1}{2}(\mathbf{h}\pm\mathbf{k})}| \qquad (4.182)$$

are all large. Under these circumstances it is known from Section 4.7.2, with high probability, that each of (4.180) and (4.181) is equal to 0. In $P\bar{1}$ every phase is 0 or π. Hence, if the four magnitudes (4.182) are large, (4.180) and (4.181) imply, by addition, that, with high probability, $\phi_{\mathbf{h}} + \phi_{\mathbf{k}} = 0$:

$$\phi_2 = \phi_{\mathbf{h}} + \phi_{\mathbf{k}} \approx 0 \qquad (4.183)$$

[†] Hauptman (1976c); Green and Hauptman (1976b).

and the larger the values of the four magnitudes (4.182) the more likely it is that (4.183) holds. The first neighborhood of $\phi_\mathbf{h} + \phi_\mathbf{k}$ is therefore defined to consist of the four magnitudes (4.182).

(b) *Conditional Probability Distribution of the Two-Phase Structure Seminvariant in* $P\bar{1}$, *Given the Four Magnitudes in its First Neighborhood.* Suppose that a crystal structure consisting of N atoms per unit cell in $P\bar{1}$ is fixed, and that the four nonnegative numbers R_1, R_2, $R_{12/2}$, and $R_{1\bar{2}/2}$ are also specified. Assume now that the ordered pair $(\mathbf{h}, \mathbf{k})$ of reciprocal vectors is a random variable which is uniformly distributed over the subset of the twofold Cartesian product $W \times W$ defined by

$$|E_\mathbf{h}| = R_1, \quad |E_\mathbf{k}| = R_2, \quad |E_{\frac{1}{2}(\mathbf{h}+\mathbf{k})}| = R_{12/2}, \quad |E_{\frac{1}{2}(\mathbf{h}-\mathbf{k})}| = R_{1\bar{2}/2} \quad (4.184)$$

and (4.178). Then the linear combination of two phases (4.177), as a function of the primitive random variables $\mathbf{h}$ and $\mathbf{k}$, is itself a random variable which, because the space group is $P\bar{1}$, takes on only the two values 0 and π. Denote by P_4^+ or P_4^- the conditional probability, given the four magnitudes (4.184), that the structure seminvariant (4.177) have the value 0 or π, respectively. Then, correct up to and including terms of order $1/N$,

$$P_4^\pm = \frac{1}{M} X_4^\pm \tag{4.185}$$

where

$$X_4^\pm = \exp\left[\mp \left(\frac{3\sigma_3{}^2 - 2\sigma_2\sigma_4}{2\sigma_2{}^3} \right) R_1 R_2 (R_{12/2}^2 + R_{1\bar{2}/2}^2) \pm \left(\frac{\sigma_3{}^2 - \sigma_2\sigma_4}{\sigma_2{}^3} \right) R_1 R_2 \right]$$
$$\times \cosh\left[\frac{\sigma_3}{\sigma_2{}^{3/2}} R_{12/2} R_{1\bar{2}/2} (R_1 \pm R_2) \right] \tag{4.186}$$

$$M = X_4^+ + X_4^- \tag{4.187}$$

σ_n is defined by (4.37), and upper and lower signs go together. It is readily verified that if all four magnitudes (4.184) are large, then $P_4^+ > \frac{1}{2}$ and $\phi_2 \approx 0$, in agreement with the heuristic argument given in Section 4.11.1(a). If, on the other hand, R_1 and R_2 are large and one of $R_{12/2}$ or $R_{1\bar{2}/2}$ is large but the other small, then $P_4^+ < \frac{1}{2}$ and $\phi_2 \approx \pi$.

4.11.2. Space Group $P2_1$

In this space group there are two sequences of nested neighborhoods of the two-phase structure seminvariant. The neighborhoods of the first sequence are particularly well suited to estimate those structure semin-variants whose values happen to lie close to 0 or π; those of the second sequence are particularly useful in estimating those seminvariants whose values are approximately $\pm\pi/2$. Because of limitations of space, only those distributions associated with the first two neighborhoods of the first sequence are described here. The reader interested in the higher neighborhoods or the second sequence of neighborhoods, which are particularly useful for enantiomorph discrimination, is referred to the literature (Green and Hauptman, 1978a,b; Hauptman and Green, 1978).

(a) *First Two Neighborhoods of the Two-Phase Structure Seminvariant in $P2_1$.* In the space group $P2_1$, the linear combination of two phases

$$\phi_2 = \phi_{h_1 k l_1} - \phi_{h_2 k l_2} \tag{4.188}$$

is a structure seminvariant if and only if

$$(h_1 - h_2, 0, l_1 - l_2) \equiv 0 \pmod{\boldsymbol{\omega}_s} \tag{4.189}$$

where $\boldsymbol{\omega}_s$, the seminvariant modulus in $P2_1$, is defined by

$$\boldsymbol{\omega}_s = (2, 0, 2) \tag{4.190}$$

In short, ϕ_2 is a structure seminvariant if and only if $h_1 - h_2$ and $l_1 - l_2$ are both even.

Assume that (4.189) holds. Construct the four-phase structure invariant

$$\phi_{h_1 k l_1} - \phi_{h_2 k l_2} - \phi_{\frac{1}{2}(h_1-h_2),q,\frac{1}{2}(l_1-l_2)} - \phi_{\frac{1}{2}(h_1-h_2),\bar{q},\frac{1}{2}(l_1-l_2)} \tag{4.191}$$

where q is an arbitrary nonzero integer. The first neighborhood of the quartet (4.191) consists of the four magnitudes

$$\left| E_{h_1 k l_1} \right|, \quad \left| E_{h_2 k l_2} \right|, \quad \left| E_{\frac{1}{2}(h_1-h_2),q,\frac{1}{2}(l_1-l_2)} \right|, \quad \left| E_{\frac{1}{2}(h_1-h_2),\bar{q},\frac{1}{2}(l_1-l_2)} \right| \tag{4.192}$$

of which, because the space group is $P2_1$, only the following three are distinct:

$$\left| E_{h_1 k l_1} \right|, \quad \left| E_{h_2 k l_2} \right|, \quad \left| E_{\frac{1}{2}(h_1-h_2),q,\frac{1}{2}(l_1-l_2)} \right| \tag{4.193}$$

It is known from comparison with Section 4.8.1.1 that if the three magnitudes (4.193) are all large the quartet (4.191) is probably close to zero. However, in $P2_1$,

$$\phi_{\frac{1}{2}(h_1-h_2),q,\frac{1}{2}(l_1-l_2)} + \phi_{\frac{1}{2}(h_1-h_2),\bar{q},\frac{1}{2}(l_1-l_2)} = \pi q \tag{4.194}$$

It follows that if the three magnitudes (4.193) are all large then

$$\phi_2 \approx \pi q \qquad (4.195)$$

that is,

$$\phi_2 \approx 0 \text{ or } \pi \qquad (4.196)$$

according as q is even or odd. The first neighborhood of the two-phase structure seminvariant ϕ_2 is therefore defined to consist of the three magnitudes (4.193). Since q is an arbitrary integer, there are many first neighborhoods.

The second neighborhood of the two-phase structure seminvariant ϕ_2 is defined to be the second neighborhood of the quartet (4.191). Reference to Section 4.8.1.2 then suggests that the second neighborhood of ϕ_2 consists of the three magnitudes (4.193) and the additional three magnitudes

$$\left| E_{\frac{1}{2}(h_1+h_2),q+k,\frac{1}{2}(l_1+l_2)} \right|, \quad \left| E_{\frac{1}{2}(h_1+h_2),q-k,\frac{1}{2}(l_1+l_2)} \right|, \quad \left| E_{h_1-h_2,0,l_1-l_2} \right| \qquad (4.197)$$

Since q is an arbitrary integer, there are many second neighborhoods. Furthermore, the theory of quartets suggests that if the six magnitudes (4.193) and (4.197) of the second neighborhood are all large, then the quartet (4.191) is probably close to zero and

$$\phi_2 = \pi q \qquad (4.198)$$

If, on the other hand, the three magnitudes (4.193) are all large and the three magnitudes (4.197) are all small, then (4.191) is probably close to π and

$$\phi_2 \approx \pi(q + 1) \qquad (4.199)$$

(b) *Conditional Probability Distribution of the Two-Phase Structure Seminvariant in* $P2_1$, *Given the Three Magnitudes in its First Neighborhood.* It is assumed that a crystal structure in $P2_1$ consisting of N atoms, not necessarily identical, in the unit cell is fixed, and that the three nonnegative numbers R_1, R_2, and $R_{1\bar{2}/10}$ are also specified. Suppose that the ordered pair $(h_1 k l_1)$ and $(h_2 k l_2)$ of reciprocal vectors is a random variable which is uniformly distributed over that subset of the twofold Cartesian product $W \times W$ defined by (4.189) and (4.190), and

$$\left| E_{h_1 k l_1} \right| = R_1, \quad \left| E_{h_2 k l_2} \right| = R_2, \quad \left| E_{\frac{1}{2}(h_1-h_2),q,\frac{1}{2}(l_1-l_2)} \right| = R_{1\bar{2}/10} \qquad (4.200)$$

The structure seminvariant ϕ_2 is then a random variable whose conditional probability distribution $P_{1|3}$, given the three magnitudes (4.193) of the first neighborhood, turns out to be, correct up to and including terms of order $1/N$,

$$P_{1|3} \approx \frac{1}{K} \exp\left\{\frac{\sigma_4}{\sigma_2{}^2}\,[2(-1)^q R_1 R_2 (R_{1\bar{2}/10}^2 - 1) \cos \Phi + R_1{}^2 R_2{}^2 \cos(2\Phi)]\right\} \tag{4.201}$$

where σ_n is defined by (4.37) and K is a normalizing parameter, dependent on the three parameters R_1, R_2, and $R_{1\bar{2}/10}$, and independent of Φ, which is readily obtained if desired but not ordinarily needed for the applications. Clearly, if R_1, R_2, and $R_{1\bar{2}/10}$ are all large, then (4.201) has a unique maximum at $\Phi = 0$ or π according as q is even or odd, in agreement with the plausible argument of Section 4.11.2(a).

(c) *Conditional Probability Distribution of the Two-Phase Structure Seminvariant in $P2_1$, Given the Six Magnitudes in Its Second Neighborhood.* As usual, suppose that a crystal structure in $P2_1$ consisting of N atoms in the unit cell is fixed and that the six nonnegative numbers R_1, R_2, $R_{1\bar{2}/10}$, $R_{12/11}$, $R_{12/1\bar{1}}$, and $R_{1\bar{2}}$ are also specified. Assume that the ordered pair $(h_1 k l_1)$ and $(h_2 k l_2)$ of reciprocal vectors is a random variable which is uniformly distributed over the subset of the twofold Cartesian product $W \times W$ defined by (4.189), (4.190), (4.200), and

$$\left| E_{\frac{1}{2}(h_1+h_2),q+k,\frac{1}{2}(l_1+l_2)} \right| = R_{12/11}$$
$$\left| E_{\frac{1}{2}(h_1+h_2),q-k,\frac{1}{2}(l_1+l_2)} \right| = R_{12/1\bar{1}} \tag{4.202}$$
$$\left| E_{h_1-h_2,0,l_1-l_2} \right| = R_{1\bar{2}}$$

Then ϕ_2, as a function of the pair $(h_1 k l_1)$ and $(h_2 k l_2)$, is a random variable and its conditional probability distribution $P_{1|6}$, given the six magnitudes (4.200) and (4.202) of its second neighborhood, turns out to be

$$P_{1|6} \approx \frac{1}{L} \exp\left\{\frac{-2(-1)^q R_1 R_2}{\sigma_2{}^3}\right.$$
$$\times [(2\sigma_3{}^2 - \sigma_2 \sigma_4) R_{1\bar{2}/10}^2 + (\sigma_3{}^2 - \sigma_2 \sigma_4)(R_{12/11}^2 + R_{12/1\bar{1}}^2) - 3(\sigma_3{}^2 - \sigma_2 \sigma_4)]$$
$$\times \cos \Phi - \left(\frac{\sigma_3{}^2 - \sigma_2 \sigma_4}{\sigma_2{}^3}\right) R_1{}^2 R_2{}^2 \cos(2\Phi)\right\}$$
$$\times \cosh\left\{\frac{\sigma_3 R_{1\bar{2}}}{\sigma_2{}^{3/2}}\,[(R_{1\bar{2}/10}^2 - 1) + 2(-1)^q R_1 R_2 \cos \Phi_1]\right\}$$
$$\times I_0\left\{\frac{2\sigma_3}{\sigma_2{}^{3/2}}\,R_{1\bar{2}/10} R_{12/11}[R_1{}^2 + R_2{}^2 + 2(-1)^q R_1 R_2 \cos \Phi]^{1/2}\right\}$$
$$\times I_0\left\{\frac{2\sigma_3}{\sigma_2{}^{3/2}}\,R_{1\bar{2}/10} R_{12/1\bar{1}}[R_1{}^2 + R_2{}^2 + 2(-1)^q R_1 R_2 \cos \Phi]^{1/2}\right\} \tag{4.203}$$

where L is a suitable normalizing parameter not needed for the present purpose. It is not difficult to verify that if all six magnitudes (4.200) and (4.202) are large then (4.203) has a unique maximum at $\Phi = 0$ or π according as q is even or odd, respectively; but if the three magnitudes (4.200) are large and the three magnitudes (4.202) are small, then (4.203) has a unique maximum at $\Phi = 0$ or π according as q is odd or even, respectively, all in accord with the heuristic argument given in Section 4.11.2(a).

4.11.3. Space Group $P2_12_12_1$

In this space group, the linear combination of two phases

$$\phi_2 = \phi_{\mathbf{h}} + \phi_{\mathbf{k}} \tag{4.204}$$

is a structure seminvariant if and only if

$$\mathbf{h} + \mathbf{k} \equiv 0 \pmod{\boldsymbol{\omega}_s} \tag{4.205}$$

where $\boldsymbol{\omega}_s$, the seminvariant modulus, is defined for $P2_12_12_1$ by means of

$$\boldsymbol{\omega}_s = (2, 2, 2) \tag{4.206}$$

In short, ϕ_2 is a structure seminvariant if and only if the three components of $\mathbf{h} + \mathbf{k}$ are even. The system of neighborhoods for the general two-phase seminvariant (4.204) is rather complicated. For this reason, only the first neighborhoods and related distributions of the restricted structure seminvariant

$$\phi_2 = \phi_{h_1 k_1 0} + \phi_{h_2 k_2 0} \tag{4.207}$$

where

$$h_1 + h_2 \equiv k_1 + k_2 \equiv 0 \pmod{2} \tag{4.208}$$

will be considered here. By symmetry, a similar analysis applies to the restricted two-phase seminvariants $\phi_{h_1 0 l_1} + \phi_{h_2 0 l_2}$ and $\phi_{0 k_1 l_1} + \phi_{0 k_2 l_2}$.

(a) *First Neighborhoods of the Restricted Two-Phase Structure Seminvariant in* $P2_12_12_1$. Let m by an arbitrary integer. If the four magnitudes

$$|E_{h_1 k_1 0}|, \quad |E_{h_2 k_2 0}|, \quad |E_{\frac{1}{2}(h_1 + h_2), \frac{1}{2}(k_1 + k_2), m}|, \quad |E_{\frac{1}{2}(h_1 - h_2), \frac{1}{2}(k_1 - k_2), m}| \tag{4.209}$$

are all large, then, from Section 4.7.1, with high probability, each of the three-phase structure invariants,

$$\phi_{h_1 k_1 0} + \phi_{\frac{1}{2}(\bar{h}_1 + \bar{h}_2), \frac{1}{2}(\bar{k}_1 + \bar{k}_2), \bar{m}} + \phi_{\frac{1}{2}(\bar{h}_1 + h_2), \frac{1}{2}(\bar{k}_1 + k_2), m} \qquad (4.210)$$

$$\phi_{h_2 k_2 0} + \phi_{\frac{1}{2}(\bar{h}_1 + \bar{h}_2), \frac{1}{2}(\bar{k}_1 + \bar{k}_2), m} + \phi_{\frac{1}{2}(h_1 + \bar{h}_2), \frac{1}{2}(k_1 + \bar{k}_2), \bar{m}} \qquad (4.211)$$

is approximately zero, so that addition of (4.210) and (4.211) yields, on account of the space-group-dependent relationships among the phases,

$$\phi_2 = \phi_{h_1 k_1 0} + \phi_{h_2 k_2 0} \approx 0 \text{ or } \pi \qquad (4.212)$$

according as

$$p = \tfrac{1}{2}(h_1 + h_2) + m \equiv 0 \text{ or } 1 \pmod{2} \qquad (4.213)$$

respectively, that is, according as the integer p is even or odd. Hence the first neighborhood of the restricted two-phase structure seminvariant (4.207) is defined to consist of the four magnitudes (4.209). Since m is an arbitrary integer, there are many first neighborhoods.

A different system of first neighborhoods is obtained by means of the four magnitudes

$$|E_{h_1 k_1 0}|, \quad |E_{h_2 k_2 0}|, \quad |E_{\frac{1}{2}(h_1 + h_2), \frac{1}{2}(k_1 - k_2), n}|, \quad |E_{\frac{1}{2}(h_1 - h_2), \frac{1}{2}(k_1 + k_2), n}| \qquad (4.214)$$

and the related three-phase structure invariants

$$\phi_{h_1 k_1 0} + \phi_{\frac{1}{2}(\bar{h}_1 + \bar{h}_2), \frac{1}{2}(\bar{k}_1 + k_2), n} + \phi_{\frac{1}{2}(\bar{h}_1 + h_2), \frac{1}{2}(\bar{k}_1 + \bar{k}_2), \bar{n}} \qquad (4.215)$$

$$\phi_{h_2 k_2 0} + \phi_{\frac{1}{2}(\bar{h}_1 + \bar{h}_2), \frac{1}{2}(k_1 + \bar{k}_2), n} + \phi_{\frac{1}{2}(h_1 + \bar{h}_2), \frac{1}{2}(\bar{k}_1 + \bar{k}_2), \bar{n}} \qquad (4.216)$$

where n is an arbitrary integer. If the four magnitudes (4.214) are large, then each of (4.215) and (4.216) is probably close to zero. In view of the space-group-dependent relationships among the phases, addition of (4.215) and (4.216) then yields

$$\phi_2 = \phi_{h_1 k_1 0} + \phi_{h_2 k_2 0} \approx 0 \text{ or } \pi \qquad (4.217)$$

according as

$$q = \tfrac{1}{2}(h_1 - h_2) + k_1 + n \equiv 0 \text{ or } 1 \pmod{2} \qquad (4.218)$$

respectively, that is, according as the integer q is even or odd. Hence the four magnitudes (4.214) constitute another first neighborhood of ϕ. Since n is an arbitrary integer, there are many first neighborhoods of this kind too.

(b) *Conditional Probability Distributions of the Restricted Two-Phase Structure Seminvariant in* $P2_12_12_1$, *Given the Four Magnitudes in Its First Neighborhood.*

(i) *Neighborhoods of the First Kind.* Here, a brief account is given of the conditional probability distribution of the two-phase structure seminvariant ϕ_2, (4.212), given the four magnitudes

$$| E_{h_1 k_1 0} |, \qquad | E_{h_2 k_2 0} |, \qquad | E_{\frac{1}{2}(h_1+h_2),\frac{1}{2}(k_1+k_2),m} |, \qquad | E_{\frac{1}{2}(h_1-h_2),\frac{1}{2}(k_1-k_2),m} |$$

where $h_1 \equiv h_2 \pmod 2$, $k_1 \equiv k_2 \pmod 2$; m is an arbitrary nonzero integer and $m h_1 k_1 h_2 k_2 (h_1 \pm h_2)(k_1 \pm k_2) \neq 0$. Using the same probabilistic background as previously described, the four magnitudes of the first neighborhood are fixed in accordance with the scheme

$$| E_{h_1 k_1 0} | = R_1, \qquad | E_{h_2 k_2 0} | = R_2 \qquad\qquad (4.219)$$

$$| E_{\frac{1}{2}(h_1+h_2),\frac{1}{2}(k_1+k_2),m} | = R_{12/2}, \qquad | E_{\frac{1}{2}(h_1-h_2),\frac{1}{2}(k_1-k_2),m} | = R_{\overline{12}/2} \qquad (4.220)$$

Denote by $P^\pm$ the conditional probability, given the four magnitudes (4.219) and (4.220), that

$$\cos \phi_2 = \pm 1 \qquad\qquad (4.221)$$

Then

$$P^\pm = \frac{1}{K} X^\pm \qquad\qquad (4.222)$$

where

$$K = X^+ + X^- \qquad\qquad (4.223)$$

$$X^\pm = \exp\left\{ \mp (-1)^p R_1 R_2 \left[\frac{2\sigma_3{}^2 - \sigma_2\sigma_4}{\sigma_2{}^3} (R_{12/2}^2 + R_{\overline{12}/2}^2) - \frac{2(\sigma_3{}^2 - \sigma_2\sigma_4)}{\sigma_2{}^3} \right] \right\}$$
$$\times I_0\left\{ \frac{2\sigma_3}{\sigma_2{}^{3/2}} R_{12/2} R_{\overline{12}/2} [R_1 \pm (-1)^p R_2] \right\} \qquad\qquad (4.224)$$

p is defined by (4.213), σ_n by (4.37), R_1, R_2, $R_{12/2}$, and $R_{\overline{12}/2}$ by (4.219), and (4.220), and upper and lower signs go together. Clearly, (4.222)–(4.224) are consistent with the heuristic argument given in Section 4.11.3(a), in that, provided that the four magnitudes R_1, R_2, $R_{12/2}$, and $R_{\overline{12}/2}$ of the first neighborhood are all large, then ϕ_2 is probably 0 or π, according as p is even or odd, respectively. If, on the other hand, R_1 and R_2 are both large, but one of $R_{12/2}$ and $R_{\overline{12}/2}$ is large and the other small, then (4.222)–

(4.224) show that ϕ_2 is probably π or 0 according as p is even or odd, respectively.

(ii) *Neighborhoods of the Second Kind.* Here a brief description is given of the conditional probability distribution of the two-phase structure seminvariant ϕ_2, (4.212), given the four magnitudes

$$| E_{h_1 k_1 0} |, \qquad | E_{h_2 k_2 0} |, \qquad | E_{\frac{1}{2}(h_1+h_2),\frac{1}{2}(k_1-k_2),n} |, \qquad | E_{\frac{1}{2}(h_1-h_2),\frac{1}{2}(k_1+k_2),n} |$$

where $h_1 \equiv h_2 \pmod 2$, $k_1 \equiv k_2 \pmod 2$, n is an arbitrary nonzero integer, and $n h_1 k_1 h_2 k_2 (h_1 \pm h_2)(k_1 \pm k_2) \neq 0$. Again, with the usual probabilistic background, the four magnitudes of the first neighborhood are fixed in accordance with the scheme

$$| E_{h_1 k_1 0} | = R_1, \qquad | E_{h_2 k_2 0} | = R_2 \tag{4.225}$$

$$| E_{\frac{1}{2}(h_1+h_2),\frac{1}{2}(k_1-k_2),n} | = R_{1\bar{2}/2}, \qquad | E_{\frac{1}{2}(h_1-h_2),\frac{1}{2}(k_1+k_2),n} | = R_{\bar{1}2/2} \tag{4.226}$$

Denote by $Q^\pm$ the conditional probability, given the four magnitudes (4.225) and (4.226), that

$$\cos \phi_2 = \pm 1 \tag{4.227}$$

Then

$$Q^\pm = \frac{1}{L} Y^\pm \tag{4.228}$$

where

$$L = Y^+ + Y^- \tag{4.229}$$

$$Y^\pm = \exp\left\{ \mp (-1)^q R_1 R_2 \left[\frac{2\sigma_3{}^2 - \sigma_2 \sigma_4}{\sigma_2{}^3} (R_{1\bar{2}/2}^2 + R_{\bar{1}2/2}^2) - \frac{2(\sigma_3{}^2 - \sigma_2 \sigma_4)}{\sigma_2{}^3} \right] \right\}$$

$$\times I_0 \left\{ \frac{2\sigma_3}{\sigma_2{}^{3/2}} R_{1\bar{2}/2} R_{\bar{1}2/2} [R_1 \pm (-1)^q R_2] \right\} \tag{4.230}$$

q is defined by (4.218), σ_n by (4.37), R_1, R_2, $R_{1\bar{2}/2}$, and $R_{\bar{1}2/2}$ by (4.225) and (4.226), and upper and lower signs go together. Clearly, (4.228)–(4.230) are in accord with the plausible reasoning of Section 4.11.3(a), that is, if the four magnitudes R_1, R_2, $R_{1\bar{2}/2}$, and $R_{\bar{1}2/2}$ of the first neighborhood are all large, then ϕ_2 is probably 0 or π according as q is even or odd, respectively. If, on the other hand, R_1 and R_2 are both large but one of $R_{1\bar{2}/2}$ and $R_{\bar{1}2/2}$ is large and the other small, then (4.228)–(4.230) imply that ϕ_2 is probably π or 0 according as q is even or odd, respectively.

4.12. Concluding Remarks

A brief account has been given of the most recent advances in the probabilistic theory of the structure seminvariants. The neighborhood principle plays the central role. For a fixed enantiomorph, the value of any structure seminvariant ϕ is primarily determined, in favorable cases, by the values of one or more small sets of appropriately chosen structure-factor magnitudes $|E|$, the neighborhoods of ϕ, and is relatively insensitive to the great bulk of remaining magnitudes. In view of the evidence presently available, a brief account of which has been given here, the neighborhood principle appears now to be firmly established. The newly acquired ability to derive the associated conditional probability distribution of a structure seminvariant, or invariant, given the magnitudes, and seminvariants when appropriate, in any of its neighborhoods, leads also to reliable estimates for large numbers of the seminvariants. Finally, the availability of reliable estimates for the structure seminvariants implies that the values of the individual phases are readily obtainable.

Acknowledgment

This work was supported by NSF grant No. CHE76-17582.

References

Cochran, W. (1955). *Acta Crystallogr.* **8**, 473.

Giacovazzo, C. (1975). *Acta Crystallogr. Sect. A* **31**, 252.

Giacovazzo, C. (1976a). *Acta Crystallogr. Sect. A* **32**, 74.

Giacovazzo, C. (1976b). *Acta Crystallogr. Sect. A* **32**, 91.

Green, E., and Hauptman, H. (1976a). *Acta Crystallogr. Sect. A* **32**, 43.

Green, E., and Hauptman, H. (1976b). *Acta Crystallogr. Sect. A* **32**, 940.

Green, E., and Hauptman, H. (1978a). *Acta Crystallogr. Sect. A* **34**, 216.

Green, E., and Hauptman, H. (1978b). *Acta Crystallogr. Sect. A* **34**, 230.

Hauptman, H. (1972). *Crystal Structure Determination: The Role of the Cosine Seminvariants*, Plenum Press, New York.

Hauptman, H. (1975a). *Acta Crystallogr. Sect. A* **31**, 671.

Hauptman, H. (1975b). *Acta Crystallogr. Sect. A* **31**, 680.

Hauptman, H. (1976a). *Some Recent Advances in the Probabilistic Theory of the Structure Invariants, Proceedings of the Prague Conference on Computing and Direct Methods in X-Ray Crystallography*, July, 1975, Munksgaard, Copenhagen.

Hauptman, H. (1976b). *Acta Crystallogr. Sect. A* **32**, 877.

Hauptman, H. (1976c). *Acta Crystallogr. Sect. A* **32**, 934.

Hauptman, H. (1977a). *Acta Crystallogr. Sect. A* **33**, 553.

Hauptman, H. (1977b). *Acta Crystallogr. Sect. A* **33**, 556.

Hauptman, H. (1977c). *Acta Crystallogr. Sect. A* **33**, 565.

Hauptman, H., and Green, E. (1976). *Acta Crystallogr. Sect. A* **32**, 45.

Hauptman, H., and Green, E. (1978). *Acta Crystallogr. Sect. A* **34**, 224.

Hauptman, H., and Karle, J. (1953). *Solution of the Phase Problem. I. The Centrosymmetric Crystal*, American Crystallographic Association Monograph No. 3, Polycrystal Book Service, Pittsburgh, Pennsylvania.

Hauptman, H., and Karle, J. (1956). *Acta Crystallogr.* **9**, 45.

Hauptman, H., and Karle, J. (1959). *Acta Crystallogr.* **12**, 93.

International Tables for X-Ray Crystallography (1962). Vol. I, The Kynoch Press, Birmingham, England.

Karle, J., and Hauptman, H. (1961). *Acta Crystallogr.* **14**, 217.

Lessinger, L., and Wondratschek, H. (1975). *Acta Crystallogr. Sect. A* **31**, 521.

Schenk, H. (1973). *Acta Crystallogr. Sect. A* **29**, 480.

Schenk, H. (1974). *Acta Crystallogr. Sect. A* **30**, 477.

Schenk, H., and de Jong, J. G. H. (1973). *Acta Crystallogr. Sect. A* **29**, 31.

Woolfson, M. (1954). *Acta Crystallogr.* **7**, 61.

Application of Calculated Cosine Invariants in Phase Determination

WILLIAM DUAX, CHARLES WEEKS, HERBERT HAUPTMAN, GEORGE DeTITTA, DAVID LANGS, EDWARD GREEN, and GERT KRUGER

5.1. Introduction

Almost all problems of real interest in crystal structure determination can be resolved if correct relative phases can be determined for the observed structure amplitudes. If three reflections have a vector sum of zero, the cosine of the sum of the three phases associated with these reflections (the universal cosine invariant) is uniquely determined by observed amplitudes. Thus if

$$\mathbf{h}_1 + \mathbf{h}_2 + \mathbf{h}_3 = 0 \tag{5.1}$$

then $\phi_{\mathbf{h}_1} + \phi_{\mathbf{h}_2} + \phi_{\mathbf{h}_3}$ is a structure invariant for every space group. The three reflections are commonly called a Σ_2 triple. It is now well understood that a knowledge of the signs and magnitudes of the universal cosine

WILLIAM DUAX, CHARLES WEEKS, HERBERT HAUPTMAN, GEORGE DeTITTA, DAVID LANGS, EDWARD GREEN, and GERT KRUGER • Medical Foundation of Buffalo, Inc., Buffalo, New York, 14203. G. Kruger was on sabbatical leave from the National Physical Research Laboratory, Pretoria, South Africa.

invariants will lead readily to correct relative phasing. It is possible, therefore, to shift the crucial feature of phase determination to one of accurate determination of the values of cosine invariants. This shift in emphasis is useful because formulas, techniques, and procedures for both calculating the values of the cosines and assessing the reliability of such calculations have been determined. The most important feature of these calculations is that the values of the cosines are derived from the observed intensities of the diffraction data. In many cases, the values of the cosine invariants can be estimated from an inspection of the intensities of the three reflections in the invariant. It has been shown that when the three reflections are very intense the sum of the three phases has a high probability of being zero. For any given combination of three reflections filling the condition of (5.1) in a real problem there is an associated probability for the sum of their phases being zero. These probability conditions have been exploited with great success in symbolic addition and multiple solution approaches to structure determination. The probability conditions related to the Σ_2 triples are also the foundation for the success of the tangent formula extension of relative phasing.

If a sufficiently large and powerful starting set of reflections having internally consistent phasing can be developed, either as a single solution or as a member of a multiple set, the tangent formula will allow correct expansion and refinement of the basis set and lead to an unambiguous structure solution. The size and character of a set of phased reflections that will lead to solution depends largely upon the nature of the crystal structure. Such factors as the number of atoms in the unit cell and the randomness of their distribution therein will affect efforts to develop an internally consistent set of phases in the early stages of either hand phasing or tangent phase extension. A crucial question then is whether the relative phases of the basis set can be correctly determined. Knowledge of the values, magnitudes, and signs, with perfect accuracy of the universal cosine invariants, would allow unambiguous phasing of all reflections and be vastly superior to a knowledge of only probable value for the cosine coupled with their associated uncertainties. While completely accurate calculation of cosines is not yet possible, current techniques provide cosines which yield information that is not provided by the more elementary probability treatment. Many internal relationships that exist among phases and invariants have been derived from a consideration of space-group-dependent symmetry relationships. Using probability methods, these internal relationships have been exploited to generate individual and relative phasing. Techniques and formulas have been developed to determine:

(a) individual phases for reflections that are themselves seminvariants —Σ_1 formulas;

(b) relative phases of two reflections whose sum is a structure seminvariant—the coincidence method and pair relationships;

(c) relationships among judiciously chosen combinations of universal cosine invariants—Diamond's determinant and quadruples.

Just as phase extension through Σ_2 relationships has relied upon interactions and indications from intense reflections, so most of the useful information obtained from symmetry considerations has been dependent upon the strongest contributors being consistent in their indications of individual and relative phasing. When indications concerning phase relationships are ambiguous, solution of structural problems may be difficult or impossible: conflicts indicate that some cosine invariants differ appreciably from probable values. The determination of which universal cosine invariants are different from their expected values is simplified if some reliable means of calculating the cosines is available.

The material presented here illustrates some of the efforts of the past ten years to use calculated cosine invariants to determine which cosines can be approximated safely by $+1$ or -1 in value, to resolve conflicts in phasing indications and to isolate cosine invariants that differ significantly from values that would be expected on the basis of elementary probability theory.

Before considering applications of the cosines, the formulas for calculation of the cosines, their relative accuracy, and the proper interpretation of the results will be described. The remainder of the chapter will be devoted to a discussion of the use of calculated cosines in improving the reliability of phasing through (a) Σ_1 evaluations, (b) pair relationships, (c) strong enantiomorph selection in space group $P2_1$, (d) NQEST, and (e) automated procedures.

5.2. Accuracy of Cosine Calculations

It is now well established that the value of the structure invariant $\phi_{h_1} + \phi_{h_2} + \phi_{h_3}$ is dependent not only on the magnitudes of E_{h_1}, E_{h_2}, and E_{h_3}, but on other magnitudes as well. The identity of the most important of these magnitudes and the precise nature of this dependence is being explored and formulas devised to provide accurate calculations of cosine invariants. Two of the earliest formulas still in use at the Medical Foundation of Buffalo are the modified triple-product formula (Hauptman *et al.*, 1969) and the MDKS formula (Fisher *et al.*, 1970; Hauptman, 1972a).

Modified Triple Product Formula (TPROD). This formula is

$$\cos(\phi_{h_1} + \phi_{h_2} + \phi_{h_3}) \approx \frac{K\psi}{|E_{h_1}E_{h_2}E_{h_3}|} + \frac{R_3}{|E_{h_1}E_{h_2}E_{h_3}|} \quad (5.2)$$

where

$$\psi = \langle(|E_k|^{1/2} - \overline{|E|^{1/2}})(|E_{h_1+k}|^{1/2} - \overline{|E|^{1/2}})$$
$$\times (|E_{k-h_3}|^{1/2} - \overline{|E|^{1/2}}), |E_k| > t\rangle_k \quad (5.3)$$

$$R_3 = \frac{\sigma_3}{4\sigma_2^{3/2}} \left[\frac{3}{2} (|E_{h_1}E_{h_3}|^2 + |E_{h_2}E_{h_3}|^2 + |E_{h_3}E_{h_1}|^2) + |E_{h_1}|^2 \right.$$
$$\left. + |E_{h_2}|^2 + |E_{h_3}|^2 - \frac{7}{2} \right] \quad (5.4)$$

$$\sigma_n = \sum_{j=1}^{N} Z_j^n, \qquad n = 2, 3 \quad (5.5)$$

$\overline{|E|^{1/2}}$ is the average value of $|E|^{1/2}$, K is a sliding scale factor, and t is a threshold value. The parameter A is defined by

$$A = \frac{2}{N^{1/2}} |E_{h_1}E_{h_2}E_{h_3}| \quad (5.6)$$

where N is the number of atoms, assumed identical, in the unit cell.

MDKS *Formula.* This formula is

$$\cos(\phi_{h_1} + \phi_{h_2} + \phi_{h_3}) = M(D - KS) \quad (5.7)$$

where

$$D = \langle(|E_{k-h_3}|^2 - 1), |E_k| > t, |E_{h_1+k}| > t\rangle_k \quad (5.8)$$

and

$$S = \langle(|E_{k-h_3}|^2 - 1, |E_k| > t\rangle_k + \langle(|E_{k-h_3}|^2 - 1), |E_{h_1+k}| > t\rangle_k \quad (5.9)$$

M and K are scaling parameters, t is a threshold parameter.

In both of these formulas cosines are evaluated through a restricted averaging over a large sample of observed intensities. The numbers of contributors in the averages are a function of the threshold values t. After extensive experimentation, it was found that the accuracy of the triple-product calculation was largely insensitive to the value of t. However, a larger value for t results in rapid, economical calculation and consequently t is generally chosen as 2.0. All reflections $\mathbf{k}$ having $|E|$ values of 2.0 or more are used in deriving the average (5.3). It can be seen from inspection that, when most of the reflections of the types $\mathbf{h}_1 + \mathbf{k}$ and $\mathbf{k} - \mathbf{h}_3$ are intense, the average will be large and the cosine calculation will be positive. In other words, if in addition to being a triple of three strong reflections, the members of the triple are also members of many other strong triples then the probability that the sum of the three phases of the first triple is nearly zero is enhanced. The scaling factor K is adjusted in such a way that, for each A value, the distribution of calculated cosine values matches the expected cosine distribution.

The MDKS formula was derived from quadruple relationships among triples and marked the first exploitation of what has subsequently come to be known as the neighborhood concept (Hauptman, 1975). Quadruples are combinations of four universal cosine invariants specifically chosen so that each of the six reflections occurs twice in the four triples (see Table 5.1).

Because each reflection and its negative are present in the quadruple, each phase and its negative are also present. Consequently, the sum of the 12 phases ($\Phi_1 + \Phi_2 + \Phi_3 + \Phi_4$) must equal zero. In centrosymmetric space groups, or in centrosymmetric projections in noncentrosymmetric space groups, the phase relationships are further simplified such that the sums of the phases of reflections in the individual triples must be 0 or π, with the cosines of the triples ± 1. Furthermore, since the sum of the 12 phases is necessarily zero, the cosines of four, two, or none of the triples in a quadruple must be $+1$ in value. If the E's of the six reflections are all large, then all four A's are of necessity large, and the probability that the

TABLE 5.1. An Example of a Quadruple

Triple			Example			Sum of phases	A
1. $\mathbf{h}_1$ $\mathbf{h}_2$		$\mathbf{h}_3$	0 03	$\bar{3}$ 01	3 0 $\bar{4}$	Φ_1	A_1
2. $-\mathbf{h}_1$ $\mathbf{k}$		$\mathbf{h}_1 - \mathbf{k}$	0 0$\bar{3}$	$\bar{1}$ $\bar{3}$5 1	1 35 2	Φ_2	A_2
3. $-\mathbf{k}$ $\mathbf{h}_3 + \mathbf{k}$		$-\mathbf{h}_3$	1 35 $\bar{1}$	2 $\bar{3}$$\bar{5}$ 3	$\bar{3}$ 0 4	Φ_3	A_3
4. $-\mathbf{h}_2$ $\mathbf{h}_1 + \mathbf{h}_2 - \mathbf{k}$		$\mathbf{k} - \mathbf{h}_1$	3 0$\bar{1}$	$\bar{2}$ 35 3	$\bar{1}$ $\bar{3}$5 $\bar{2}$	Φ_4	A_4

cosine of any one of the four triples is $+1$ in value is enhanced by its highly specific association with the others. If, on the other hand, the magnitudes of the six reflections are such that three of the A's are very small, then the probability that the fourth triple is $+1$ in value is called into question. A series of quadruples having one triple in common can be constructed by varying $\mathbf{k}$. Conditional information about the probability that the cosine of a particular invariant is $+1$ in value can be gained by examining the A values of triples with which it combines to form quadruples.

Expressed in another way, if the E's of the three main terms $\mathbf{h}_1$, $\mathbf{h}_2$, and $\mathbf{h}_3$ are large, and if the E's of two of the remaining six terms in a quadruple, $\mathbf{k}$ and $\mathbf{h}_1 + \mathbf{k}$, are also large, then the probability that $\cos(\phi_{\mathbf{h}_1} + \phi_{\mathbf{h}_2} + \phi_{\mathbf{h}_3})$ is $+1$ will be dependent upon the magnitude of the sixth term or, more importantly, on the average value for the sixth terms in a series of quadruples in which the E's of $\mathbf{k}$ and $\mathbf{h}_1 + \mathbf{k}$ are strong. It is this average that is evaluated in the MDKS formula where the magnitudes in the restricted averages D and S can be seen to be those of the reflections in the quadruple.

The scaling parameters M and K are again determined in such a way that the resulting distribution of calculated cosines agrees with the theoretical distribution. There is nothing in either formula that restricts the calculated values to the range -1 or $+1$. Cosines that are calculated to be greater than 1 are presumed to have a high probability of being near $+1$ in value. Individual calculated cosines may be completely accurate, but they are often inaccurate and occasionally totally wrong. Most successful applications have come through use of combinations of calculated cosines: a least-squares analysis of the calculated cosines was used in the determinations of four steroid structures. Subsequent to the solution of two of these structures, a thorough analysis was made of the accuracy of cosines calculated by the modified triple-product formula (Hauptman *et al.*, 1971). The following are some of the important conclusions based upon that analysis:

(a) Invariants involving certain reflections may be computed less accurately than a general set of invariants, and they may be subject to a nonrandom error.

(b) Invariants computed after imposing restrictions on $|E_k|$ are, on average, as accurate as those computed when $\mathbf{k}$ is allowed to range over all reflections, but computing time is substantially less. The condition $|E_\mathbf{k}| > 2$ is suggested for general use.

(c) The scaling parameter K is largely independent of A when the range of $\mathbf{k}$ is severely restricted. When the range of $\mathbf{k}$ is unrestricted, K

is a function of A, and in the case of these structures, the dependence is linear.

(d) Cosine invariants whose true values are large are computed more accurately than those that are, in reality, small. Cosines that are small show a relatively large nonrandom error since their calculated or estimated values are usually larger than their true values.

(e) Cosines with large A appear to be computed much more accurately than those with small A, but this is largely a manifestation of the lower accuracy of the computation of cosines whose observed values are relatively small, because such cosines are more frequent at low A.

The relative accuracy with which cosine invariants, $\cos(\phi_{h_1} + \phi_{h_2} + \phi_{h_3})$, are calculated by the modified triple-product and the MDKS formulas was analyzed for the structure of a steroid complex (Duax et al., 1972). The pertinent results of that analysis were as follows:

(a) The MDKS formula was more effective than TPROD in identifying the negative invariants in the most important high-A range.

(b) Cosines calculated by the MDKS formula had a higher proportion of negatively indicated values than could reasonably be expected.

(c) By the MDKS method, a smaller total number of Σ_2 triples were eligible for routine use in the phase extension process; that is, a smaller number of triples could reliably be assumed to have $+1$ values.

Another example of the use of calculated cosines in detecting both those cosines that are nearest $+1$ in value and those that are farthest from expected values is provided in an analysis of the distribution of cosines in the subset used to phase the structure of valinomycin.

Cosines were calculated by the modified triple-product formula. Sixty-two cosine invariants that were directly involved in the initial phasing procedure had large positive calculated values and were assigned $+1$ values. The three most negatively calculated cosines, all less than -0.5, were assigned -1 values. Internally consistent phasing of 61 reflections based upon their cosine assignments led to successful tangent formula phase extension and structure solution. The true cosines of the 65 invariants used in the phasing procedure are plotted in Fig. 5.1, together with the curve of the expected values of the cosines as a function of A alone.

It can be seen that the true average value of the positive invariants is much closer to $+1$ (average deviation, 0.05) than to the theoretical distribution (average deviation, 0.25). Thus the accurate identification, by means

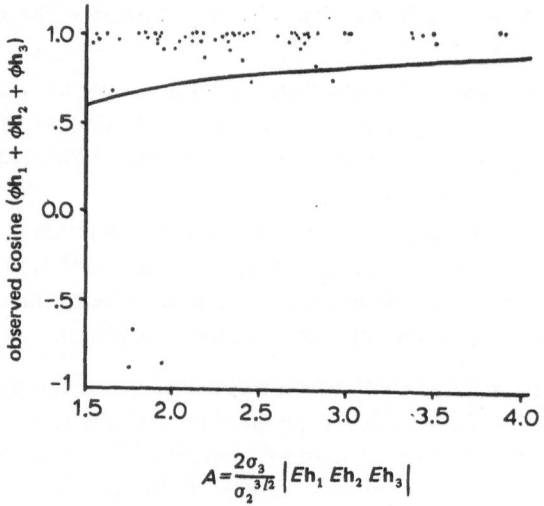

$$A = \frac{2\sigma_3}{\sigma_2^{3/2}} \left| E_{h_1} E_{h_2} E_{h_3} \right|$$

FIGURE 5.1. True value for 65 cosines used in the solution of the valinomycin structure together with the curve of expected values. They show the improvement which results from the ability both to identify *a priori* and to estimate reliably those cosines whose true values are approximately ±1.

of the calculated cosine values, of those cosines that are very close to +1 played a very important part in the process of phase determination. However, the correct identification of the three cosines in the neighborhood of −1 was crucial.

5.3. Quartets, Quintets, and Triplets

The quartets are the four-phase cosine invariants $\cos(\phi_{h_1} + \phi_{h_2} + \phi_{h_3} + \phi_{h_4})$, where $h_1 + h_2 + h_3 + h_4 = 0$. Hauptman (1974a,b) has shown that, for a sufficiently large $B = (2/N) \left| E_{h_1} E_{h_2} E_{h_3} E_{h_4} \right|$, where N is the number of assumed identical atoms in the unit cell,

$$(\phi_{h_1} + \phi_{h_2} + \phi_{h_3} + \phi_{h_4})$$

$$\approx \begin{cases} +1 & \text{if } \left| E_{h_1+h_2} \right|, \left| E_{h_1+h_3} \right|, \left| E_{h_1+h_4} \right| \text{ are all large} \quad (5.10) \\ -1 & \text{if } \left| E_{h_1+h_2} \right|, \left| E_{h_1+h_3} \right|, \left| E_{h_1+h_4} \right| \text{ are all small} \quad (5.11) \end{cases}$$

Two aspects of this result are remarkable. First, the sum of four phases is shown to be dependent not only upon the intensity of the four reflections the sum of whose indices is zero but also upon the intensities of three other reflections of specific identity. These seven reflections represent the second neighborhood of the quartet, and the contribution of more data to the evaluation of the cosine facilitates greater accuracy of cosine determination. Second, cosine invariants of value -1 can be predicted quite reliably, although still not as reliably as can values of $+1$. This capability has proven to be of significance in permitting the accurate determination of relative phases and successful identification of correct phase sets in multiple solution phase extension.

Similarly, quintets, or five-phase invariants, $\phi_{h_1} + \phi_{h_2} + \phi_{h_3} + \phi_{h_4} + \phi_{h_5}$ where $h_1 + h_2 + h_3 + h_4 + h_5 = 0$, can be shown to be primarily dependent upon the intensities of 15 reflections. If one considers a special quintet in which a single reflection appears twice (as h_4 and $\bar{h}_4$) the remaining members of the quintet are a triplet and the cosine of this triplet is shown to be dependent upon ten reflections. The term triplet is used to distinguish it from a triple because the probable phase of the triplet can be defined explicitly in terms of the ten-member neighborhood, whereas previous evaluation of the same triples has been a function of the intensities of only the three reflections in the triple, evaluation of the degree to which these strong reflections interact with other strong reflections, or exploration of memberships in six reflection neighborhoods, or quadruples.

Numerous ten-member neighborhoods can be found for each triplet by varying the redundant reflection in the quintet. The triplet formula presently in development was derived for space group $P1$. The success of the formula in evaluating cosines of projection data in $P2_1$ is illustrated in Tables 5.2 and 5.3. In Table 5.2, the strongest $h0l$ triplets in the structure of tetrahymanol are sorted as a function of A. The crystal structure of tetrahymanol has 62 nonhydrogen atoms in the asymmetric unit.

In Table 5.3, the same triplets are sorted on values of modified A, a function of the intensities of the reflections in the ten-magnitude neighborhoods which determines the distribution of values of the cosine. Not only are the largest values of modified A extremely well correlated with those cosines whose true values are $+1$, but, as Table 5.3 shows, there is also a good correlation between the most negative values of modified A and those cosines whose true values are -1. Comparison of Tables 5.2 and 5.3 shows clearly the superiority of modified A over A in estimating the cosine values of the triplet and confirms the expectation that the ability to use the ten magnitudes of the second neighborhoods, rather than only the three magnitudes of the first neighborhood, increases the number of triples whose values can be reliably estimated.

TABLE 5.2. The 50 *h0l* Triples in the Structure of Tetrahymanol Having Largest
A Values, Their True Cosines, and Modified *A* Values Based upon
Triplet Calculations

Triple serial number	*A*	True cosine	Modified *A* from triplet calculation [a]
1	5.31	1.0	190.03
2	4.88	1.0	84.50
3	4.56	1.0	25.75
4	3.89	1.0	35.49
5	3.77	1.0	—
6	3.44	1.0	12.69
7	3.34	−1.0	—
8	3.20	1.0	84.54
9	3.04	1.0	43.65
10	2.88	−1.0	−22.44
11	2.87	1.0	14.95
12	2.80	−1.0	—
13	2.63	1.0	55.39
14	2.41	1.0	9.83
15	2.28	−1.0	0.45
16	2.28	1.0	5.47
17	2.23	1.0	42.30
18	2.08	1.0	7.64
19	2.04	1.0	4.33
20	2.00	1.0	17.12
21	1.99	1.0	4.49
22	1.96	−1.0	0.62
23	1.93	1.0	7.71
24	1.93	1.0	6.05
25	1.72	1.0	0.21

TABLE 5.2 (*continued*)

Triple serial number	A	True cosine	Modified A from triplet calculation [a]
26	1.71	1.0	17.56
27	1.70	1.0	6.73
28	1.69	1.0	5.77
29	1.67	1.0	64.48
30	1.66	1.0	8.67
31	1.64	1.0	25.53
32	1.63	1.0	14.19
33	1.51	1.0	3.17
34	1.51	1.0	2.17
35	1.49	1.0	8.52
36	1.48	−1.0	—
37	1.48	1.0	4.52
38	1.45	−1.0	−0.39
39	1.43	1.0	8.60
40	1.35	1.0	1.25
41	1.28	−1.0	−0.05
42	1.19	1.0	5.70
43	1.13	−1.0	1.48
44	1.09	1.0	2.74
45	1.06	−1.0	−0.47
46	1.05	1.0	0.03
47	0.99	1.0	−0.07
48	0.89	1.0	2.69
49	0.84	1.0	0.80
50	0.75	1.0	2.17

[a] The dash (—) denotes no triplet calculation possible because $|E|$'s satisfying required conditions were not available.

TABLE 5.3. The 46 Triples from Table 5.2 for Which Triplet Evaluation was Possible Sorted on Modified A

Triple serial number	A	True cosine	Modified A from triplet calculation
1	5.31	1.0	190.03
2	3.20	1.0	84.54
3	4.88	1.0	84.50
4	1.67	1.0	64.48
5	2.63	1.0	55.39
6	3.04	1.0	43.65
7	2.23	1.0	42.30
8	3.89	1.0	35.49
9	4.56	1.0	25.75
10	1.64	1.0	25.53
11	1.71	1.0	17.56
12	2.00	1.0	17.12
13	2.87	1.0	14.95
14	1.63	1.0	14.19
15	3.44	1.0	12.69
16	2.41	1.0	9.83
17	1.66	1.0	8.67
18	1.43	1.0	8.60
19	1.49	1.0	8.52
20	1.93	1.0	7.71
21	2.08	1.0	7.64
22	1.70	1.0	6.73
23	1.93	1.0	6.05
24	1.69	1.0	5.77
25	1.19	1.0	5.70
26	2.28	1.0	5.47
27	1.48	1.0	4.52
28	1.99	1.0	4.49
29	2.04	1.0	4.33
30	1.51	1.0	3.17
31	1.09	1.0	2.74
32	0.89	1.0	2.69
33	1.51	1.0	2.17

TABLE 5.3 (*continued*)

Triple serial number	A	True cosine	Modified A from triplet calculation
34	0.75	1.0	2.17
35	1.13	−1.0	1.48
36	1.35	1.0	1.25
37	0.84	1.0	0.80
38	1.96	−1.0	0.62
39	2.28	−1.0	0.45
40	1.72	1.0	0.21
41	1.05	1.0	0.03
42	1.28	−1.0	−0.05
43	0.99	1.0	−0.07
44	1.45	−1.0	−0.39
45	1.06	−1.0	−0.47
46	2.88	−1.0	−22.44

5.4. Σ_1 Cosines

Phases whose values are independent of space-group-dependent permissible origins are called structure seminvariants. Space-group-dependent formulas for the evaluation of the probable phases of many such reflections have been derived (Hauptman and Karle, 1953; Weeks and Hauptman, 1970). These are commonly called Σ_1 formulas and Σ_1 reflections. The formulas are scaled averages of appropriately chosen contributors of magnitudes $|E|$, and the predicted phases of the Σ_1 reflections are governed by the relative intensities of the individual contributors. When strong contributors conflict, the phase indication is ambiguous.

The cause of this ambiguity can be understood and resolved by analysis of the cosines associated with the contributors to the Σ_1 averages. As an example, one of the Σ_1 formulas in the space group $P2_12_12_1$ is

$$E_{02k,2l} = N^{1/2}\langle(-1)^{h+k}(|E_{hkl}|^2 - 1)\rangle_h \qquad (5.12)$$

To evaluate E_{022}, the average is taken over reflections 011, 111, 211, ..., $h11$, with the signs of contributors alternating. Clearly, if reflections with *h* odd have the largest $|E|$'s the average will be negative, and if reflections with *h* even are strongest the average will be positive. Negative or positive calculations of E_{022} indicate a phase for ϕ_{022} of π or 0, respectively. If reflections with *h* even and odd are equally intense there is no strong indication of phase.

Each contributing reflection can be combined with the 022 reflection to create a universal cosine invariant, as shown in Table 5.4. Space group conditions require the following:

(a) The 022 reflection must have a phase of 0 or π.

(b) A phase relationship exists between the symmetry forms in columns **B** and **C**:

   (1) If *h* is even, $\phi_C = \pi - \phi_B$, $\phi_B + \phi_C = \pi$.
   (2) If *h* is odd, $\phi_C = -\phi_B$, $\phi_B + \phi_C = 0$.

(c) If $\phi_{022} = 0$,

   (1) $\cosine(\phi_A + \phi_B + \phi_C)$ in 1, 3, 5, 7, ... = -1;
   (2) $\cosine(\phi_A + \phi_B + \phi_C)$ in 2, 4, 6, ... = $+1$.

(d) If $\phi_{022} = \pi$,

   (1) $\cosine(\phi_A + \phi_B + \phi_C)$ in 1, 3, 5, 7, ... = $+1$;
   (2) $\cosine(\phi_A + \phi_B + \phi_C)$ in 2, 4, 6, ... = -1.

Obviously, half of the contributors to a Σ_1 summation must have associated universal cosine invariants that are $+1$ in value and half that are -1 in value. If strong contributors are in conflict, the calculated cosines should differentiate between the positive and the negative values.

A typical example of calculated cosines for Σ_1 contributors is provided

TABLE 5.4. Universal Cosine Invariants that Correspond to Contributors to the Σ_1 Summation for Reflection 022

Triple	A	B	C
1	022	$0\bar{1}\bar{1}$	$0\bar{1}\bar{1}$
2	022	$\bar{1}\bar{1}\bar{1}$	$1\bar{1}\bar{1}$
3	022	$\bar{2}\bar{1}\bar{1}$	$2\bar{1}\bar{1}$
4	022	$\bar{3}\bar{1}\bar{1}$	$3\bar{1}\bar{1}$
⋮	⋮	⋮	⋮

TABLE 5.5. Calculated Cosines for the Σ_1 Contributors for the 020 Reflection

Contributor $0kl$	A^*	Calculated cosines	
		MDKS	TPROD
011	1.5	0.38	1.2
012	3.2	1.09	1.06
013	1.0	0.35	0.97
014	1.5	1.72	1.33

by the case of the 020 reflection in urea-estradiol (Table 5.5). Two of the cosines must be -1 and two must be $+1$ in value. Although the TPROD calculation is ambiguous the MDKS evaluation correctly indicates the negative cosines. Thus, although the Σ_1 calculation was inconclusive, the further calculation and analysis of the cosine invariants associated with the Σ_1 contributors led to the correct value of zero for ϕ_{020}.

5.5. Pair Relationships

The coincidence method of Grant *et al.* (1957) provided a means of relating the signs of pairs of structure factors in centrosymmetric space groups by utilizing suitable combinations of triple-product sign relationships, each of which was known with high probability. For example, if

$$S(\mathbf{h})S(\mathbf{h'}) \simeq S(\mathbf{h} + \mathbf{h'}) \tag{5.13}$$

with probability P_1, and

$$S(\mathbf{h})S(\mathbf{h'}) \simeq S(\mathbf{h} - \mathbf{h'}) \tag{5.14}$$

with probability P_2, then

$$S(\mathbf{h} + \mathbf{h'}) \simeq S(\mathbf{h} - \mathbf{h'}) \tag{5.15}$$

with probability

$$P = 2P_1P_2 - P_1 - P_2 + 1 \tag{5.16}$$

and no knowledge of $S(\mathbf{h})$ or $S(\mathbf{h'})$ is required. Recently (Hauptman, 1972a,b) it has been found that the relationships among pairs of phases that are related as in equation (5.15) may be expressed in terms of the space-group-dependent cosine seminvariants, $\cos(\phi_{\mathbf{h}_1} + \phi_{\mathbf{h}_2})$. In these so-called pair formulas, the cosine seminvariants are computed in terms of a summation product of two ($|E|^2 - 1$)'s for all pairs of reflections that are related to each of the paired reflections in Σ_2 triples. The resulting summations provide a stronger measure of the relationship among such phases than can be obtained from the single pair of relationships given in equations (5.13) and (5.14). In space group $P2_12_12_1$, two types of formulas exist which allow the cosine seminvariants, $\cos(\phi_{\mathbf{h}_1} + \phi_{\mathbf{h}_2})$, where the phases $\phi_{\mathbf{h}_1}$ and $\phi_{\mathbf{h}_2}$ are both restricted to be either 0 or π or both restricted to be $\pm\pi/2$, to be computed from normalized structure-factor magnitudes alone. The detailed derivation of such formulas has been presented elsewhere (Hauptman, 1972b), and only the results are summarized here. The first type of formula, in the form applicable to $0kl$ reflections, is

$$
\begin{aligned}
E_{0k_1l_1}E_{0k_2l_2} &= |E_{0k_1l_1}E_{0k_2l_2}| \cos(\phi_{0k_1l_1} + \phi_{0k_2l_2}) \\
&\simeq \tfrac{1}{2}N\langle(-1)^{h+\frac{1}{2}(k_1+k_2)}(|E_{h,\frac{1}{2}(k_1+k_2),\frac{1}{2}(l_1+l_2)}|^2 - 1) \\
&\quad \times (|E_{h,\frac{1}{2}(k_1-k_2),\frac{1}{2}(l_1-l_2)}|^2 - 1)\rangle_h \\
&\simeq \tfrac{1}{2}N\langle(-1)^{h+\frac{1}{2}(k_1-k_2)+l_1}(|E_{h,\frac{1}{2}(k_1+k_2),\frac{1}{2}(l_1-l_2)}|^2 - 1) \\
&\quad \times (|E_{h,\frac{1}{2}(k_1-k_2),\frac{1}{2}(l_1+l_2)}|^2 - 1)\rangle_h
\end{aligned}
\tag{5.17}
$$

where $k_1 \pm k_2$ and $l_1 \pm l_2$ are even, that is, k_1 and k_2 have the same parity and l_1 and l_2 have the same parity, and N is the number of atoms, assumed identical, in the unit cell. Clearly, the contributions from the two parts of this formula may be combined. Analogous formulas for the $h0l$ and $hk0$ reflections may be found through a cyclic permutation of the indices. Because of the cumbersome notation occurring in formulas of this sort, it is convenient to introduce the notation $\mathbf{h}_1 = 0k_1l_1$, $\mathbf{h}_2 = 0k_2l_2$, $\mathbf{h}_3 = h, \frac{1}{2}(k_1 + k_2), \frac{1}{2}(l_1 + l_2)$ or $h, \frac{1}{2}(k_1 + k_2), \frac{1}{2}(l_1 - l_2)$, $\mathbf{h}_4 = h, \frac{1}{2}(k_1 - k_2)$, $\frac{1}{2}(l_1 - l_2)$ or $h, \frac{1}{2}(k_1 - k_2), \frac{1}{2}(l_1 + l_2)$, and $E_{\mathbf{h}_1} = E_1$, etc. In the type I formulas, (5.17), reflections $\mathbf{h}_3$ and $\mathbf{h}_4$ may be, and generally are, three-dimensional, but in the type II formulas, of which

$$
\begin{aligned}
E_{0k_1l_1}E_{0k_1l_2} &= |E_{0k_1l_1}E_{0k_1l_2}| \cos(\phi_{0k_1l_1} + \phi_{0k_1l_2}) \\
&\simeq \tfrac{1}{2}N\langle(-1)^{k+\frac{1}{2}(l_1-l_2)}(|E_{0,k,\frac{1}{2}(l_1+l_2)}|^2 - 1)(|E_{0,k+k_1,\frac{1}{2}(l_1-l_2)}|^2 - 1)\rangle_k
\end{aligned}
\tag{5.18}
$$

is an example, reflections $\mathbf{h}_3 = 0, k, \frac{1}{2}(l_1 + l_2)$, and $\mathbf{h}_4 = 0, k + k_1, \frac{1}{2}(l_1 - l_2)$, not to be confused with the $\mathbf{h}_3$ and $\mathbf{h}_4$ in (5.17), always have at least one zero index. It is apparent that formulas of the second type are applicable only to pairs $\mathbf{h}_1$ and $\mathbf{h}_2$ having a common nonzero component.

For example, in 9α-fluorocortisol the scaled average of the products of $|E|^2 - 1$ for the pairs of $h17,2$ and $h10$ reflections and the pairs of $h17,0$ and $h12$ reflections yielded -144 as the calculated value of $E_{016,2}E_{018,2}$, indicating that 016,2 and 018,2 have different phases. The quantity $|E_{\mathbf{h}_1}E_{\mathbf{h}_2}|$ $\times \cos(\phi_{\mathbf{h}_1} - \phi_{\mathbf{h}_2})$ is tabulated rather than $E_{\mathbf{h}_1}E_{\mathbf{h}_2} = |E_{\mathbf{h}_1}E_{\mathbf{h}_2}|\cos(\phi_{\mathbf{h}_1} + \phi_{\mathbf{h}_2})$ of (5.17) and (5.18), because the relationship between the two phases is immediately apparent from the sign of the former. If this sign is positive, the phases have the same value, but they differ by π radians if the sign is negative. The magnitude of the calculated product of the normalized structure-factor amplitudes is quite different from the observed value. However, this is not a serious problem because it is known that the phase of each of the paired reflections is restricted to one of two possible values, and the pair formula is used only to determine whether these values are the same or differ by π radians. The probability that the correct relationship between the phases is that indicated increases as the absolute value of the calculated product increases, and one measure of the reliability of the relative phase indication is given by the significance level,

$$s = 2n^{1/2}(\text{obs } |E_{\mathbf{h}_1}E_{\mathbf{h}_2}| + \text{calc } |E_{\mathbf{h}_1}E_{\mathbf{h}_2}|)/N \qquad (5.19)$$

where n is the number of contributors to the average in (5.17) or (5.18), and N is the number of atoms in the unit cell.

The relationship of the pair formulas to the coincidence method of Grant *et al.* (1957) may be made clearer by examining the nature of the individual contributors to the pair summations, and an analysis of this type also helps one to understand how and when the pair formulas will give incorrect relative phase indications. The case of 111,0 and 121,0 will be considered as an illustration. The type I summation for these two reflections is over the products of $(|E|^2 - 1)$'s for the pairs of $15l$ and $016,l$ reflections and the pairs of $05l$ and $116,l$ reflections. The paired reflections 111,0 and 121,0 each form a triple of the Σ_2 type with some symmetry variant of each $15l$ and $016,l$ and with each $05l$ and $116,l$. The largest contributor to the summation occurs when $\mathbf{h}_3 = 152$ and $\mathbf{h}_4 = 016,2$ and the Σ_2 triples corresponding to this contributor are given in Table 5.6.

It is convenient to introduce the notation $\cos(134) = \cos(\phi_{\mathbf{h}_1} + \phi_{\mathbf{h}_3*}$ $+ \phi_{\mathbf{h}_4*})$ and $\cos(234) = \cos(\phi_{\mathbf{h}_2} + \phi_{\mathbf{h}_3*} + \phi_{\mathbf{h}_4*})$, where the asterisk indicates that symmetry variant of the associated reflection needed to yield a structure invariant. Owing to the space-group symmetries, $\cos(134) = \pm\cos(234)$.

TABLE 5.6. Σ_2 Triples Corresponding to a Large Contributor to the Summation for the Pair $(111,0; 121,0)$

h (h_1 or h_2)	k (h_3 variant)	$-h - k$ (h_4 variant)	A^*
1 21 1	$\bar{1}\bar{3}2$	0 $\bar{1}\bar{6}$ 2	2.17
1 11 0	$\bar{1}52$	0 $\bar{1}\bar{6}$ $\bar{2}$	2.86

If $\phi_{\bar{1}5\bar{2}} = \alpha$, $\phi_{0\bar{1}\bar{6},2} = \beta$, $\phi_{111,0} = \phi_{h_1}$, and $\phi_{121,0} = \phi_{h_2}$, then

$$\Phi_1 = \phi_{111,0} + \phi_{\bar{1}52} + \phi_{0\bar{1}\bar{6},\bar{2}} = \phi_{h_1} + \alpha + \beta \qquad (5.20)$$

and

$$\Phi_2 = \phi_{121,0} + \phi_{\bar{1}5\bar{2}} + \phi_{0\bar{1}\bar{6},2} = \phi_{h_2} + \alpha + \beta \qquad (5.21)$$

It follows that

$$\Phi_1 = \Phi_2 \quad \text{if } \phi_{h_1} = \phi_{h_2} \qquad (5.22)$$

and

$$\Phi_1 = \pi + \Phi_2 \quad \text{if } \phi_{h_1} = \pi + \phi_{h_2} \qquad (5.23)$$

Hence $\phi_{h_1} = \phi_{h_2}$ or $\phi_{h_1} = \pi + \phi_{h_2}$ according as $\cos \Phi_1 = \cos \Phi_2$ or $\cos \Phi_1 = -\cos \Phi_2$. If $\cos \Phi_1 = \pm \cos \Phi_2 \simeq 0$, no information is available. If

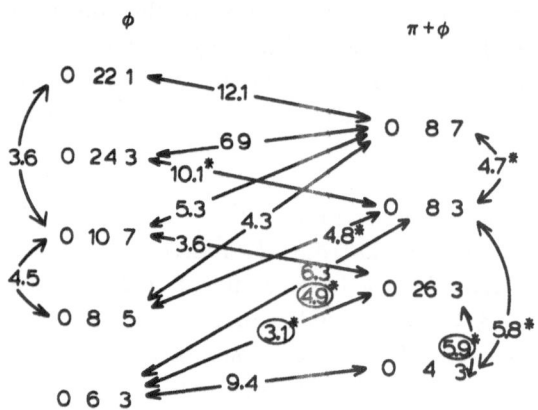

FIGURE 5.2. Pairings among the 0*gu* reflections of 9α-fluorocortisol reflections. Value of *s*, (5.19), are indicated on the arrows, and an asterisk indicates that a pair relationship of type II was used. In those cases where the pair formula gave the opposite relationship from the one shown, the *s* values are circled.

TABLE 5.7. Σ_2 Triples for Large Contributors (Both $A > 1$) to Some Selected Pairs in 9α-Fluorocortisol

| h_1 | h_2 | h_3 | h_4 | $A(134)$ | cos(134) | | $A(234)$ | cos(234) | | Calculated |
					TPROD	TRUE		TPROD	TRUE	$\mid E_{h_1}E_{h_2}\mid \cos(\phi_{h_1}-\phi_{h_2})$
063	0 26 3	5 16 3	5 10 0	1.22	0.79	0.89	1.35	0.70	0.89	−285
		0 16 4	0 10 7	1.05	1.77	1.00	1.16	1.38	1.00	−185
043	0 26 3	0 15 4	0 11 1	2.68	1.09	1.00	2.73	0.70	−1.00	−1631
		8 15 3	8 11 0	2.04	0.55	−0.96	2.08	0.81	0.96	−409
		1 15 3	1 11 0	1.71	0.85	0.99	1.75	0.87	0.99	534
063	0 24 3	0 15 4	0 9 1	1.84	0.75	1.00	1.26	0.56	−1.00	−645
063	0 8 3	6 7 3	6 10	2.43	0.67	0.95	1.78	0.68	0.95	−1391
		0 7 3	0 16	1.95	1.14	1.00	1.43	0.43	−1.00	1002

the A values for the two triples are both large, as is true in this example, then it is probable that both cosine seminvariants will have positive values and that the phase relationship as indicated by (5.22), which agrees with the overall indication from equation (5.17), will be correct. Consistent with the pair (5.17) and (5.18) is the assumption that the two cosine seminvariants related to a single dominant contributor have the same sign. It is apparent then that the pair relationship will fail in those cases where there is a single dominant contributor, and that contributor has one positive and one negative associated cosine seminvariant. Thus, erroneous conclusions based on the pair formulas can be avoided in large part by examining the large contributors and checking the calculated values of their cosines which are predicted by the modified triple-product and MDKS formulas.

Some of the strongest pair interactions in the 0gu reflections in 9α-fluorocortisol are shown in Fig. 5.2. The circled s values indicate points of conflict. The two largest contributors to the mildly discrepant pair (063; 026,3) disagreed with the overall sum, and the positive calculated cosines associated with these contributors, listed in Table 5.7, clearly gave the correct relationship between these phases. In addition to the three pair averages which are incorrect, two large contributors to the (043; 026,3) pair and a large contributor to each of the pairs (063; 024,3) and (063; 083) are wrong, and these contributors are also shown in Table 5.7. It can be seen that, in each case where a cosine seminvariant should have been negative, the calculated value, although positive, was small, so that there was in fact some doubt that the indicated pair relationship (5.17) or (5.18) was valid. Furthermore, the evidence for the relative 0gu phases being as shown in Fig. 5.2 was quite strong, and any changes in these phases would have resulted in a greater number of violations.

5.6. Strong Enantiomorph Selection

In fixing the enantiomorph one attempts to employ a structure invariant L whose value is approximately equal to $\pm\pi/2$ in order to ensure strong enantiomorph discrimination. Even so, for complex structures, a single structure invariant constitutes a very narrow foundation on which to base a technique for phase determination. For this reason one has often found, in practice, that enantiomorph specification has not been decisively made and has led to an initial E map that contains fragments of both enantiomorphs and presents great difficulties of interpretation. In the present section a technique for enantiomorph specification is described that employs, instead of just a single invariant, a class of several structure invariants, each member of which is approximately equal to $\pm\pi/2$. In this way the base of the phase determination process is considerably broadened with resultant unambiguous enantiomorph discrimination (Hauptman and Duax, 1972; Duax and Hauptman, 1972).

If the integer k is fixed, then, as h and l range over those integers for which the normalized structure factor magnitudes $|E_{hkl}|$ are large, it is known that the corresponding phases ϕ_{hkl} will, in general, provided that $|E_{02k,0}|$ is not too large, take on values distributed over the range 0–2π. If the origin is fixed so that $\phi_{02k,0} = 0$, which may always be done in $P2_1$, then some of the ϕ_{hkl}'s are expected to have values in the neighborhood of 0 or π and others in the neighborhood of $\pm\pi/2$. In view of the relationship

$$\phi_{\bar{h}k\bar{l}} = \phi_{h\bar{k}l} \qquad \text{if } k \text{ is even} \tag{5.24}$$

or

$$\phi_{\bar{h}k\bar{l}} = \pi + \phi_{h\bar{k}l} \qquad \text{if } k \text{ is odd} \tag{5.25}$$

it follows that, if k is even,

$$\cos(\phi_{h\bar{k}l} + \phi_{\bar{h}k\bar{l}} + \phi_{02k,0}) = \cos 2\phi_{h\bar{k}l} \tag{5.26}$$

$$\simeq \pm 1 \tag{5.27}$$

according as

$$\phi_{h\bar{k}l} \simeq 0 \text{ or } \pi \tag{5.28}$$

or

$$\phi_{h\bar{k}l} \simeq \pm\pi/2 \tag{5.29}$$

respectively. If, on the other hand, k is odd, then

$$\cos(\phi_{h\bar{k}l} + \phi_{\bar{h}k\bar{l}} + \phi_{02k,0}) = -\cos 2\phi_{h\bar{k}l} \tag{5.30}$$

$$\simeq \mp 1 \tag{5.31}$$

according as

$$\phi_{hkl} \simeq 0 \text{ or } \pi \tag{5.32}$$

or

$$\phi_{h\bar{k}l} \simeq \pm\pi/2 \tag{5.33}$$

respectively. In other words, the special cosine invariants (5.26) and (5.30), by doubling the phase $\phi_{h\bar{k}l}$, exaggerate, for fixed k, the deviations in the values of the ϕ_{hkl}. Thus differences of $\pi/2$ among the ϕ_{hkl} (permitting strong enantiomorph discrimination) imply, from (5.26)–(5.33), that the values of some cosine invariants are approximately -1, so that the identification of these cosines by means of their calculated values is feasible. This observation is the motivation for the method of strong enantiomorph discrimination that is described here, and already shows clearly the strong dependence of the method on the calculated cosine invariants, in particular those calculated to be negative.

In summary, the basic idea is to find an integer k and two "orthogonal" classes, I and II, of phases ϕ_{hkl} having the following properties:

1. $|E_{02k,0}|$ is moderately large, say $\simeq 2$;
2. every $|E_{hkl}|$ corresponding to any phase ϕ_{hkl} in class I or II is large, say > 1;
3. any two phases in class I differ from each other by 0 or π, approximately;
4. any two phases in class II differ from each other by 0 or π, approximately;
5. any phase in class I differs from any phase in class II by $\pi/2$ approximately.

For example, in the determination of the structure of valinomycin, two orthogonal classes, I and II, of phases ϕ_{h4l} were tentatively identified. First, $|E_{080}|$ was observed to be moderately large (1.921). Second, using the modified triple-product procedure (Hauptman *et al.*, 1969), the values of all cosine invariants

$$\cos(\phi_{h4l} + \phi_{\bar{h}\bar{4}\bar{l}} + \phi_{080}) \tag{5.34}$$

with

$$A = 0.1586 \, |E_{h4l}^2 E_{080}| > 1.5 \tag{5.35}$$

were calculated. In accordance with the method of strong enantiomorph discrimination (Hauptman and Duax, 1972), those cosines that were calculated to be greater than unity led to four phases ϕ_{h4l} which were placed unambiguously in class I. Those cosines calculated to be most negative led to two phases ϕ_{h4l} which were placed unambiguously in class II. The results are summarized in Table 5.8 which lists also, for comparison, the true values of the cosine invariants as obtained from the refined structure.

Next, the membership of classes I and II was confirmed and the classes themselves extended by calculating appropriate cosine invariants,

$$\cos(\phi_{h4l} + \phi_{h'4l'} + \phi_{-h-h',0,-l-l'}) \tag{5.36}$$

TABLE 5.8. Tentative Definition of the Orthogonal Classes I and II of Phases ϕ_{h4l} via Calculated Cosine Invariants, $\cos(\phi_{h\overline{3}l} + \phi_{\overline{h}\overline{3}l} + \phi_{080})$, $|E_{080}| = 1.921$

Class	h4l	$\lvert E \rvert$	A	$\cos(\phi_{h\overline{3}l} + \phi_{\overline{h}\overline{3}l} + \phi_{080})$ Calculated	True
I	11 4 $\overline{1}$	2.506	1.914	+1.44	+0.983
	17 4 $\overline{7}$	2.428	1.796	+1.06	+0.993
	10 4 $\overline{10}$	2.320	1.640	+1.61	+0.675
	16 4 $\overline{5}$	2.265	1.562	+2.14	+0.963
II	21 4 5	2.518	1.931	−0.85	−0.860
	5 4 $\overline{3}$	2.392	1.743	−0.23	−0.888

in accordance with the procedure (Hauptman and Duax, 1972). Thus,

$$\cos_{\text{calc}}(\phi_{11,\overline{41}} + \phi_{10,4\overline{10}} + \phi_{109}) = +2.02, \qquad A = 1.544 \qquad (5.37)$$

confirms the presence of $\phi_{11,4\overline{1}}$ and $\phi_{10,4\overline{10}}$ in the same class (I). Again, since $\phi_{16,4\overline{5}}$ and $\phi_{11,4\overline{1}}$ are in class I, the values of the calculated cosines,

$$\cos_{\text{calc}}(\phi_{14\overline{2}} + \phi_{16,\overline{45}} + \phi_{\overline{17},011}) = +1.36, \qquad A = 1.568 \qquad (5.38)$$

$$\cos_{\text{calc}}(\phi_{14\overline{2}} + \phi_{11,\overline{41}} + \phi_{10,01}) = +1.03, \qquad A = 1.618 \qquad (5.39)$$

require that $\phi_{14\overline{2}}$ be placed in class I, too. Next, since $\phi_{16,4\overline{5}}$ and $\phi_{14\overline{2}}$ are in class I,

$$\cos_{\text{calc}}(\phi_{19,47} + \phi_{\overline{15},\overline{49}} + \phi_{\overline{30}\overline{2}}) = +1.43, \qquad A = 2.672 \qquad (5.40)$$

and

$$\cos_{\text{calc}}(\phi_{19,47} + \phi_{14\overline{2}} + \phi_{\overline{35},09}) = +1.87, \qquad A = 1.694 \qquad (5.41)$$

imply that $\phi_{19,47}$ likewise belongs to class I. Continuing in this way, the membership of class I was extended and firmly established. The final membership of class II was firmly established in a similar way. The ten phases ϕ_{h4l} finally placed in class I and the six phases ϕ_{h4l} placed in class II are shown in Table 5.9.

Only one cosine invariant directly connecting the classes I and II had a sufficiently large A value to warrant calculating the cosine. This,

$$\cos_{\text{calc}}(\phi_{443} + \phi_{9,\overline{41}\overline{0}} + \phi_{\overline{15},07}) = +0.30, \qquad A = 1.572 \qquad (5.42)$$

TABLE 5.9. Final Definition of the Orthogonal Classes I and II of Phases
ϕ_{h4l} Using Table 5.8 and Calculated Cosine Invariants,
$$\cos(\phi_{h4l} + \phi_{h'4l'} + \phi_{-h-h',0,-l-l'})$$

Class	h4l	$|E|$	ϕ_{true} (deg)
I	11 4 $\bar{1}$	2.506	5
	4 4 3	2.497	−152
	17 4 $\bar{7}$	2.428	7
	10 4 $\bar{10}$	2.320	−15
	1 4 $\bar{2}$	2.289	−178
	16 4 $\bar{5}$	2.265	−168
	14 4 $\bar{5}$	2.096	11
	19 4 $\bar{7}$	1.993	−168
	5 4 13	1.945	20
	11 4 $\overline{11}$	1.818	27
II	24 4 4	3.627	89
	21 4 5	2.518	90
	9 4 $\bar{10}$	2.406	71
	5 4 $\bar{3}$	2.392	−95
	14 4 $\bar{8}$	2.143	−141
	4 4 $\overline{12}$	2.064	−62

was sufficiently close to zero to serve as additional confirmation of orthogonality.

Having firmly established the orthogonal classes I and II of phases ϕ_{h4l}, it could be inferred, in view of the theoretical basis, that

$$\phi_{h_1 4 l_1} + \phi_{h_2 4 l_2} + \phi_{-h_1-h_2,0,-l_1-l_2} \approx \pm 90° \qquad (5.43)$$

$$\phi_{h_1 4 l_1} + \phi_{\bar{h}_2 4 l_2} + \phi_{-h_1+h_2,0,-l_1+l_2} \approx \pm 90° \qquad (5.44)$$

where $\phi_{h_1 4 l_1}$ is an arbitrary phase in class I and $\phi_{h_2 4 l_2}$ is an arbitrary phase in class II. In short, the invariants (5.43) and (5.44) constitute a class of structure invariants each member of which is approximately equal to $\pm 90°$ and, therefore, suitable for decisive enantiomorph selection. The corresponding cosines constitute a class of 65 cosine invariants whose values are approximately 0 and for which the average value of A is 0.50.

The relationship between the ten phases ϕ_{h4l} of class I and the origin fixing reflections is such that the phases of class I must be approximately equal to 0 or 180° (Duax and Hauptman, 1972). Hence the value of the six phases ϕ_{h4l} of class II must be equal to $\pm 90°$. As it turned out, values could be unambiguously assigned to only six phases of class I and to four phases of class II. However, the true values of all ten phases in class I and all six phases in class II are shown in the last column of Table 5.9, and, with minor exceptions, they show good agreement with the expected values of 0 or 180° for class I and $\pm 90°$ for class II.

5.7. NQEST

In multiple solution techniques a number of ambiguities are introduced at the outset, they are assigned initial numerical values, and tangent extension and refinement are performed. If a large number of phase sets must be tested, the calculation and inspection of the resulting E maps may present a real problem in terms of computing time and possible inspection errors. Ideally one would like to know *a priori* whether a plausible phase set is likely to yield a solution without calculating and inspecting its resulting E map. Various figures of merit, such as the "absolute figure of merit" (Germain *et al.*, 1971), the Ψ_0 test (Cocharan and Douglas, 1955), and the residual R_K (Karle and Karle, 1966) have been proposed to rank phase sets in the order of their plausibility. However, it has been found by experience that none of these figures of merit discriminates consistently against hopelessly incorrect phase sets, nor can they be depended upon to indicate the presence of a correct phase set. It would be clearly desirable to construct a figure of merit that is reliable, absolute, and easily calculated. Recently secured estimates for the cosine invariants, $\cos(\phi_{h_1} + \phi_{h_2} + \phi_{h_3} + \phi_{h_4})$, make it possible to propose such a figure of merit (DeTitta *et al.*, 1975a).

A PQ (positive quartet) is defined as a quartet with "cross terms," $|E_{h_1+h_2}|$, $|E_{h_1+h_3}|$, $|E_{h_1+h_4}|$, all large and a NQ (negative quartet) to be one with cross terms all small, (5.10), (5.11). Note that a PQ is not necessarily positive nor is a NQ necessarily negative.

Our subsequent attention will be directed to the NQs because they form the basis of the required figure of merit, but it should be apparent that the PQS are equally important in their own right. The PQS are not, however, suitable as the basis for the required figure of merit because they are insensitive to the integrity of a phase set derived from tangent formula procedures. The reason for this is that tangent procedures generally ensure that cosine invariants $\cos(\phi_{h_1} + \phi_{h_2} + \phi_{h_3} + \phi_{h_4}) \simeq +1$, for large $A = (2/N^{1/2})|E_{h_1}E_{h_2}E_{h_1-h_2}|$; that is, the strongest phase relationships are generally not violated by the tangent formula procedure. It can be easily shown that, for a given quartet invariant, if any one of the conditions

$$\cos(\phi_{h_1} + \phi_{h_3} + \phi_{-h_1-h_3}) = +1 \text{ and } \cos(\phi_{h_3} + \phi_{h_4} + \phi_{h_1+h_3}) = 1$$

or

$$\cos(\phi_{h_1} + \phi_{h_3} + \phi_{-h_1-h_3}) = +1 \text{ and } \cos(\phi_{h_2} + \phi_{h_4} + \phi_{h_1+h_3}) = 1 \quad (5.45)$$

or

$$\cos(\phi_{h_1} + \phi_{h_4} + \phi_{-h_1-h_4}) = +1 \text{ and } \cos(\phi_{h_2} + \phi_{h_3} + \phi_{h_1+h_4}) = 1$$

is fulfilled, then necessarily $\cos(\phi_{h_1} + \phi_{h_2} + \phi_{h_3} + \phi_{h_4}) = +1$.

The cosine invariants of (5.45) are the six cosine triples associated with each quartet, the associated triples. For PQ the A values of the associated triples are large, by (5.10), and for NQ the A values of the associated triples are small, (5.11). Therefore, we expect the phase sets generated by tangent formula procedures to identify the PQ well, regardless of the correctness of the set. However, the tangent refinement procedure is ordinarily designed to avoid the associated triples of the NQ and, therefore, to avoid biasing the proposed figure of merit.

For some specific values of B_{min} and E_{cross},[†] construct n quartets such that $B^{\ddagger} \geq B_{min}$ and $\{|E_{h_1+h_2}|, |E_{h_1+h_3}|, |E_{h_1+h_4}|\} \leq E_{cross}$. In addition, for a specific value of E_{main},[†] ensure that only quartets of interest will be constructed by imposing the condition that the "main terms" satisfy $\{|E_{h_1}|, |E_{h_2}|, |E_{h_3}|, |E_{h_4}|\} \geq E_{main}$. The latter condition limits the quartets generated to those with each main term greater than or equal to E_{main}, the smallest normalized structure factor phased in the tangent refinement procedure. Then an estimate, NQEST, of the integrity of the phase set with respect to the NQ is defined by

$$\text{NQEST} = \sum_{h_1,h_2,h_3}^{n} B \cos(\phi_{h_1} + \phi_{h_3} + \phi_{h_3} + \phi_{h_4}) \bigg/ \sum_{h_1,h_2,h_3}^{n} B \quad (5.46)$$

Clearly, NQEST ranges from -1 to $+1$ and, in view of (5.11), is expected to be negative for a correct set of phases provided that E_{cross} is sufficiently small and B_{min} is sufficiently large.

[†] E_{cross} is defined as the least upper bound of the cross terms and E_{main} as the greatest lower bound of the main terms.

[‡] See (4.48) and Section 5.3.

TABLE 5.10. Figures of Merit for PGA$_1$M

Set	ABSFOM	$10^{-4}\Psi_0$	RESID	NQEST
1	1.035	1.87	41.04	+0.674
2	1.049	1.03	38.98	+0.648
3	1.055	1.04	30.82	+0.602
4	0.967	0.98	40.64	−0.499
5	1.030	1.02	39.74	+0.591
6	1.047	1.03	40.42	+0.669
7	1.039	1.03	40.65	+0.670
8	0.959	0.98	40.55	−0.507
9	1.085	1.05	37.34	+0.674
10	1.087	1.05	37.44	+0.675
11	1.082	1.05	37.14	+0.671
12	0.969	0.98	40.15	−0.527
13	1.066	1.05	36.93	+0.657
14	1.090	1.05	37.45	+0.675
15	1.088	1.05	37.42	+0.675
16	0.978	0.98	39.68	−0.533

In the determination of the crystal structure of prostaglandin A$_1$ (PGA$_1$M), MULTAN was permitted to generate 16 phase sets and phase extension was carried out to $|E| = 1.43$. NQ were generated with $B_{min} = 1.0$, $E_{main} = 1.75$, $E_{cross} = 0.7$, and NQEST calculated over the 123 NQ for each of the 16 sets, Table 5.10. All 16 E maps were calculated and inspected in order of increasing NQEST, that is, from most negative to most positive. Phase sets 16 and 12, having the most favorable values of NQEST, immediately revealed the structure, with the strongest 19 peaks in each map corresponding to correct atomic positions. The two solutions are related by a trivial translation along the polar axis. Phase sets 8 and 4, having the next most favorable values of NQEST and also related to each other by a translation along b, revealed essentially a complete molecular structure misplaced along the direction of the C(1)–C(8) hydrocarbon chain of the prostaglandin molecule (see DeTitta et al., 1975b). The remaining 12 phase sets revealed only the direction of the hydrocarbon chain which ran continuously across the cell in each map.

5.8. Automated Procedures

As the accuracy of the estimates of cosine values improves, the automatic utilization of this information in phase extension and refinement becomes feasible. A least-squares procedure for phase extension based upon cosines calculated by the triple-product formula was first used successfully in the determination of the structures of estriol (Hauptman *et al.*, 1969).

More recently in QTAN, a global, fixed point multisolution phase-refinement procedure, phase extension is based upon only those triples whose cosines are reliably indicated to be $+1$ by a statistical analysis based on the joint conditional probability distribution of the quadruple phase relationship (Langs and DeTitta, 1975), and the concurrent use of NQEST in an active mode. Because triplet analysis results in the calculation of a modified A value, the use in the tangent formula of triplets ordered on modified A proves to be a very effective method for introducing calculated cosine information into phase-extension procedures. Thus QTAN not only avoids the weak links in a tangent formula phase extension process, but, by cumulatively rejecting incorrect phase sets via NQEST, permits one to survey efficiently enormous numbers of potential solution sets, in this way increasing the likelihood that the correct solution is covered.

5.9. Exercises

1. Assume the space group to be $P2_12_12_1$. If $\phi_{440} = 0$, what are the values of the cosine invariants, $\cos(\phi_h + \phi_k + \phi_{-h-k})$, for the triples given in Table 5.11? The $P2_12_12_1$ phase shifts are given in Table 5.12.

TABLE 5.11. Universal Cosine Invariants that Correspond to Contributors to the Σ_1 Summation for Reflection 440

h	k	−h − k
440	$\bar{2}\bar{2}0$	$\bar{2}\bar{2}0$
440	$\bar{2}\bar{2}\bar{1}$	$\bar{2}\bar{2}1$
440	$\bar{2}\bar{2}\bar{2}$	$\bar{2}\bar{2}2$
440	$\bar{2}\bar{2}\bar{3}$	$\bar{2}\bar{2}3$

TABLE 5.12. Phase Relationships Appropriate to Space Group $P2_12_12_1$

hkl parity	hkl	$\bar{h}kl$	$h\bar{k}l$	$hk\bar{l}$
ggg uuu	α	$-\alpha$	$-\alpha$	$-\alpha$
ggu uug	α	$-\alpha$	$\pi - \alpha$	$\pi - \alpha$
guu ugg	α	$\pi - \alpha$	$-\alpha$	$\pi - \alpha$
gug ugu	α	$\pi - \alpha$	$\pi - \alpha$	$-\alpha$

2. Given $\phi_{20,10} = 0$ and the two Σ_2 triples in Table 5.13, what is ϕ_{810} if the space group is $P2_12_12_1$?

3. Assume the space group to be $P2_12_12_1$. (a) Is it possible for the cosine invariants $(\phi_h + \phi_k + \phi_{-h-k})$, of all the triples given in Table 5.14 to have positive values? Explain. (b) If your answer to (a) is no, then indicate an internally consistent set of signs for these cosine invariants.

4. Assume a structure in space group $P2_1$ with a disproportionate number of $h1l$ reflections that are very intense. If the 020 reflection is also very intense what might you expect the relative phases of the strongest $h1l$ reflections to be? Constructing Σ_2 triples of the form $h1l$, $\bar{h}1\bar{l}$, $0\bar{2}0$ and calculating cosines yields the results shown in Table 5.15. Assume that the triples having calculated cosines of 1.0 or greater are in fact +1 in value, and assign a phase of 0 to 111 reflection. What is the phase of 020? What do you know about the phases of the other $h1l$ reflections? How would you determine their individual phases? If the cosine that is calculated to be -0.8 is in fact -1.0 in value, what are the two possible values for the phase of the 211 reflection? If the $h1l$ reflections

TABLE 5.13. Σ_2 Triple for Exercise 2

h_1	h_2	h_3	A	$\cos(\phi_1 + \phi_2 + \phi_3)$
20 1 0	$\bar{14}$ 5 0	$\bar{6}\bar{6}$0	7.02	1.0
8 1 0	$\bar{14}$ 5 0	6$\bar{6}$0	2.61	1.0

TABLE 5.14. Cosine Invariants for Exercise 3

Triple	h	k	$-h-k$	A
1	023	$\bar{1}\bar{2}\bar{8}$	105	3.42
2	0213	$\bar{1}\bar{2}\bar{8}$	$10\bar{5}$	2.26
3	023	$03\bar{8}$	$0\bar{5}5$	2.49
4	0213	$03\bar{8}$	$0\bar{5}\bar{5}$	1.65

were strong but the 020 reflection was very weak, speculate about the relative phases of the *hl* reflections. How would you proceed to use Σ_2 relationships to test the merit of your speculations?

5. Consider the set of four triples and associated structure seminvariants $(\phi_h + \phi_k + \phi_{-h-k})$ given in Table 5.16.

   (a) What is the sum of the phases of the 12 reflections occurring in these triples?

   (b) Assume the space group to be $P2_12_12_1$. What combinations of values for the four structure seminvariants are possible?

   (c) What is the general rule concerning the number of triples in a centrosymmetric quadruple that may have cosine seminvariants $\cos(\phi_h + \phi_k + \phi_{-h-k})$ with values of -1?

TABLE 5.15. Cosine Invariants and Calculated Cosines for Exercise 4

hkl	$\bar{h}k\bar{l}$	02*k*0	A	Calculated cosine
111	$\bar{1}1\bar{1}$	$0\bar{2}0$	5.0	1.2
214	$\bar{2}1\bar{4}$	$0\bar{2}0$	4.8	1.3
417	$\bar{4}1\bar{7}$	$0\bar{2}0$	4.7	1.1
313	$\bar{3}1\bar{3}$	$0\bar{2}0$	4.6	1.2
412	$\bar{4}1\bar{2}$	$0\bar{2}0$	4.5	1.0
211	$\bar{2}1\bar{1}$	$0\bar{2}0$	4.3	-0.8
316	$\bar{3}1\bar{6}$	$0\bar{2}0$	4.3	1.5
311	$\bar{3}1\bar{1}$	$0\bar{2}0$	4.2	1.4

TABLE 5.16. Cosine Invariant for Exercise 5

Triple	h	k	$-h - k$
1	21 2 0	$\bar{1}\bar{1}$ 5 0	$\bar{10}$ 7 0
2	$\bar{21}\,\bar{2}$ 0	6 $\bar{6}$ 0	15 8 0
3	11 $\bar{3}$ 0	$\bar{6}$ 6 0	$\bar{3}\,\bar{1}$ 0
4	10 7 0	$\bar{13}$ 8 0	5 1 0

Acknowledgments

The authors wish to express their appreciation to Ms. Deanna Hefner, Ms. Brenda Giacchi, and Ms. Estelle Robel for their dedicated assistance in preparing this manuscript. Research was supported in part by grant No. CA-10906 from the National Cancer Institute, grant No. HL-15378 from the National Heart, Lung, and Blood Institute, grant No. GM-19684 from the National Institute of General Medical Sciences, DHEW, and grant No. CHE76-17582 from the National Science Foundation.

References

Cochran, W., and Douglas, A. S. (1955). *Proc. R. Soc. London Ser. A* **227**, 486.

DeTitta, G. T., Edmonds, J. W., and Duax, W. L. (1975a). *Prostaglandins* **9**, 659.

DeTitta, G. T., Edmonds, J. W., Langs, D. A., and Hauptman, H. (1975b). *Acta Crystallogr. Sect. A* **31**, 472.

Duax, W. L., and Hauptman, H. (1972). *Acta Crystallogr. Sect. B* **28**, 2912.

Duax, W., Weeks, C., and Hauptman, H. (1972). *Acta Crystallogr. Sect. B* **28**, 1857.

Fisher, J., Hancock, H., and Hauptman, H. (1970). Naval Research Laboratory report No. 7132.

Germain, G., Main, P., and Woolfson, M. M. (1971). *Acta Crystallogr. Sect. A* **27**, 368.

Grant, D., Howells, R., and Rogers, D. (1957). *Acta Crystallogr.* **10**, 489.

Hauptman, H. (1972a). *Crystal Structure Determination: The Role of the Cosine Seminvariants*, Plenum Press, New York.

Hauptman, H. (1972b). *Acta Crystallogr. Sect. B* **28**, 2337.

Hauptman, H. (1974a). *Acta Crystallogr. Sect. A* **30**, 472.

Hauptman, H. (1974b). *Acta Crystallogr. Sect. A* **30**, 822.

Hauptman, H. (1975). *Acta Crystallogr. Sect. A* **31**, 680.

Hauptman, H., and Duax, W. (1972). *Acta Crystallogr. Sect. A* **28**, 393.

Hauptman, H., Fisher, J., Hancock, H., and Norton, D. (1969). *Acta Crystallogr. Sect. B* **25**, 811.

Hauptman, H., and Karle, J. (1953). *Solution of the Phase Problem. I. The Centrosymmetric Crystal*, American Crystallographic Association Monograph No. 3, Polycrystal Book Service, Pittsburgh, Pennsylvania.

Karle, J., and Karle, I. (1966). *Acta Crystallogr.* **21**, 849.

Langs, D. A., and DeTitta, G. T. (1975). *Acta Crystallogr. Sect. A* **31**, 516.

Weeks, C., and Hauptman, H. (1970). *Z. Kristallogr.* **131**, 437.

Phase Correlation with Calculated Cosine Invariants for Routine Structure Analysis

PAUL T. BEURSKENS
and TH. E. M. VAN DEN HARK

6.1. Phase Correlation Procedure

The sign correlation procedure, one of the oldest computer-oriented applications of the Σ_2 relationship, was developed in the early sixties (Beurskens, 1964). At that time, the procedure was applicable only to centrosymmetric structures. The name "sign correlation procedure" will be used now only when referring explicitly to centrosymmetric structures; otherwise, the more general term "phase correlation procedure" will be used. The sign correlation procedure was programmed for the IBM 1620 computer (Beurskens, 1963) and applied successfully to several centrosymmetric structures (Sax *et al.*, 1965). Thereafter, an automatic program was written in ALGOL-60 (Beurskens, 1965).

While in the symbolic addition and multiple-solution procedures single phase indications often have to be trusted in the initial stage of the phase determination process, in the phase correlation procedure all emphasis is laid on avoiding single phase indications. As a consequence, it is necessary to bring into the calculations as many reflections as possible right from the beginning of the procedure. Usually a starting set of 20 or more reflec-

PAUL T. BEURSKENS and TH. E. M. VAN DEN HARK • Crystallography Laboratory, Toernooiveld, Nijmegen, The Netherlands.

tions is chosen, the phases of which are represented by letter symbols. During the phase determination procedure, relations between the letter symbols are found and used to eliminate symbols. Thus, while the number of reflections that take part in the analysis is increased, the number of unknown symbols is gradually reduced.

In the past few years, new insights have been obtained into several essential points of the sign correlation procedure. A new automatic computer program for the execution of the sign correlation procedure has been written in FORTRAN IV, and the applicability of the phase correlation procedure to phase determination for noncentrosymmetric structures has been investigated. The procedure has proved to be a powerful one, particularly in combination with cosine-invariant calculations, as will be discussed in the following sections.

6.1.1. Definitions

H is the set of reflections h for which the phases are known, either absolutely or in terms of one or more letter symbols. The set includes all symmetry-related reflections.

Hi ($i = 1, 2, 3, \ldots$) signifies reflections that have been assigned a letter symbol Pi to represent the phase. These initial choices form a subset of the set H.

The symbol Si denotes the sign of a reflection Hi in the sign correlation procedure.

$H + H'$ denotes any possible vector sum $h + h'$, using also symmetry-related reflections.

K is a selected set of reflections $k = h + h'$ that are to be used as temporarily accepted reflections, but certainly not all possible reflections $h + h'$.

L is the set of all reflections $l = h + k$ or $k + k'$, and all reflections $h + h'$; thus K is a subset of L.

6.1.2. Selection of the Starting Set

Some 10–30 reflections Hi are chosen, with $|E|$ values as large as possible, such that there are no Σ_2 interactions (1.8) between these reflections. Origin- and, where appropriate, enantiomorph-determining reflections may also be added to the starting set.

Frequently, however, the origin (and enantiomorph) will be fixed at a later stage of the phase determination; some letter symbols are eliminated by giving them an arbitrarily chosen phase. Other known phases, such as Σ_1 results, may also be added.

6.1.3. Phase Generation

All possible pairs of the reflections $\mathbf{h}$ are considered, so as to obtain phases, absolute or symbolic, for the reflections $\mathbf{h} + \mathbf{h}'$. The set K is selected from reflections with large $|E|$ values for which the phases have been found with high reliability [large $\alpha_\mathbf{h}$, see (3.6)]. Where different symbols are found for the same reflection, the most reliable phase indication is chosen; the phases for these reflections are "temporarily accepted." The phase generation is continued so as to obtain phases for the reflections $\mathbf{h} + \mathbf{k}$ and $\mathbf{k} + \mathbf{k}'$. The resulting reflection set L is now analyzed in order to find new acceptable phases and relations, or correlation equations, among the letter symbols. Some reflections in L may have been found, from many independent phase indications, to have the same phase. This means that even though some of the temporarily accepted phases are incorrect, the phases of the reflections of set L are unlikely to be incorrect. These reflections are now "definitely accepted" and added to set H. The majority of the newly accepted reflections were also present in set K. They are now found with a much higher reliability; some reflections of set K are not accepted. Some altogether new reflections, those reflections that were not accepted temporarily, may also be accepted. A new cycle of phase correlation (see Fig. 6.1) with the expanded set H may now be started by recalculation of sets K and L.

6.1.4. Correlation Equations

In each cycle of the phase correlation procedure, the sets of reflections H and L are screened to find relations between the symbols. The phases of many reflections will be expressed by more than one symbol giving relations such as those in Table 6.2. A relation that has been found with a high reli-

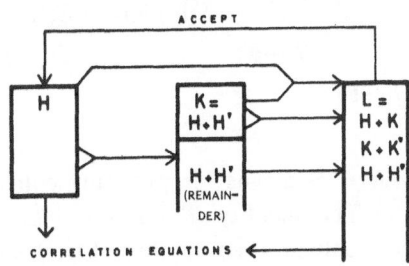

FIGURE 6.1. Diagram showing the cyclic phase correlation procedure.

ability, and very frequently with no contradictions, is selected and may be accepted to be true. This result is called a correlation equation. The correlation equation is used to eliminate one of the unknown symbols *Pi*. If more correlation equations become acceptable, more symbols can be eliminated.

6.1.5. End of the Phase Correlation Procedure

The iterative procedure of expanding set *H* and eliminating symbols *Pi* in an alternative sequence usually leads to a point where no more correlation equations or reliable phases from set *L* can be accepted. Sometimes symbols are used to express the phases of only a few reflections; they are rejected, that is, withdrawn from the procedure without expressing them in terms of other symbols, or eliminated by using less reliable correlation equations. When the origin is not completely fixed, one or more symbols may be eliminated to completely fix the origin. Where some symbols remain unknown, a multiple solution to the phase problem is obtained. Sometimes it makes sense to complete the phase generation for each of the solutions separately: all phases expressed in terms of letter symbols are replaced by numerical values before applying the tangent formula.

6.1.6. Quadruples and Quartets in Relation to the Phase Correlation Procedure

In the phase correlation procedure large numbers of quadruples and quartets are used implicitly.

A quadruple (correlation equation; de Vries, 1965) is a relationship between four triples and can be generated from six different reflections (de Vries, 1963, 1965; Viterbo and Woolfson, 1973). The simple form of a quadruple for a centrosymmetric structure is shown below:

$$
\begin{aligned}
S(\mathbf{h}) \quad \cdot S(\mathbf{k}) \quad \cdot S(-\mathbf{h}-\mathbf{k}) \qquad\qquad\qquad\qquad\qquad &= T_1 \approx + \\
S(\mathbf{h}+\mathbf{k}) \quad \cdot S(\mathbf{l}) \quad \cdot S(-\mathbf{h}-\mathbf{k}-\mathbf{l}) \qquad\qquad &= T_2 \approx + \\
S(-\mathbf{k}) \qquad\qquad \cdot S(-\mathbf{l}) \qquad\qquad\qquad \cdot S(\mathbf{k}+\mathbf{l}) \quad &= T_3 \approx + \\
S(-\mathbf{h}) \qquad\qquad\qquad\qquad\quad \cdot S(\mathbf{h}+\mathbf{k}+\mathbf{l}) \quad \cdot S(-\mathbf{k}-\mathbf{l}) &= T_4 \approx + \\
\rightarrow T_1 \cdot T_2 \cdot T_3 \cdot T_4 &= +
\end{aligned}
$$

where $\approx$ means "is probably equal to."

The identity $T_1 \cdot T_2 \cdot T_3 \cdot T_4 = +$ is always satisfied. This means that only an even number of triples in a quadruple can be incorrect.

Quadruples occur in the phase correlation procedure if, for example, two reflections of set H and n couples of two temporarily accepted reflections lead to the same phase, absolute or symbolic, for a reflection of set L. In this case, $3n + 1$ triples have been used of which n quadruples can be constructed, all with one and the same triple in common.

A quartet is a relationship between the phases of four reflections of which the sum of the indices equals zero. It has been called a Σ_5 relationship by Hauptman and Karle (1953), and a coincidence of the second kind by de Vries (1965). Where the $|E|$ values of the four participating reflections *and* of the three cross terms are large, the sum of the four phases probably equals a value near 0 rather than π; this is the strengthened quartet of Schenk (1973).

The general form of a strengthened quartet, for centrosymmetric structures, is

$$S(\mathbf{h}) \cdot S(\mathbf{k}) \cdot S(\mathbf{l}) \cdot S(-\mathbf{h} - \mathbf{k} - \mathbf{l}) \approx +$$

provided that $|E(\mathbf{h})|$, $|E(\mathbf{k})|$, $|E(\mathbf{l})|$, $|E(\mathbf{h} + \mathbf{k} + \mathbf{l})|$, $|E(\mathbf{h} + \mathbf{k})|$, $|E(\mathbf{k} + \mathbf{l})|$, and $|E(\mathbf{h} + \mathbf{l})|$ are large. Recently, it was observed that if the $|E|$ values of the three cross terms are very small, the sum of the four phases tends to be π instead of 0, and is termed a negative quartet (Hauptman, 1974a,b, 1975; Schenk, 1974).

The quartets occurring in the phase correlation procedure have two cross terms with large $|E|$ values; the third cross term is not checked. Examples of quadruples and quartets are given below.

6.1.7. Dinaphtho[1,2-*a*; 1′,2′-*h*]anthracene[†]

The following example is based on the structure of dinaphtho[*a,h*] anthracene.

Crystal data: monoclinic, $a = 8.167(3)$, $b = 15.000(5)$, $c = 15.641(5)$ Å; $\beta = 90.43(2)°$. Space group $P2_1/c$, $Z = 4$; $C_{30}H_{18}$.

[†] Hummelink-Peters *et al.*, 1975.

About 2000 symmetry-independent intensities were measured. The structure was solved by direct methods using the program MULTAN of Germain, Main, and Woolfson (1971). The most probable of the eight solutions was correct.

In this section, we report some numerical results, obtained by the sign correlation program SCOR, in order to illustrate the procedure and also to show the relationship of the procedure with quadruples and quartets. SCOR was applied to 388 reflections with $|E| > 1.3$. Thus the triples used in the procedure have a triple product of at least 2.2, or a probability that they satisfy the Σ_2 relationship of at least 59.7%.

6.1.7.1. Starting Set

A starting set of 26 reflections Hi was chosen from the strongest $|E|$ values, such that there exist no triples $Hi + Hj = Hk$; these initial reflections were given symbols Si. During the selection of this starting set, all reflections $(Hi + Hj)$ are calculated. The algorithm used is explained by its results, as shown in Table 6.1. It should be noted that the reflection $71\bar{1}$ is *not* given a new symbol, because the sign of this reflection depends on products $Hi \cdot Hj$.

6.1.7.2. Sign Generation and Search for Correlation Equations

From the signs of all reflections $(Hi + Hj)$, that is, from the signs calculated from the starting set, the most probable new signs are selected and denoted Ki, as shown in Table 6.1. Preset control parameters led to the selection of 134 reflections **k**, which were used to calculate symbolic signs for more reflections. Inspection of the results for the reflections of set L showed that some correlation equations occurred very frequently with few or no contradictions, and with high probabilities. Six correlation equations were accepted and used to eliminate six symbols. To continue the sign generation, a symbolic sign of a reflection of set L was definitely accepted when it was found by at least 10 consistent triples with sufficiently high probabilities. In this way, 18 reflections were accepted and added to the set of initial choices, H.

The results for *one* reflection of set L are given in Table 6.2. The reflection $71\bar{1}$ has been found eight times with the symbolic sign $S1 \cdot S2$: once directly from starting set reflections, and seven times from temporarily accepted reflections. In determining the same symbolic sign for the reflection

TABLE 6.1. Signs for the 20 Strongest Reflections [a]

Reflection	Symbolic sign
4 8 $\bar{6}$ H1	S1
3 9 5 H2	S2
3 12 1 H3	S3
7 1 $\bar{1}$ K1	$S1 \cdot S2$, $-S3 \cdot S4$, $S6 \cdot S14$, $S10 \cdot S13$
4 11 $\bar{2}$ H4	S4
2 12 5 H5	S5
4 8 8 H6	S6
3 8 8 H7	S7
4 8 $\bar{7}$ H8	S8
1 8 12 H9	S9
0 2 3 H10	S10
7 0 2 K2	$S1 \cdot S7$
2 7 10 H11	S11
0 2 4 H12	S12
7 1 2 H13	S13
3 7 9 K3	$S2 \cdot S12$, $S8 \cdot S13$
3 7 $\bar{9}$ H14	S14
3 11 2 K4	$-S2 \cdot S10$
4 13 0 H15	S15
5 7 $\bar{6}$ H16	S16

[a] Among these reflections there are 16 starting-set reflections, denoted by Hi ($i = 1, 2, \ldots, 16$), and four reflections from set K, denoted by Kj ($j = 1, 2, \ldots, 4$). The symbolic signs for the Kj reflections are given in order of decreasing probability.

$71\bar{1}$ eight times, 22 triples have been used. If the sign of the reflection $71\bar{1}$ is *in*correctly indicated by the symbolic sign $S1 \cdot S2$, then at least 8 of the 22 triples must be incorrect. This is unlikely to be true, and the symbolic sign $S1 \cdot S2$ for the reflection $71\bar{1}$ may be trusted. Nevertheless, our preset limit was 10 consistent Σ_2 indications. After accepting the six correlation equations, however, the sign $S1 \cdot S2$ for the reflection $71\bar{1}$ was found 20 times consistently and was definitely accepted.

6.1.7.3. Second Cycle

The basic set for the second cycle is formed by the 26 initial reflections and the 18 newly accepted reflections, such as $71\bar{1}$, formerly denoted $K1$ and now denoted $H27$. Further intermediate results are given in Table 6.3.

TABLE 6.2. Sign Correlation Results for Reflection $71\bar{1}$

(a) Reflections used to give the sign $S1 \cdot S2$ to $71\bar{1}$ eight times

Reflections Ki	Sign	Result	Sign
7 1 $\bar{1}$	$S1 \cdot S2$	7 1 $\bar{1}$	$S1 \cdot S2$
3 11 2 4 10 3	$-S2 \cdot S10$ $S1 \cdot S10$	7 1 $\bar{1}$	$S1 \cdot S2$
2 1 $\bar{7}$ 5 0 6	$S2 \cdot S9$ $S1 \cdot S9$	7 1 $\bar{1}$	$S1 \cdot S2$
7 2 3 0 3 4	$S2 \cdot S4$ $S1 \cdot S4$	7 1 $\bar{1}$	$S1 \cdot S2$
1 2 $\bar{5}$ 6 1 4	$S2 \cdot S11$ $-S1 \cdot S11$	7 1 $\bar{1}$	$S1 \cdot S2$
3 7 9 4 6 $\overline{10}$	$S2 \cdot S12$ $S1 \cdot S12$	7 1 $\bar{1}$	$S1 \cdot S2$
6 3 $\bar{5}$ 1 2 4	$S2 \cdot S19$ $S1 \cdot S19$	7 1 $\bar{1}$	$S1 \cdot S2$
5 3 10 2 4 $\overline{11}$	$S2 \cdot S5$ $-S1 \cdot S5$	7 1 $\bar{1}$	$S1 \cdot S2$

(b) Other symbolic signs for reflection $71\bar{1}$

$-S3 \cdot S4$	6 indications
$-S8 \cdot S17$	4 indications
$S6 \cdot S14$	3 indications
$S10 \cdot S13$	3 indications
$S7 \cdot S23$	2 indications

plus 7 symbolic signs indicated only once

(c) Symbolic relations (given in order of decreasing probability) that follow from the symbolic signs in (a) and (b) above

$$S1 \cdot S2 = -S3 \cdot S4$$
$$S1 \cdot S2 = -S8 \cdot S17$$
$$S1 \cdot S2 = S6 \cdot S14$$
$$S1 \cdot S2 = S10 \cdot S13$$
$$S3 \cdot S4 = S8 \cdot S17$$
etc.

TABLE 6.3. Intermediate Results of the Sign Correlation Procedure

Cycle number	Number of reflections H	Number of reflections K	Eliminated symbols	Number of definitely accepted reflections
1	26	134	$S4, S8, S12, S13, S14, S19$	+18
2	44	196	$S9, S11, S15, S16, S17, S18, S20, S21, S24, S26$	+58
3	99	146	$S6, S7$	+55
4	154	180	—	—

6.1.7.4. End of the Sign Correlation Procedure

The origin was fixed by giving to the symbols $S2$, $S3$, and $S10$ a plus sign. Three symbols, $S22$, $S23$, and $S25$, were rejected because only a small number of reflections had symbolic signs dependent on these symbols. The sign correlation procedure was stopped when set H contained 154 reflections: for the large majority of the remaining reflections, either probable symbols or absolute signs were calculated. Two symbols, $S1$ and $S5$, remained unknown. The correlation equations found at the end of the procedure were not reliable enough to be accepted, so a fourfold solution to the phase problem was obtained; one of these solutions was, of course, correct.

6.1.7.5. Quadruples

The generation of symbolic signs, right from the initiation of the procedure, is directly related to the existence of quadruples among the triples that are used. As an example, the reflection $71\bar{1}$, temporarily used as $K1$, was found seven times with the same symbolic sign, and, together with the 3×7 triples involved, it is possible to construct seven quadruples. Two quadruples are shown in Table 6.4. Many of the reflections used in the initial stage of the sign correlation procedure are strong, thus giving rise to strong and reliable quadruples.

TABLE 6.4. Some Quadruples [a]

$$Q1: \quad S(48\overline{6}) \cdot S(395) \cdot S(7\overline{1}1) \quad\quad \approx S1 \cdot S2 \cdot (S1 \cdot S2)$$

$$S(7\overline{1}1) \cdot S(3\,\overline{11}\,\overline{2}) \cdot S(\overline{4}\,10\,3) \approx (S1 \cdot S2) \cdot (-S2 \cdot S10) \cdot (-S1 \cdot S10)$$

$$S(\overline{3}\overline{9}\overline{5}) \cdot S(0\overline{2}3) \cdot S(3\,11\,2) \quad \approx S2 \cdot (-S10) \cdot (-S2 \cdot S10)$$

$$S(\overline{4}86) \cdot S(02\overline{3}) \cdot S(4\,\overline{10}\,\overline{3}) \approx S1 \cdot (-S10) \cdot (-S1 \cdot S10)$$

$$Q2: \quad S(48\overline{6}) \cdot S(395) \cdot S(7\overline{1}1) \quad\quad \approx S1 \cdot S2 \cdot (S1 \cdot S2)$$

$$S(7\overline{1}1) \cdot S(\overline{2}\overline{1}7) \cdot S(\overline{5}0\overline{6}) \quad \approx (S1 \cdot S2) \cdot (S2 \cdot S9) \cdot (S1 \cdot S9)$$

$$S(\overline{3}\overline{9}\overline{5}) \cdot S(1\,8\,12) \cdot S(21\overline{7}) \quad \approx S2 \cdot S9 \cdot (S2 \cdot S9)$$

$$S(\overline{4}86) \cdot S(\overline{1}\,8\,\overline{12}) \cdot S(506) \quad \approx S1 \cdot S9 \cdot (S1 \cdot S9)$$

[a] Only the starting set reflections from Table 6.1 and temporarily accepted reflections from Table 6.2 are used.

TABLE 6.5. Strengthened Quartets Occurring in Quadruple $Q1$ of Table 6.4

$S(h)$ $\cdot$ $S(k)$ $\cdot$ $S(l)$ $\cdot$ $S(-h-k-l) \approx +$		
$S(4\,\bar{8}\,\bar{6}) \cdot S(3\,9\,5) \cdot S(\bar{3}\,\bar{11}\,\bar{2}) \cdot S(\bar{4}\,10\,3)$ $\approx +$	$h+k \equiv 7\,1\,\bar{1}$	$k+l \equiv 0\,\bar{2}\,3$
$S(4\,\bar{8}\,\bar{6}) \cdot S(\bar{7}\,\bar{1}\,1) \cdot S(0\,\bar{2}\,3) \cdot S(3\,11\,2)$ $\approx +$	$h+k \equiv \bar{3}\,\bar{9}\,5$	$h+l \equiv 4\,\bar{10}\,\bar{3}$
$S(3\,9\,5) \cdot S(\bar{7}\,\bar{1}\,1) \cdot S(0\,2\,\bar{3}) \cdot S(4\,\bar{10}\,\bar{3})$ $\approx +$	$h+k \equiv \bar{4}\,8\,6$	$h+l \equiv 3\,11\,2$

6.1.7.6. Quartets

As a consequence of the application of the correlation procedure, strengthened quartets play an important role in the initial stages of the sign-generation procedure. Some strengthened quartets are given as an example in Table 6.5.

6.2. Simple Cosine-Invariant Calculations

The Σ_2 procedures, based on the application of the sum of angles and analogous formulas (see Chapter 1) often lead to the correct solution of the phase problem. Sometimes they fail, perhaps because of incorrect triples that have been used during the initial steps of the phase determination. By variations in the execution parameters, such as selection criteria, origin, and ambiguity choices, these incorrect triples may accidently be avoided and the correct solution of the problem obtained.

On the other hand, the use of these incorrect triples may be avoided by cosine-invariant calculations. In Chapter 5, several application of cosine-invariant calculations have been considered. The techniques which make the most effective and safest use of the calculated cosine invariants were designed to tackle difficult problems. Unfortunately, the accuracy with which the cosine invariants can be calculated is still poor.

In this section, we shall describe relatively simple and inexpensive applications of cosine-invariant calculations which are useful for routine structure analyses. For simplicity, we have used so far the $B3,0$ formula (Karle and Hauptman, 1958). We define

$$\Phi_{hk} = \phi(\mathbf{h}) + \phi(\mathbf{k}) + \phi(-\mathbf{h} - \mathbf{k})$$

$$A_{hk} = 2N^{-1/2} \,|\, E(\mathbf{h}) \,|\,|\, E(\mathbf{k}) \,|\,|\, E(-\mathbf{h} - \mathbf{k}) \,|$$

The $B3,0$ formula, in its simplest form, then becomes

$$\tfrac{1}{2}N^{1/2}A_{hk}\cos\Phi_{hk} = AB_{hk} + C \tag{6.1}$$

where A is a known positive constant, C is a small positive correction term, and B_{hk} is given by

$$B_{hk} = \langle[(|\,E(\mathbf{l})\,|^2 - 1)(|\,E(\mathbf{h}+\mathbf{l})\,|^2 - 1)(|\,E(\mathbf{h}+\mathbf{k}+\mathbf{l})\,|^2 - 1)]\rangle_{\mathbf{l}}$$

The left-hand side of (6.1) is called a *triple invariant*. In the first structure determination based upon this formula (Kanters *et al.*, 1966) the numerical results were used to find those triples that contradict the Σ_2 relationship, that is, incorrect triples with $\cos\Phi_{hk} \approx -1$.

The calculation of even a small number of invariants gives useful results. Accurate cosine values are not required: the results of the invariant calculations are merely used to reject the triples that have a large chance of being incorrect, and to select the triples that should be used in the early stages of the phasing process. Application of a Σ_2 procedure to the triples, from which the majority of incorrect triples have been withdrawn, increases the chances of finding the correct solution to the phase problem. When a relatively large number, say 100, of the strongest reflections have been phased in this way, the phase determination may be continued by routine Σ_2 procedures.

The numerical results of the calculations with the $B3,0$ formula on different structures show a wide range of $(AB_{hk} + C)$ values; the results for triples with large A_{hk} values can be interpreted in the following way.

(a) $AB_{hk} + C \gg 0$: Very large values are reliable indications for $\cos\Phi_{hk} = +1$, thus affirming the validity of the Σ_2 relationship. Medium to large $(AB_{hk} + C)$ values are still good affirmations of the validity of the Σ_2 relationship.

(b) $AB_{hk} + C \approx 0$: Weak results are unreliable. For noncentrosymmetric structures these weak results are certainly not good indications, because $\cos\Phi_{hk} \approx 0$. The theoretical distribution curve of $\cos\Phi_{hk}$ does have a minimum where the experimental $(AB_{hk} + C)$ distribution shows a maximum. For centrosymmetric structures, $\cos\Phi_{hk}$ can have only the values ± 1. Many of the weak results correspond to $\cos\Phi_{hk} = +1$.

(c) $AB_{hk} + C \ll 0$: Very large negative results are strong indications that the Σ_2 relationship is not satisfied. The possible implication $\cos\Phi_{hk} = -1$ is of limited importance; that is, the reliability of the indication $\cos\Phi_{hk} = -1$ is not large enough to be of practical use.

It is clear that the triples with the highest $(AB_{hk} + C)$ values will be used in the initial steps of the phase determination with a Σ_2 procedure. As the phase determination progresses, more and more triples with lower $(AB_{hk} + C)$ values may enter into the procedure. Despite the fact that triples with very high $(AB_{hk} + C)$ values are very reliable, incorrect triples may occur occasionally. Therefore, we prefer to use the phase correlation procedure to determine the phases of the individual reflections.

6.2.1. Aminomalonic Acid[†]

Crystal data: orthorhombic, $a = 4.990$, $b = 8.507$, $c = 10.854$ Å; space group $P2_1cn$, $Z = 4$; $CHNH_2(CO_2H)_2$. Intensity data: 560 independent reflections, including 105 $0kl$ reflections.

At that time we had no routine programs for noncentrosymmetric direct methods, and we first solved the centrosymmetric $0kl$ projection; thereafter, we had no problems in finding the three-dimensional structure. For this projection the $B3,0$ formula (6.1) reduces to either a relation between the signs of three centric projection reflections:

$$E_h E_k E_{h+k} = AB_{hk} + C \qquad \text{for } \mathbf{h} = 0kl, \quad \mathbf{k} = 0k'l' \qquad (6.2)$$

or an expression for the sign of one centric projection reflection, a Σ_1-type interaction:

$$(-1)^h \mid E_h \mid^2 E_{h+k} = AB_{hk} + C \qquad \text{for } \mathbf{h} = hkl, \quad \mathbf{k} = \bar{h}kl \qquad (6.3)$$

We calculated 148 relations of types (6.2) and (6.3). The numerical results were used as follows. The 49 results with $(AB + C) > 9.0$ determined the signs of most of the 45 $0kl$ reflections with $\mid E \mid \geq 0.9$; none of these results was incorrect. About 18 results in the range $9.0 > \mid AB + C \mid > 7.0$ were used to find the remaining signs, only two of which were incorrect.

[†] Kanters *et al.* (1966).

6.2.2. Scaling of Calculated Triple Invariants

We will now discuss a weighting scheme from which the reliability of calculated triple invariants can be estimated, relative to the reliability of the Σ_2 relationship (van den Hark, 1976).

As noted before, we calculate triple invariants for a limited number of triples in order to save computer time. Thus, it is necessary that the scheme, from which the reliability of a triple is estimated, permits the simultaneous use both of triples for which triple invariants have been calculated and of triples for which this has not been done.

In order to define the weighting scheme, we will introduce KAB_{hk} values which are related to $(AB_{hk} + C)$ values. We will first define KAB_{hk} values and summarize some characteristics.

6.2.3. Definition of KAB_{hk} Values

The $(AB_{hk} + C)$ values, as they are calculated with the $B3,0$ formula, show a wide range, usually quite different for different structures. Therefore, comparison of the results of the $B3,0$ formula in terms of $(AB_{hk} + C)$ values is difficult. We define

$$KAB_{hk} = K(AB_{hk} + C + X) \qquad (6.4)$$

where K is a scaling factor and X a constant, both of which are adjusted to obtain the following characteristics.

(a) The order of the triples is the same for sorting on $(AB_{hk} + C)$ values as for sorting on KAB_{hk} values.

(b) Negative KAB_{hk} values indicate unreliable Σ_2 interactions.

(c) Positive KAB_{hk} values have magnitudes that are comparable with A_{hk} values. This criterion permits the estimation of the reliability: if $KAB_{hk} > A_{hk}$, the reliability of the triple is larger than that of the corresponding Σ_2 interaction.

(d) KAB_{hk} values for triples from different structures are comparable, because of (c).

The constant X is added to the calculated $(AB_{hk} + C)$ so that triples with a positive KAB_{hk} value have at least a 50% probability of being correct, that is, $|\Phi_{hk}| < \pi/2$. The constant X can be chosen such that, for fixed A_{hk}, the fraction of triples with negative KAB_{hk} values agrees with the fraction of negative triple invariants, denoted by f^-, that is expected theoretically. For triples from centric structures or centric projections of noncentric structures, f^- is given by

$$f^- = 0.5 - 0.5 \tanh 0.5 A_{hk}$$

For triples from general reflections in noncentrosymmetric structures, values of f^- for different A_{hk} values have been tabulated by Hauptman (1972).

Hence, for a given structure, the experimental values for X are approximately independent of A_{hk}. Therefore, we use a constant value for X, for all calculated triple invariants.

The scaling factor K is chosen such that the positive KAB_{hk} values are comparable with the A_{hk} values. For triples from centrosymmetric structures or from centric projections of noncentrosymmetric structures, with fixed A_{hk} values, the theoretical expectation value of $A_{hk} \cos \Phi_{hk}$ is given by $A_{hk}(1.0 - 2f^-)$.

For triples from general reflections of noncentrosymmetric structures an expression, as a function of A_{hk}, is given by Hauptman (1972).

We calculate the scale K for about 100 triples with the largest A_{hk} values. This constant value is used for the calculation of KAB_{hk} values. Thus, the reliability of a calculated triple, with a given result for KAB_{hk}, is still dependent on the A_{hk} value of this triple.

The weights w_{hk} are assigned to all triples that are to be used for phase determination, that is, both to triples for which triple invariants have been calculated and to triples for which such calculations have not been done. We define a weighting scheme such that all triples can be used simultaneously for phase determination. The reliability of a triple can be estimated with existing formulas, such as (1.18); we define the weights such that in these equations A_{hk} can be replaced by w_{hk}.

Triples for which no triple invariants have been calculated are therefore weighted by

$$w_{hk} = A_{hk} \tag{6.5}$$

Triples for which triple invariant calculations resulted in negative KAB_{hk} values will not be used for phase determination and, therefore,

$$w_{hk} = 0 \tag{6.6}$$

Triples for which triple invariant calculations resulted in positive KAB_{hk} values are weighted by

$$w_{hk} = 2(A_{hk} \cdot KAB_{hk})^{1/2} \tag{6.7}$$

The reasons for this definition are as follows.

(a) As a consequence of the scaling procedure, the triples with the largest A_{hk} values will have KAB_{hk} values which, on average, are comparable with their A_{hk} values for predicting the reliability of the triples.

(b) For triples with $A_{hk} = KAB_{hk}$, (6.7) leads to double the weight given by (6.5).

(c) For the strongest triples, the highest results for KAB_{hk} will have maximum weight; extremely large KAB_{hk} values will be reduced by the square root in (6.7).

(d) Triples with smaller A_{hk} values will, on average, have less reliable KAB_{hk} values; the KAB_{hk} values for smaller A_{hk} values overestimate the weight; therefore, the product of A_{hk} and KAB_{hk} appears in (6.7).

It may be noted that numerical results (van den Hark, 1976) show that the reliability of triples with large KAB_{hk} values is much better than those with large A_{hk} values. Therefore, (6.7) is a rather conservative application of the $B3,0$ formula. There are, however, more triples with large KAB_{hk} values than with large A_{hk} values, so the reliability of the initial steps of the phasing procedure is greatly improved.

6.3. Phase Correlation with Calculated Triple Invariants

It is, in general, much more difficult to solve a noncentrosymmetric than a centrosymmetric structure. For centrosymmetric structures, cos Φ_{hk} can have only the values ± 1; for noncentrosymmetric structures, cos Φ_{hk} may, in general, have any value between $+1$ and -1. Where a noncentrosymmetric structure has one or more centric projections, it will be of great help to have the phases of a number of centric projection reflections before extending the phase determination to general reflections. A centric projection can be solved relatively easily: a projection may be considered to be solved when the signs of the centric reflections are expressed in terms of a few unknown letter symbols. In solving a projection, not only triples among centric reflections are used, but also triples in which a centric reflection and two symmetry-related three-dimensional reflections occur; we will call these triples Σ_1-type interactions.

These considerations lead to the classification of space groups in four categories: (a) Centrosymmetric space groups; (b) noncentrosymmetric space groups with three centric projections, for example, $P2_12_12_1$; (c) noncentrosymmetric space groups with one centric projection, for example

$P2_1$; (d) noncentrosymmetric space groups without a centric projection; space groups such as $P1$ and Pm, which have only centric axial reflections also belong to this category. Procedures for (a), (b), and (c) will be described; our procedures have never been tried for space groups of category (d).

6.3.1. Centrosymmetric Space Groups

In this case triple invariants are calculated with the $B3,0$ formula for all triples that can be generated from all strong reflections, say, with $|E|$ ≥ 1.7. Thereafter, the sign correlation procedure is applied to the triples using the weighting scheme (6.5)–(6.7).

Of practical importance is the number of triples that is to be calculated. If the lower limit of $|E|$ is slightly decreased, the number of triples to be calculated is greatly increased. For medium to strong reflections, say, with $1.4 < |E| < 1.7$, we used a reflection selection procedure based upon the $B2,0$ formula (Karle and Hauptman, 1959):

$$| E_{\mathbf{h}}' |^2 = 1 + D\langle(| E_{\mathbf{k}} |^2 - 1)(| E_{\mathbf{k+h}} |^2 - 1)\rangle_{\mathbf{k}} \qquad (6.8)$$

E' is the normalized structure factor of the squared structure. It has been shown (van den Hark et al., 1974a) that reflections with $|E'| > |E|$ tend to give more reliable $B3,0$ results, so the $B2,0$ formula is useful in saving computer time.

Photodimer of o-distyrylbenzene[†]

Crystal data: monoclinic, $a = 28.047$, $b = 9.504$, $c = 12.600$ Å; $\beta = 103.4°$; $C2/c$, $Z = 4$; $C_{44}H_{36}$. The molecule appeared to be situated on a twofold rotation axis. Reflection data: 1519; 970 "observed."

[†] 5,6,11,12-Tetraphenyldibenzo(2,3,8,9)tricyclo(8.2.0.0^{4,7})dodeca-2,8-diene (van den Hark et al., 1974a).

The $B3,0$ formula (6.1) was used to calculate 460 triple invariants for 114 reflections with $|E| \geq 1.7$. The $B2,0$ formula (6.8) was used to calculate $|E'|$ values for 90 reflections with $1.4 < |E| < 1.7$. For 26 reflections, we found $|E'|$ to be greater than 2.5. Inclusion of these reflections extended the $B3,0$ calculations to 332 additional triples.

The sign correlation procedure was applied to the triples with the largest $(AB_{hk} + C)$ values: 514 signs were obtained. The most probable of two remaining solutions revealed the structure clearly.

It was found *a posteriori* that the four strongest reflections, all entering into many Σ_2 relations, are involved in eight invalid Σ_2 interactions with reflections having $|E| \geq 1.7$. Only one of these interactions is present in the upper half of the table of $B3,0$ results. For further details the reader is referred to the original paper.

6.3.2. Noncentrosymmetric Space Groups with Three Centric Projections

The $B3,0$ formula is applied to all triples among centric projection reflections with, say, $|E| > 1.0$: triples between reflections from different projections are also included, but both triples with, by symmetry, $\Phi_{hk} = \pm\pi/2$ and triples of the Σ_1 type are omitted. The centric reflections are then expressed in terms of a few letter symbols, thus constituting a very good starting set for further phase generation, of three-dimensional reflections, by routine applications of the sum of angles related formulas. The use of the phase correlation procedure followed by tangent-formula refinement is advised. Generally, no problems are expected for these space groups.

2-Di-(*p*-anisyl)methyl-1,3-dioxolan[†]

Crystal data: Orthorhombic, $a = 15.815$, $b = 16.90$, $c = 5.789$ Å; space group $P2_12_12_1$, $Z = 4$; $C_{18}H_{20}O_4$. Intensity data: approximately 1500 reflections.

[†] van den Hark *et al.*, 1974b.

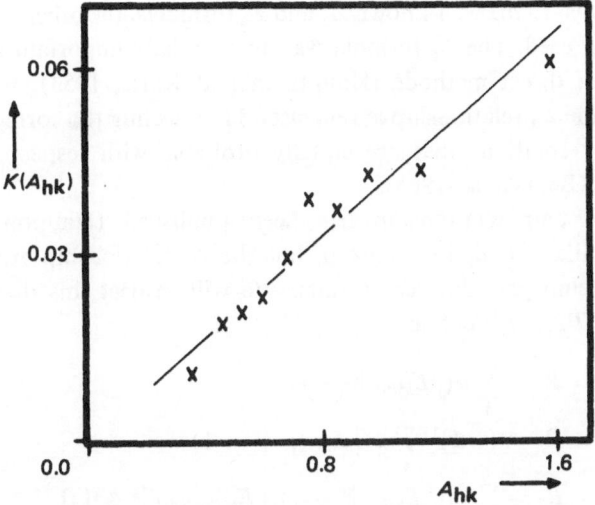

FIGURE 6.2. The sliding scaling factor as a function of A_{hk}.

In Fig. 6.2 we show the sliding scaling factor $K(A_{hk})$, that is, the ratio between the expectation value of $A_{hk} \cos \Phi_{hk}$ and the experimental average of $(AB_{hk} + C + X)$ for a given range of A_{hk}. The actual scale, used for the calculation of KAB_{hk} values (6.4), was taken from the higher A_{hk} values ($K = 0.06$); the rest of the curve is taken into account by introducing the factor A_{hk} in (6.7).

6.3.3. Noncentrosymmetric Space Groups with One Centric Projection

The centric projection may be solved in the same way as described in Section 6.3.2; in this case, it makes sense also to use information from other sources, such as from the Σ_3 and pair relationships described below. Phase extension to general reflections, using phase correlation on results from the $B3,0$ formula, requires an enantiomorph-fixing procedure. This usually is achieved by adapting a multisolution procedure. Very strong enantiomorph discrimination is possible if many reflections take part in the origin-fixing procedure.

A *pair relationship* is an expression for $\phi(\mathbf{h}_1) + \phi(\mathbf{h}_2)$. It is necessary that the reflections $\mathbf{h}_1$ and $\mathbf{h}_2$ be chosen in such a way that $\phi(\mathbf{h}_1) + \phi(\mathbf{h}_2)$ is a structure seminvariant. For the space group $P2_1$, for example, $\mathbf{h}_1$ and $\mathbf{h}_2$ are chosen such that $\mathbf{h}_1 + \mathbf{h}_2 = 2h\ 0\ 2l$.

In contrast to the well-known Σ_1 and Σ_2 formulas, the other Σ_n formulas are not often used. The Σ_3 formula was an especially important tool in the early days of direct methods (Hauptman and Karle, 1953). It has been shown that the Σ_3 relationship is very useful in selecting the correct solution out of many solutions that are equally probable with respect to the Σ_2 relationship (Beurskens, 1965).

Another pair relationship has been published (Hauptman, 1972) that is very like the Σ_3 relationship, but the Σ_3 relationship and the new pair relationship give different results. We will restrict this discussion to space group $P2_1$. We define

$$R_1 = \sum_k \alpha(|E_{h'kl'}|^2 - 1)$$

$$R_2 = (-1)^{k_1} \sum_k \alpha(|E_{h''kl''}|^2 - 1) \qquad (6.9)$$

$$R_3 = \sum_k \alpha[(|E_{h'kl'}|^2 - 1)(|E_{h''k+k_1 l''}|^2 - 1)]$$

where $\alpha = (-1)^k$, and define further $2h' = h_1 + h_2$, $2l' = l_1 + l_2$, $2h'' = h_1 - h_2$ and $2l'' = l_1 - l_2$.

The following Σ_3 interactions are defined:

$$\phi_{h_1 k_1 l_1} + \phi_{h_2 \bar{k}_1 l_2} \approx \phi(R_1)$$

$$\phi_{h_1 k_1 l_1} + \phi_{h_2 \bar{k}_1 l_2} \approx \phi(R_2) \qquad (6.10)$$

In these formulas, derived for centrosymmetric space groups, large positive or negative results for R_1 and R_2 are strong indications that the seminvariant $\phi_{h_1 k_1 l_1} + \phi_{h_2 \bar{k}_1 l_1}$ is approximately equal to 0 or π, respectively. The expression R_3 is part of the pair relationship of Hauptman (1972).

We have found that the results are far more reliable if, for any one pair, R_1, R_2, and R_3 are all large and have the same sign. We want to calculate a weight, w_{123}, for each pair and give its reliability. Pairs with strong and consistent indications R_1, R_2, and R_3 must take the largest weights. We define n_1, n_2, and n_3 as the number of reflections that contribute to the summations in R_1, R_2, and R_3, and define the weight by

$$w_i = R_i n_i^{-1/2} \qquad (i = 1, 2, 3) \qquad (6.11)$$

The absolute value of w_3 has approximately the same magnitude as the significance level defined by Hauptman (1972).

Next, we combine the two Σ_3 indications (6.11) giving w_{12}:

$$w_{12} = 0.5S(w_1 + w_2)[S(w_1)|w_1|^{1/2} + S(w_2)|w_2|^{1/2}]^2 \qquad (6.12)$$

TABLE 6.6. Some Numerical Examples of the *smr* Formula (6.12) [a]

w_1	w_2	w_{12}
100	100	200
150	50	187
200	0	100
250	−50	38
300	−100	27

[a] For all entries $w_1 + w_2 = 200$.

We will call w_{12} the *square-mean-root* (*smr*) value of w_1 and w_2. Some results of (6.12) are shown in Table 6.6. It is clear that w_{12} has the largest value when w_1 and w_2 have comparable values and that w_{12} drops very rapidly when the difference between w_1 and w_2 increases. Combination of w_{12} and w_3 is done after scaling the w_{12} values such that the strongest results for w_{12} and for w_3 become comparable. The combination of w_{12} and w_3, giving w_{123}, is also done with the *smr* formula (6.12).

In general, there are only a limited number of reliable pairs; when going from the top 50 to the next 50 pairs, the reliability drops significantly. For simple structures, such as centrosymmetric structures or noncentrosymmetric structures with three centric projections, the results of the pair relationship are unimportant, but for difficult problems, such as structures in space group $P2_1$, the results are very useful. It is important to realize that the information from the pair relationship is quite different from the information that is obtained from the Σ_2 relationship; the weak reflections play an important role in the pair relationship. We never just use the results of the pair relationship, but always in combination with calculated triple invariants.

Heptahelicene[†]

Crystal data: monoclinic, $a = 14.022$, $b = 15.094$, $c = 9.221$ Å; $\beta = 93.20°$; space group $P2_1$, $Z = 4$ (two symmetry-independent molecules); $C_{30}H_{18}$. Intensity data: 3565 symmetry-independent reflections.

[†] Beurskens *et al.*, 1976.

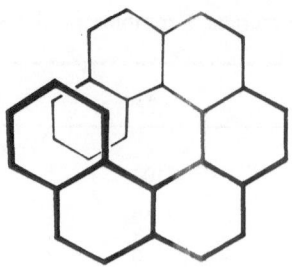

In this crystal structure determination only a few strong $h1l$ reflections were available, but there were many strong $h2l$ and $h3l$ data. We shall describe the way in which the $h2l$ and $h3l$ reflections, in addition to the centric reflections, were used to fix the origin and enantiomorph. Instead of single $h2l$ and $h3l$ reflections, we used sets of reflections for which the phases were correlated. The correlation of the phases was carried out with calculated triple invariants, having large w_{hk} values, and in which one centric reflection, with a known phase, was involved in combination with pairs having large $|w_{123}|$ values. As an example, the results for the $h2l$ reflections are shown in Fig. 6.3. All reflections in this figure are involved in at least three strong pair or triple relationships; all phases are either α or $\alpha + \pi$. Similarly, we had a set of seven $h3l$ reflections with phases β or $\beta + \pi$.

Let us use the set of $h3l$ reflections for origin fixing by defining $\beta = 0$. Then the origin is not fixed unambiguously: moving the origin an integer multiple of 1/3 along the twofold screw axis does not affect the phases of the $h3l$ reflections. In space group $P2_1$ an origin shift by a fractional distance Δy changes the phases of hkl reflections by a value $\Delta\phi_{hkl}$, where $\Delta\phi_{hkl} = 2\pi k\,\Delta y$. As a consequence, all reflections hkl with $k \neq 3n$ will have a threefold phase ambiguity. When an $h2l$ reflection (see Fig. 6.3) has a phase α with respect to a fixed origin on the twofold screw axis, the phase of this reflection with respect to the origin at $\Delta y = 1/3$ and $\Delta y = 2/3$ will be $\alpha - \pi/3$ and $\alpha + 2\pi/3$, respectively. The value of α is unknown, but it is clear that restricting the values of α to 1/3 of the phase circle completes the origin fixing: for example, α may have any value between $\pm\pi/3$.

The enantiomorph can be fixed by further restricting the possible values of α. Restricting α to, for example, positive values does fix the enantiomorph. We decided to continue the phase determination with a value of $\pi/6$ for α; then the true value of α deviates not more than $\pi/6$ from this assignment. At this stage of the phase determination, we had 23 centric projection reflections and 16 general reflections with numerical phases or with phases expressed in terms of two letter symbols; both

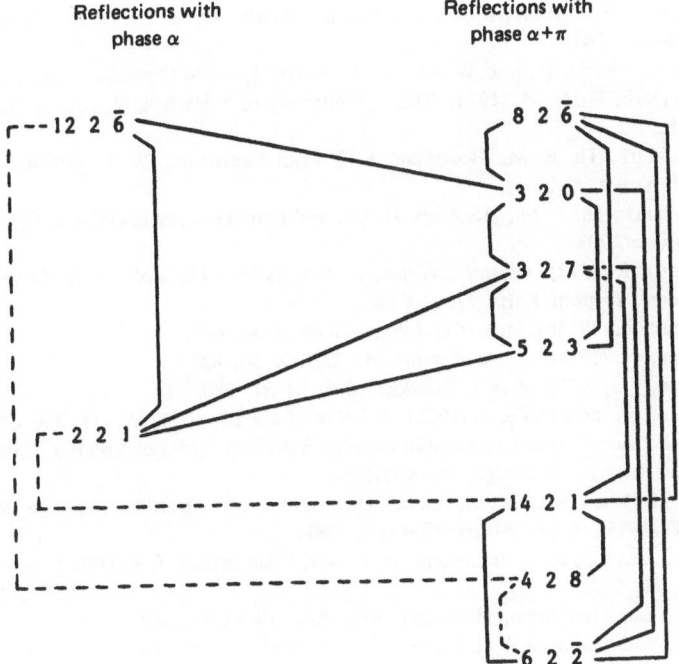

FIGURE 6.3. The correlation of phases of *h2l* reflections. The solid line indicates a triple in which a projection reflection is involved; the dashed line indicates a pair relationship.

symbols were assigned to centric reflections and represent, therefore, phases of 0 or π.

Despite the fact that the phase determination was performed very carefully, none of the four solutions revealed the correct crystal structure. In one of the Fourier maps, two heptahelicene molecules were found which proved to have the correct orientation but an incorrect position in the unit cell. Both molecules appeared to be displaced by a vector perpendicular to the twofold screw axis (see Chapter 7).

References

Beurskens, P. T. (1963). "Technical Report on Sign Correlation by the Sayre Equation," Crystallography Laboratory, University of Pittsburgh, Pittsburgh, Pennsylvania.
Beurskens, P. T. (1964). *Acta Crystallogr.* **17**, 462.
Beurskens, P. T. (1965). Thesis, University of Utrecht, Utrecht, The Netherlands.

Beurskens, P. T., Beurskens, G., and van den Hark, Th. E. M. (1976). *Cryst. Struct. Commun.*, 241.

Germain, G., Main, P., and Woolfson, M. M. (1971). *Acta Crystallogr. Sect. A* **27**, 368.

van den Hark, Th. E. M. (1976). Thesis, University of Nijmegen, Nijmegen, The Netherlands.

van den Hark, Th. E. M., Beurskens, P. T., and Laarhoven, W. H. (1974a). *J. Cryst. Mol. Struct.* **4**, 227.

van den Hark, Th. E. M., Hendriks, H. M., and Beurskens, P. T. (1974b). *Cryst. Struct. Commun.*, 703.

Hauptman, H. (1972). *Crystal Structure Determination: The Role of the Cosine Seminvariants*, Plenum Press, New York.

Hauptman, H. (1974a). *Acta Crystallogr. Sect. A* **30**, 472.

Hauptman, H. (1974b). *Acta Crystallogr. Sect. A* **30**, 822.

Hauptman, H. (1975). *Acta Crystallogr. Sect. A* **31**, 680.

Hauptman, H., and Karle, J. (1953). *Solution of the Phase Problem. I. The Centrosymmetric Crystal*, American Crystallographic Association Monograph No. 3, Polycrystal Book Service, Pittsburgh, Pennsylvania.

Hummelink-Peters, B. G. M. C., van den Hark, Th. E. M., Noordik, J. H., and Beurskens, P. T. (1975). *Cryst. Struct. Commun.*, 281.

Kanters, J. A., Kroon, J., Beurskens, P. T., and Vliegenthart, J. A. (1966). *Acta Crystallogr.* **21**, 990.

Karle, J., and Hauptman, H. (1958). *Acta Crystallogr.* **11**, 264.

Karle, J., and Hauptman, H. (1959). *Acta Crystallogr.* **12**, 404.

Sax, M., Beurskens, P. T., and Chu, S. (1965). *Acta Crystallogr.* **18**, 252.

Schenk, H. (1973). *Acta Crystallogr. Sect. A* **29**, 77.

Schenk, H. (1974). *Acta Crystallogr. Sect. A* **30**, 477.

Viterbo, D., and Woolfson, M. M. (1973). *Acta Crystallogr. Sect. A* **29**, 205.

de Vries, A. (1963). Thesis, University of Utrecht, Utrecht, The Netherlands.

de Vries, A. (1965). *Acta Crystallogr.* **18**, 473.

Application of Direct Methods to Difference Structure Factors

PAUL T. BEURSKENS
and TH. E. M. VAN DEN HARK

7.1. Introduction

This chapter discusses the application of direct methods to the solution of a structure where a part of it is already known. Often the known part of the structure consists of one or more heavy atoms, located by means of Patterson techniques. The known part may also be a fragment of a molecule, found by direct methods.

In these procedures, called DIRDIF, *dir*ect methods are applied to the *dif*ference structure factors. The procedures are initiated by subtracting the known (heavy-atom) contribution from the observed structure factor, assuming that the observed and calculated structure factors have the same phase, to obtain the magnitude and phase of the contribution from the rest of the structure (light atoms). The Σ_2 phase relationship (tangent formula) is used to recalculate the phases of the light-atom contributions, and, subsequently, to recalculate the magnitude of this contribution. An iterative procedure is used to optimize the phases and amplitudes before a difference Fourier map is calculated. The computer programs (Beurskens *et al.*, 1978) are most useful when the known atoms do not fix the phases of the difference structure factors well, either because they are in special or pseudospecial positions, or because they are not sufficiently heavy relative to the other atoms.

PAUL T. BEURSKENS and TH. E. M. VAN DEN HARK • Crystallography Laboratory, Toernooiveld, Nijmegen, The Netherlands.

At present we distinguish the following procedures:

DIRDIF.A: the special case for centrosymmetric structures
DIRDIF.B: the general case for centrosymmetric structures
DIRDIF.C: the special case for noncentrosymmetric structures
DIRDIF.D: the general case for noncentrosymmetric structures

The special case is defined where the known atoms are on special or pseudospecial positions, such that the origin (and/or enantiomorph) is not completely fixed by the known atoms. When the origin is not completely fixed, the known atoms do not contribute to some parity classes of reflections; the phases of these reflections are, thus, completely undetermined. The phases of one or more reflections can be chosen for complete origin specification. When the known atoms are in a centrosymmetric arrangement in a noncentrosymmetric structure, they do not fix the enantiomorph. After shifting the origin to a center of symmetry all calculated phases are restricted to the values 0 or π. One reflection that is expected to have a phase of $\pm\pi/2$ is used for enantiomorph specification. Noncentrosymmetric structures in which the known atoms fix the origin but not the enantiomorph also belong to the special case.

Solving the remaining unknown part of the structure in the special case with the conventional heavy-atom technique is not straightforward. In a difference Fourier synthesis several structures are found mixed up, and the identification of the atoms belonging to one and the same structure may be difficult.

The general case is defined where the known atoms contribute to all classes of reflections, such that the origin (and enantiomorph) are completely fixed. Thus the phases of all reflections are, in principle, determined, and it should be possible to find the complete structure by successive difference Fourier syntheses. However, direct methods can be used to minimize the errors in the phases of the difference Fourier coefficients, and to correct the amplitudes, before a synthesis is calculated.

7.1.1. Some Applications of DIRDIF Procedures

(a) DIRDIF.A   Space group $P\bar{1}$: two symmetry-independent positions $0, 0, 0$ and $0, \frac{1}{2}, 0$ determine signs for reflections hkl with $k = 2n$; one reflection hkl with $k = 2n + 1$ must be used to fix the origin completely.

(b) DIRDIF.A   Space group $P2_1/c$: one atom in the asymmetric unit of the unit cell on the pseudospecial position $x, \frac{1}{4}, z$ determines signs for reflections hkl with $l = 2n$; one reflection hkl with $l = 2n + 1$ must be used to fix the origin completely.

(c) DIRDIF.B   Space group $P\bar{1}$: one atom on $0, 0, 0$.

(d) DIRDIF.B   Space group $P\bar{1}$: two symmetry-dependent atoms in general positions.

(e) DIRDIF.C   Space group $P1$: one atom on $0, 0, 0$; all calculated phases are 0; select reflections that may have a phase of $\pm\pi/2$ and accept one phase for enantiomorph specification.

(f) DIRDIF.C   Space group $P1$: two independent atoms on $0, \frac{1}{4}, 0$ and $0, \frac{3}{4}, 0$; origin and enantiomorph specifications as in examples (a) and (e), respectively.

(g) DIRDIF.D   Space group $P1$: two different or any nonsymmetrical arrangement of three or more atoms.

(h) DIRDIF.D   Space group $P2_12_12_1$: one atom per asymmetric unit, on a general (not pseudospecial) position.

The procedures DIRDIF.A, DIRDIF.B, and DIRDIF.D have been studied and applied to several crystal structure determinations; DIRDIF.C is in a preliminary stage of development (Prick *et al.*, 1978).

The DIRDIF procedures were originally developed for heavy-atom structures (Beurskens and Noordik, 1971; Gould *et al.*, 1975; van den Hark *et al.*, 1976). The procedures can equally well be applied to partially known equal-atom structures (Beurskens *et al.*, 1976). We retain the subscripts H (heavy) and L (light) for the known part and the remaining part of the structure, respectively.

7.1.2. Two-Dimensional Wilson Plot

A well designed Wilson-plot routine is to be used to obtain the best possible estimates for the difference structure factors. Usually, the temperature factors of the heavy atoms differ from the overall-temperature factor of the rest structure, and this will then be of great importance for the calculation of difference structure factors.

It is assumed that the reflections are distributed over a two-dimensional array in ranges of s and $|E_H|$. We define for a reflection $\mathbf{h}$:

$|F_o|$ Observed structure amplitude, on absolute scale

I Observed intensity on relative scale K

s $2\lambda^{-1} \sin \theta$

K Scale factor

F_H Calculated structure factor on absolute scale for the known part (heavy atom) of the structure with an overall temperature factor B_H

g_H, g_L, g Temperature-corrected atomic scattering factors for a heavy atom, a light atom, and any atom, respectively

B_H, B_L, B_{ov} Overall temperature factors used for the heavy atoms, the unknown part of the structure, and the whole structure, respectively

$\Delta B_H, \Delta B_L$ Corrections to be applied to B_H and B_L

$\sum_H, \sum_L, \sum$ Summation over the heavy atoms, light atoms, and all atoms of a unit cell, respectively

S_H, S_L, S_{ov} Normalizing function for the heavy atoms, the light atoms, and/or the whole structure, respectively; for example,

$$S_H = \varepsilon \sum_H g_H^2 \qquad (7.1)$$

E_H, E_L Normalized structure factors for a structure consisting of the heavy atoms or the light atoms, respectively; for example,

$$|E_H|^2 = |F_H|^2/S_H \qquad (7.2)$$

$G_o = \langle I/S_{ov} \rangle_{\mathbf{h}}$

$G_H = K\langle |F_o|^2 \exp(-\Delta B_H \cdot \tfrac{1}{2}s^2)/S_{ov} \rangle_{\mathbf{h}}$

$G_L = K\langle S_L \exp(-\Delta B_L \cdot \tfrac{1}{2}s^2)/S_{ov} \rangle_{\mathbf{h}}$

where the averaging is done over the appropriate range of s and $|E_H|$. The quantity to be minimized is then

$$\sum_{\text{ranges}} (G_o - G_H - G_L)^2 \qquad (7.3)$$

and the parameters to be refined are K, ΔB_H, and ΔB_L.

In the special case, where the known atoms do not contribute to some parity classes of reflections, ΔB_L and K are obtained at once from the reflections of these parity classes and ΔB_H then is obtained from the remaining reflections.

7.2. Difference Structure Factors

The difference structure factor is defined for the structure minus the known part of the structure. The phases and the magnitudes of the difference structure factors are unknown, except for some extreme cases described below.

In our DIRDIF procedures, the phases are determined or refined by direct methods and the magnitudes of the difference structure factors can then be calculated.

$|F_o|$, F_H, E_H, and E_L and the normalizing function S_L have been defined in Section 7.1. In addition, we define for a reflection $\mathbf{h}$

F_L The contribution of the unknown part of the structure (light atoms), or the most probable estimate for this contribution. Similarly, E_L is the most probable value for the normalized structure factor of the unknown part of the structure:

$$E_L = F_L/S_L^{1/2} \qquad (7.4)$$

ϕ_H Phase of F_H

ϕ_L Phase of F_L

F_o A *phased* value for the observed structure factor, defined by

$$F_o = F_L + F_H \qquad (7.5)$$

and the normalized difference structure factor (7.4) now becomes

$$E_L = (F_o - F_H)/S_L^{1/2} \qquad (7.6)$$

In conventional Fourier procedures, the difference Fourier coefficients

$$\Delta F_1 = (|F_o| - |F_H|)\exp i\phi_H \qquad (7.7)$$

are calculated and accepted as an estimate for F_L. Only in very favorable circumstances can the complete light-atom structure be deduced unambiguously from the Fourier synthesis based on these coefficients. In the present procedure, a Σ_2 (tangent formula) refinement procedure is used to convert input ΔF_1 values to more probable F_L values. This procedure depends on a probability estimate for ΔF_1, relative to the extreme opposite possibility, ΔF_2:

$$\Delta F_2 = (-|F_o| - |F_H|)\exp i\phi_H \qquad (7.8)$$

where now F_o is completely out of phase with F_H. It should be noted that

$|\Delta F_1| \le |F_L| \le |\Delta F_2|$. Although conventionally ΔF_1 is used as a difference Fourier coefficient, ΔF_2 will be the more probable value for relatively small terms in noncentrosymmetric structures (see below).

In the present procedure use is made of those reflections where ΔF_1 is far more probable than ΔF_2, and ΔF_1 is used as a first estimate of F_L. The application of direct methods then leads to new phases ϕ_L, and the magnitude $|F_L|$ then also is to be recalculated, using (7.5), which may be written as

$$|F_o| = ||F_L| \exp(i\phi_L) + F_H| \qquad (7.9)$$

For the application of this equation we have to distinguish two cases, depending on whether $|F_H|$ is less than or greater than $|F_o|$. These two cases are illustrated in Figs. 7.1 and 7.2.

For centrosymmetric structures phases are either 0 or π; F_L is either ΔF_1 or ΔF_2, and (7.7) and (7.8) reduce to

$$\Delta F_1 = S(F_H) \cdot |F_o| - F_H$$
$$\Delta F_2 = -S(F_H) \cdot |F_o| - F_H$$

The normalized difference structure factor E_L is defined by (7.6). The ΔF_1 and ΔF_2 values defined above are brought on the same scale by defining

$$E_1 = \Delta F_1 / S_L^{1/2} \quad \text{and} \quad E_2 = \Delta F_2 / S_L^{1/2} \qquad (7.10)$$

The initially unknown E_L values correspond to the normalized structure factors of a structure consisting of only the unknown atoms. Conventional direct methods may, in principle, be applied to E_L values where they are

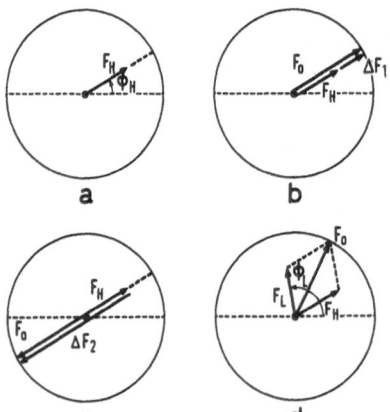

a   b   c   d

FIGURE 7.1. Illustrations of the definition of ΔF_1 and ΔF_2, and construction of F_L for case 1: $|F_H| < |F_o|$. (a) $|F_o|$ circle, with calculated F_H. (b) Definition of ΔF_1; F_H and F_o are in phase. (c) Definition of ΔF_2; F_H and F_o are out of phase by π. (d) General case; ϕ_L is assumed to be known.

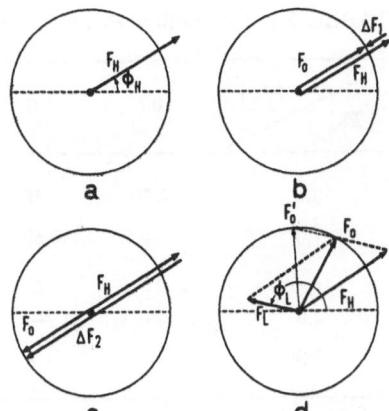

FIGURE 7.2. Illustration of the definition of ΔF_1 and ΔF_2, and construction of F_L for case 2: $|F_H| > |F_o|$; (a)–(d) as in Fig. 7.1.

known. For reflections where E_1 is far more probable than E_2, we use E_1 as an initial estimate for E_L and then use these results to initiate a tangent-formula refinement procedure.

The original $|E_1|$ values do not form a normalized set of structure factors; the average of squares of $|E_1|$ is less than unity. During the refinement of phases, the magnitudes of $|E_L|$ are, on average, being increased and the $|E_L|$ values approach the normalized theoretical distribution.

7.2.1. Probability Considerations

For centrosymmetric structures, we define P_1 as the probability that the *sign* of E_1 is correct. For case-1 reflections ($|F_H| < |F_o|$), the probability that the *magnitude* of E_1 is correct is also given by P_1 (Woolfson, 1956). That is,

$$P_1 = P(|E_1|)/[P(|E_1|) + P(|E_2|)] \tag{7.11}$$

where $P(|E|)$ is the distribution of $|E|$ values in centrosymmetric space groups. Some numerical results for P_1 are given in Table 7.1.

For noncentrosymmetric structures, the standard deviation for the phase of E_1 can be calculated (van den Hark *et al.*, 1976; Beurskens *et al.*, 1979); some numerical results are given in Table 7.2. This standard deviation should be used as a basis for the calculation of the weight of the phase of E_1. For practical reasons, however, we prefer to use P_1 as a measure for the relative probability for the phase of E_1: P_1 is defined by (7.11), where $P(|E|)$ is now the distribution of $|E|$ values in noncentrosymmetric space groups (Fig. 7.3).

TABLE 7.1. Probability P_1 for Dual E_L Values in Centrosymmetric Structures

$\|E_1\|$ \ $\|E_2\|$	0.6	1.0	1.5	2.0	3.0	4.0
0.6	0.50	0.58	0.72	0.86	0.987	0.9995
1.0	—	0.50	0.65	0.82	0.982	0.9993
1.5	—	—	0.50	0.71	0.97	0.9990
2.0	—	—	—	0.50	0.92	0.998
3.0	—	—	—	—	0.50	0.971

TABLE 7.2. Standard Deviation (deg) for the Phase of E_1 for Dual E_L Values
in Noncentrosymmetric Structures

$\|E_1\|$ \ $\|E_2\|$	0.6	1.0	1.4	2.0	3.0	4.0
0.6	104	102	92	75	59	54
1.0	—	104	94	72	50	44
1.4	—	—	104	78	47	38
2.0	—	—	—	104	50	35

The weight for the phase of E_1 is chosen to be

$$w_1 = (2P_1 - 1)^2 \tag{7.12}$$

for $P_1 \geq \frac{1}{2}$, where P_1 is given by (7.11) with $P(\|E\|)$ chosen appropriately. The case $P_1 < \frac{1}{2}$ is discussed below. The weight w is used to initiate the phase refinement procedure.

7.2.2. Classification of Reflections

The distribution function for acentric reflections has a maximum at $\|E\| = \frac{1}{2}\sqrt{2} = 0.7$. The occurrence of this maximum forced us to consider three distinct cases (Fig. 7.3).

In case (a), $\|E_1\| > 0.7$, many reflections will have both $\|E_1\|$ and $\|E_2\|$ greater than 0.7. The number of reflections that belong to this category

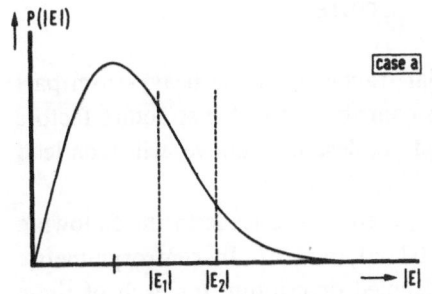

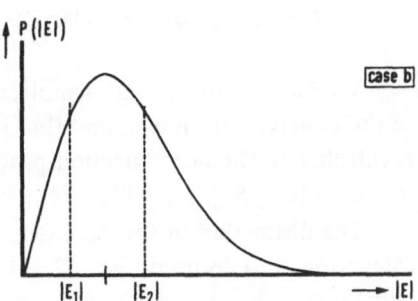

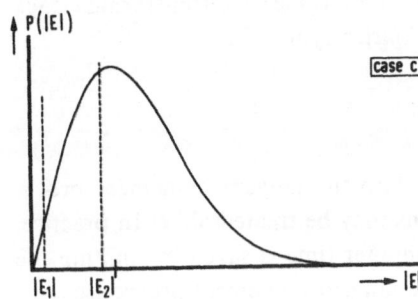

FIGURE 7.3. Classification of reflections depending on $|E_1|$ and $|E_2|$ pairs (non-centrosymmetric structures). Case a: $0.7 < |E_1| < |E_2|$. Case b: $|E_1| < 0.7 < |E_2|$. Case c: $|E_1| < |E_2| < 0.7$.

depends on the known fraction of the scattering power; usually it is about half the total number of reflections, or less.

This case is the most important, as $|E_1|$, as well as any possible value for $|E_2|$, is large enough to be of importance in a Fourier synthesis, and to be useful in the application of the tangent formula. For these reflections $|E_1|$ is more probable than $|E_2|$, and E_1 may be selected and used as a first estimate for E_L, where the probability for $|E_1|$ is significantly greater than the probability for $|E_2|$. In case (a), $P_1 \geq \frac{1}{2}$ and $0 \leq w_1 < 1$.

For case (b) $|E_1| < 0.7 < |E_2|$, and for case (c), $|E_2| < 0.7$, the value of $|E_1|$ is *not* the most probable value. The tangent formula cannot determine the phase of such a reflection; nevertheless E_1 or E_2 for the reflection may be used as a Fourier coefficient if justified by either its relative probability or its standard deviation. In this class, if $P_1 < \frac{1}{2}$, then $w_1 \equiv 0$.

For centrosymmetric structures, or special reflections having a centric distribution of $|E|$ values, E_1 is more probable than E_2 (unless $F_H = 0$): all reflections with $|E_1|$ greater than some minimum value will be treated as case-(a) reflections. The reflections with smaller $|E_1|$ values will not be subjected to the Σ_2 calculations.

7.3. Description of the Procedure

We have assumed that a molecular fragment, or the heavy-atom part of the structure, is known, and that its contribution to the structure factors is calculated. The normalization procedure, described above, will then lead to values for $|F_o|$, F_H, ΔF_1, ΔF_2, E_1, E_2, P_1, and w_1.

The distinction in several cases, as given above, leads to the following categories of reflections: a1, a2, b1, b2, c1, and c2. In centrosymmetric structures, only a1 and a2 apply. A detailed description for each of these categories is given below.

At the end of the procedure the final E_L values are transformed back to F_L values and used in a weighted Fourier synthesis.

7.3.1. Tangent Refinement

The case-(a) reflections may enter into the tangent refinement procedure. In principle, all of these reflections may be treated alike. In practice, however, a considerable amount of computer time is saved by limiting the number of reflections that enter into the tangent refinement procedure.

As input to the tangent formula, we use those reflections where $|E_1|$ exceeds a given minimum value greater than 0.7, say 1.2. We then use this formula to calculate phases for all reflections where $|E_1|$ exceeds another minimum greater than 0.7, say 0.9.

The tangent formula may be given as

$$\phi_{L,t} = \text{phase of } (E_L)_\mathbf{h} \approx \text{phase of } \Sigma_2 \qquad (7.13)$$

where

$$\Sigma'_2 = \sum_\mathbf{k} w_\mathbf{k} w_{\mathbf{h}-\mathbf{k}} E_\mathbf{k} E_{\mathbf{h}-\mathbf{k}}$$

and $E_\mathbf{k}$ is the most probable E_L value for the reflection $\mathbf{k}$. Analogous to the centrosymmetric formula, and to the formulas (7.11) and (7.12), we use the following simple expression for the relative probability and the corresponding weight:

$$P_t = \tfrac{1}{2} + \tfrac{1}{2} \tanh \sigma_3 (\sigma_2)^{-3/2} |E_\mathbf{h}| \Sigma_2 \qquad (7.14)$$

$$w_t = (2P_t - 1)^2 \qquad (7.15)$$

the weighted tangent formula is described by Germain et al. (1971).

In the first cycle, we have only E_1 values as first estimates for E_L for the reflections $\mathbf{k}$ and $(\mathbf{h} - \mathbf{k})$. Whether or not the output phases ϕ_L are accepted depends on the corresponding weights. If $w_t > w_1$ then the new phase ϕ_L is accepted with weight w_t. If $w_t < w_1$, we have to consider the results in some more detail. We define

$$\Delta\phi = \phi_{L,t} - \phi(E_1) \tag{7.16}$$

That is, $\Delta\phi$ is the correction to the original phase of E_1. If $w_t < w_1$ and $|\Delta\phi| > \pi/2$, then the result of the tangent formula for this reflection is rejected. If $w_t < w_1$ and $|\Delta\phi| < \pi/2$, then the calculated $\phi_{L,t}$ value is only partially accepted:

$$\phi_L(\text{new}) = \phi(E_1) + \frac{w_t}{w_1} \Delta\phi \tag{7.17}$$

In addition, there is a limitation on $\Delta\phi$ for reflections of category a2, as shown below, and ϕ_L may be set accordingly. A new ϕ_L value is used to calculate a new value for E_L by (7.9), which may then be used as input for the next tangent refinement cycle.

The weighting scheme, defined by (7.11) and (7.12), gives acceptable results for the first input phases for the tangent formula. As a consequence of the definitions for w_1 and w_t, given by (7.12) and (7.15), on average, a slightly overestimated w_1, relative to w_t results, but the original phases $\phi(E_1)$ still play an important role in the second cycle of refinement. The weights for the following cycles of the refinement will be largely determined by the results of the tangent formula.

Category a1

For reflections with $|F_H| < |F_o|$ we have, from Fig. 7.1

$$\phi(E_1) = \phi_H \tag{7.18}$$

The strongest of these (noncentrosymmetric reflections) or all (centrosymmetric reflections) form the category a1. This set of reflections is of most importance. Many of them will enter into the tangent formula calculations and will have, after four to six cycles of tangent refinement, a calculated E_L value. If this calculated value is unreliable, then the original E_1 value will be used.

The remaining reflections are those with $|E_1| < E_{\min}$, say 0.9; for these reflections the E_1 value will be used, with its proper weight w_1.

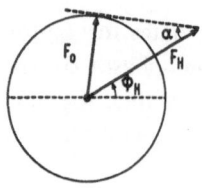

FIGURE 7.4. Definition of the limiting angle for reflections with $|F_H| > |F_o|$; $\alpha \equiv \Delta\phi_{max}$.

Category a2

For reflections with $|F_H| > |F_o|$ we have, from Fig. 7.2,

$$\phi(E_1) = \phi_H + \pi \qquad (7.19)$$

As can be seen from Fig. 7.2d, there are two possible $|E_L|$ values for a given phase ϕ_L; the smallest $|E_L|$ value is the most probable one and, naturally, this is the value to be used in our procedure. For a given $|F_o|$ and F_H there is a restriction on the possible phase values of ϕ_L (see Fig. 7.4). Consequently, the phase correction, $\Delta\phi$, in (7.16), is restricted; the maximum value for $|\Delta\phi|$ is given by

$$\sin|\Delta\phi_{max}| = |F_o|/|F_H| \qquad (7.20)$$

For $|E_1| > 0.7$ and, most likely, $|E_2| < 4.0$, we have $\Delta\phi_{max} = 44.6°$ as the largest possible value for $\Delta\phi_{max}$. Thus, the few reflections in this category have well-determined phases and will, therefore, be given unit weight ($w_1 = 1$) in all calculations. If the $\Delta\phi$, calculated by the tangent formula (7.16), exceeds $\Delta\phi_{max}$ for the given reflection, then $\Delta\phi_{max}$ is substituted for $\Delta\phi$. In the last cycle, however, the calculated $\Delta\phi$ is accepted to allow for possible experimental errors in $|F_o|$ and model errors in F_H.

Category b1

Because of the low $|E_1|$ value, the tangent formula may lead to incorrect results, and a change in phase, leading to larger $|E_L|$ values cannot be trusted; hence, the tangent formula is not used at all. Although the reflections in this category have rather low $|E_1|$ values, this category is not unimportant because of its large number of reflections. E_1 values will be used, with their proper weights w_1, only where $w_1 > w_2$. Reflections with $w_1 < w_2$ are rejected for several reasons (see also category c1).

Category b2

The reflections belonging to this category have reliable phases $\phi(E_1)$ ($w_1 = 1$, see category a2), and are, therefore, useful for Fourier synthesis;

ϕ_L is taken as $\phi(E_1)$, and the absolute value $|E_L|$ is taken as its expectation value:

$$\int_1^2 |E| P(|E|) d|E| \Big/ \int_1^2 P(|E|) d|E| \qquad (7.21)$$

Category c1

For these reflections, E_2 is a more probable estimate for E_L than E_1, in contrast to the conventional definition of difference-Fourier coefficients. It is certainly *not* useful to put these reflections as E_1 into a Fourier synthesis. To use E_2 as Fourier coefficients, on the other hand, may easily lead to an increased Fourier noise level because of the uncertainties in $|F_o|$ and F_H. At present we reject these reflections.

Category c2

This small set of reflections is treated as that in category b2.

7.3.2. Origin Specification

The special case requires origin and/or enantiomorph specification. The enantiomorph specification is, in principle, possible if one uses a multisolution technique. As we have not yet finished work on this case, we will restrict our present discussion to the centrosymmetric special case, DIRDIF.A.

The heavy atoms are in known special or pseudospecial positions, such that the origin is not fixed and the phase problem is not solved.

We then have two classes of reflections. There are the relatively strong reflections where, *on average*, all the atoms of the structure may contribute to the structure factors, and there are the relatively weak reflections, having no contribution from the known heavy atoms; these weak reflections belong to parity classes where only the light atoms contribute to the structure factors.

For the strong reflections, the values of E_1 and w_1 are calculated, and these reflections will be treated as in the general case.

For the weak reflections, $F_H = 0$, $w_1 = 0$, and $|E_L| = |E_1| = |E_2|$, for all reflections of the corresponding parity classes. These reflections are brought into the calculations by assigning origin choices. The signs of many reflections in these parity classes are found by the Σ_2 relationship, using the correlation technique described in Chapter 6 in order to avoid weak sign indications. When the known atoms do not fix the phases of the strong reflections well, some additional reflections of the weak parity classes are given symbolic signs, or letter symbols, and correlation equations will be found to solve the symbol ambiguity.

Thus, the DIRDIF.A procedure is the same as DIRDIF.B but applied more carefully to difference Fourier coefficients *and* origin choices.

In the references cited above, examples are given where the DIRDIF procedures proved to be useful in routine structure analysis of heavy-atom compounds, as well as for partially known equal-atom structures. Two examples are given here.

7.3.3. Trichloro-bis(triethylphosphine)cobalt(III) [†]

Crystal data: triclinic, $a = 7.775$, $b = 28.26$, $c = 18.019$ Å; $\alpha = 90.97$, $\beta = 97.88$, $\gamma = 90.19°$. Space group $P\bar{1}$ with $Z = 8$; $Co(P(C_2H_5)_3)_2Cl_3$. Intensity data: 7430 independent reflections (3630 above background).

A superficial study of the Patterson synthesis revealed the four independent cobalt atoms to be at pseudospecial positions: $0, 0, \frac{1}{4}$; $0, \frac{1}{2}, \frac{1}{4}$; $\frac{1}{2}, \frac{1}{4}, 0$; $\frac{1}{2}, \frac{1}{4}, \frac{1}{2}$; forming a subcell of $\frac{1}{8}$ the unit cell.

Only reflections of the parity class *ggg* should have contributions from Co on these positions. DIRDIF.A could have been used, but then three origin choices are needed. As the Patterson Co–Co peaks appear around the special positions, troubles with DIRDIF.A can be expected, and more accurate Co positions were sought from the Patterson; the resulting positions were

$$-0.02, 0.01, \tfrac{1}{4}; \qquad 0.02, 0.49, \tfrac{1}{4}; \qquad 0.44, 0.26, 0; \qquad 0.56, 0.24, \tfrac{1}{2}$$

This set of positions, although not special or pseudospecial, shows a special arrangement: there is a center of symmetry at $0, 0, \frac{1}{4}$ and the cobalt atoms form an *A*-centered unit cell (subcell of $\frac{1}{2}$ the unit cell). Because of heavy overlap, including vectors from the Cl and P atoms, a more detailed solution was not possible, and we decided to put this solution into the DIRDIF.A procedure.

The class of strong reflections is formed by the *ggg*, *ugg*, *guu*, and *uuu* reflections; the phases of the *ggg* are well determined by the cobalt atoms and, although the cobalt contribution to the *ugg*, *guu*, and *uuu* is weak, on average, a sufficient number of reflections had acceptable weights w_1. Of the remaining four parity classes, one reflection was chosen to further

[†] Van Enckevort *et al.*, 1977.

specify the origin. The automatic run of the program DIRDIF.A, followed by a Fourier synthesis located correctly all of the P and Cl atoms, but one of the cobalt atoms was misplaced. The remaining carbon atoms were found after application of the DIRDIF.B procedure.

7.3.4. Heptahelicene

This structure determination was used as an example also in 6.3.3. At the end of the example, it was stated that two misplaced heptahelicene molecules were found from the Fourier synthesis. The troubles in the phase determination process are caused by the geometrical periodicities in the structure, giving systematically and consistently incorrect Σ_2 interactions. The molecules are found in the Fourier synthesis with correct orientation. So if one uses only one molecule, and reduces the symmetry to space group $P1$, this molecule can be considered as a known part of the structure.

The monoclinic reflection data set is expanded using $|F_o(h\bar{k}l)| = |F_o(hkl)|$. The partial structure factors F_H are calculated for the one molecule and the DIRDIF.D program applied to the difference structure factors to find the most probable values of phases *and* amplitudes of the structure factors for the remaining three molecules.

The electron density map showed all four molecules, that is, the input molecule as well as three new molecules. The twofold screw axis was easily found: the structure was shifted to bring the twofold screw axis to its proper position, and the coordinates of symmetry-dependent atoms were averaged. The structure then refined rapidly in the proper space group $P2_1$.

7.4. Some Observations

The input molecule of the preceding example was also found on the electron density map; this situation is caused by errors in the scale factor and the atomic positions. The peaks in a conventional difference Fourier map will be enhanced by the application of direct methods. From some trial runs, it was learned that if more and more atoms are known, these atoms will disappear, and the remaining atoms will have higher peaks on the Fourier synthesis.

There is an important difference between the tangent recycling technique of Karle (1968) and the present procedure. In the former method, the same reflections and Σ_2 interactions which were used in the initial phasing procedure and resulted in the recognition of a structural fragment are used again for the refinement of phases. In contrast, quite different reflections and Σ_2 interactions are used in our procedure. Weak reflections, for example, may have a large contribution from the known structural fragment and may be important in our procedure.

Acknowledgments

We thank Drs. R. O. Gould and Gezina Beurskens for their participation in this research. One of us (Th. E. M. van den Hark) acknowledges support of the Dutch Foundation for Pure Research, ZWO/FOMRE.

References

Beurskens, P. T., and Noordik, J. H. (1971). *Acta Crystallogr. Sect. A* **27**, 187.
Beurskens, P. T., van den Hark, Th. E. M., and Beurskens, G. (1976). *Acta Crystallogr. Sect. A* **32**, 821.
Beurskens, P. T., Bosman, W. P. J. H., Gould, R. O., van den Hark, Th. E. M., and Prick, P. A. J. (1978). Technical Report *1978/1*, Crystallography Laboratory, Toernooiveld, Nijmegen, The Netherlands.
Beurskens, P. T., Prick, P. A. J., and Doesburg, H. M. (1979). *Acta Crystallogr. Sect. A*, to be published.
van Enckevort, W. J. P., Hendriks, H. M., and Beurskens, P. T. (1977). *Cryst. Struct. Commun.*, 531.
Germain, G., Main, P., and Woolfson, M. M. (1971). *Acta Crystallogr. Sect. A* **27**, 368.
Gould, R. O., van den Hark, Th. E. M., and Beurskens, P. T. (1975). *Acta Crystallogr. Sect. A* **31**, 813.
van den Hark, Th. E. M., Prick, P., and Beurskens, P. T. (1976). *Acta Crystallogr. Sect. A* **32**, 816.
Karle, J. (1968). *Acta Crystallogr. Sect. B* **24**, 182.
Prick, P. A. J., Beurskens, P. T., and Gould, R. O. (1978). *Acta Crystallogr. Sect. A* **34**, S42.
Woolfson, M. M. (1956). *Acta Crystallogr.* **9**, 804.

Phase Extension and Refinement Using Convolutional and Related Equation Systems

DAVID SAYRE

8.1. Introduction

The mathematical solution of a set of relations among the phases of the structure factors of a crystal may be attempted either *ab initio*, or by starting from some partial or approximate information concerning the desired solution. In the first case one speaks of a direct method of phase determination, in the second case of a direct method of phase extension or refinement. The preceding chapters have dealt principally with the first case, although phase extension and refinement have been mentioned at some length as .components of methods whose overall purpose is *ab initio* phasing. Starting with this chapter, attention will shift to the second case. Mathematically the shift will involve a change in the type of relation system solved. Chapter 8 will discuss methods of phase extension and refinement based on the solution of convolutional and related systems of equations, and Chapters 9 and 10 will deal, respectively, with methods involving the maximization of Karle–Hauptman determinants and the solution of relation systems arising out of structural redundancy. The shift will involve also a change in solution technique. In *ab initio* phasing the methods tend to emphasize the rapid classification of solution space into regions which are thought to be worth, or not worth, visiting, together with special techniques,

DAVID SAYRE • Department of Mathematical Sciences, IBM Thomas J. Watson Research Center, Yorktown Heights, New York 10598.

such as the use of magic integers (White and Woolfson, 1975) for exploring rapidly large regions of solution space. In solving from a starting set, the methods consist essentially in iteratively computing an increment to one's current position in solution space.

Practically the shift involves a change in the range of application to include large and very large structures. It arises because the systems solved in phase extension and refinement—unlike the systems expressing the expected value of a structure invariant or seminvariant, which are usually chosen for *ab initio* solution—retain their accuracy independently of the size of structure concerned. Applications are therefore fairly frequent in protein and even (Chapter 10) virus crystallography, and these applications will be stressed in this and the following two chapters. In these areas of crystallography the required initial phase sets are usually available through heavy-atom phasing or from knowledge of a related structure, and direct methods then offer the possibility of fast and convenient phase extension and improvement.

In the applications to large structures it is necessary to work with lower-resolution data than has been generally assumed in the earlier portions of the book. This procedure represents a challenge for the methods of Chapter 8, perhaps less so for those of Chapter 9, but is not a problem for those of Chapter 10.

8.2. Outline of the Convolutional Equation Systems[†]

After the demonstration by Karle and Hauptman (1950) that F_0F_h lies near F_kF_{h-k} when $|F_h|$, $|F_k|$, and $|F_{h-k}|$ are large, it became necessary (a) to develop *ab initio* structure-determining methods from this simple local type of relationship and to find ways of enhancing the rather moderate accuracy of such relationships, and (b) to find additional relationships of considerably higher accuracy, if possible. As for (a), the story has been told in the preceding chapters. Several lines of development have occurred for (b), and they form the subjects of this and the following chapters.

One possibility for (b) was to combine a large number of indications of the above type for F_h. A method for doing this was provided by the observation (Sayre, 1952) that if F is a set of structure factors that yields

---

[†] We use = to denote a relation of equality when the relation is derived nonprobabilistically, $\approx$ when the relation is derived probabilistically, and $\sim$ when it is derived probabilistically and with additional approximations.

as its Fourier sum a density function ϱ which consists of equal and fully resolved atoms, then

$$F * F \overset{(r,e)}{=} [(f * f)/f]F \tag{8.1}$$

or

$$\sum F_k F_{h-k} \overset{(r,e)}{=} [(\sum f_k f_{h-k})/f_h]F_h \tag{8.2}$$

a relationship of the desired type. Here an asterisk denotes convolution, and we have placed r and e in parentheses above the relation to remind us of the assumptions of atomic resolution and equality of atoms. The proof of (8.1) is simple. Under the assumption of resolution

$$\varrho^2 \overset{(r)}{=} \varrho^{sa} \tag{8.3}$$

where ϱ^2 is the result of squaring ϱ and ϱ^{sa} is the result of replacing each atom in ϱ by the square of that atom; atom equality is not required for this step. Then, under the assumption of atom equality, the right-hand side of (8.3) may be Fourier transformed into the right-hand side of (8.1); resolution is not required for this step. Finally, the left-hand side of (8.3) may always be transformed into the left-hand side of (8.1), by virtue of the convolution theorem of Fourier analysis. The entire scheme is as follows, omitting a factor of $1/V$ from both sides of the lower line:

$$
\begin{array}{ccc}
\varrho^2 & \overset{(r)}{=} & \varrho^{sa} \\
\updownarrow & & \updownarrow {\scriptstyle (e)} \\
F * F & \overset{(r,e)}{=} & [(f * f)/f]F
\end{array} \tag{8.4}
$$

It should be noted that the argument is not probabilistic, as has sometimes been stated (Karle, 1964, 1969; Stout and Jensen, 1968; Carpenter, 1969).

A particularly simple form of (8.1) results from noting that $F = fT$, where T is the trigonometrical part of the structure factor, whereupon

$$(fT) * (fT) \overset{(r,e)}{=} (f * f)T \tag{8.5}$$

Finally, a form that is convenient for comparison with the probabilistic systems to be discussed next is obtained by noting that in the equiatom case $T = N^{1/2}E$, whereupon

$$(fE) * (fE) \overset{(r,e)}{=} N^{-1/2}(f * f)E \tag{8.6}$$

I am indebted to Professor Georges Tsoucaris (private communication) for suggesting the form $[(f * f)/f]$, in which the coefficient Vf^{sa}/f of the right-hand side of (8.1) is given here. The coefficient is given in explicit form in (8.2). In cases where the atom resolution is not complete Tsoucaris's formulation bases ϱ^{sa} on the squared band-limited atom rather than on the band-limited squared atom as was done by Sayre (1952). Tsoucaris' formulation is more resistant to errors arising from band limitation than is the earlier one and makes unnecessary, or less necessary, the empirical correction for band limitation introduced by Sayre (1972): the same formulation has been noted by Rothbauer (1976). I am indebted also to Robert Greene, who calculated the f curves used in the numerical examples at the end of this section. These correspond to narrowest band-limited atoms in three dimensions: see Slepian and Pollack (1961) for a discussion of narrowest band-limited pulses in one dimension. This method gives somewhat better results than the Gaussian atoms used by Sayre (1952). A significant generalization of (8.1), allowing the condition e to be dropped, has been given by Rothbauer (1976).

Probabilistic arguments yielding systems similar but not identical to (8.1) have been given by several authors. Hughes (1953) gave the result

$$\langle E_k E_{h-k} \rangle_k \stackrel{(r',e)}{\approx} N^{-1/2} E_h \tag{8.7}$$

where r' represents a weaker resolution assumption expressed in terms of the vanishing of certain sums under the averaging over $\mathbf{k}$.

Karle and Hauptman (1956) gave

$$\langle E_k E_{h-k} \rangle_k \stackrel{(r')}{\sim} \sigma_2^{-3/2} \sigma_3 E_h \tag{8.8}$$

a more approximate system but one in which the assumption of equal atoms has been dropped. When the atoms are equal, (8.8) reduces to (8.7).

Karle and Karle (1966) gave

$$\langle E_k E_{h-k} \rangle_k \stackrel{(r'')}{\sim} \frac{\langle | E_k | | E_{h-k} | \rangle_k}{\langle I_0(K)/I_1(K) \rangle_k | E_h |} E_h \tag{8.9}$$

where $K = 2\sigma_2^{-3/2} \sigma_3 | E_h | | E_k | | E_{h-k} |$ and where r'' is a condition weaker than r' in that any $\mathbf{k}$ may be omitted for which $| E_k |$ or $| E_{h-k} |$ is small. Here I_0 and I_1 denote modified Bessel functions of the first kind.

Finally, Hauptman (1971) gave

$$\langle E_k E_{h-k} \rangle_k \stackrel{(r'',e)}{\approx} \left[\left\langle \frac{I_1(K) | E_k | | | E_{h-k} |}{I_0(K)} \right\rangle_k \Big/ | E_h | \right] E_h \tag{8.10}$$

TABLE 8.1. Accuracy of Equations (8.1), (8.7), and (8.10) in a Typical Case [a]

Resolution (Å)	Equation (8.1)		Equation (8.7)		Equation (8.10)		Equation (8.10) restricted	
	$\langle\Delta\phi\rangle$ (deg)	R	$\langle\Delta\phi\rangle$ (deg)	R	$\langle\Delta\phi\rangle$ (deg)	R	$\langle\Delta\phi\rangle$ (deg)	R
0.77	1.0	0.008	3.6	0.039	—	—	3.4	0.038*
1.0	2.4	0.025	5.4	0.075	—	—	4.5	0.055*
1.2	4.6	0.054	7.9	0.096	7.9	0.090*	6.2	0.067*
1.4	7.7	0.106	11.4	0.171	11.4	0.141*	9.0	0.082*
1.6	12.5	0.106*	13.3	0.144*	13.3	0.144*	11.9	0.104*
1.8	13.6	0.149*	16.2	0.184*	16.2	0.171*	8.8	0.124*
2.0	10.8	0.144*	12.3	0.177*	12.3	0.146*	6.9	0.096*
2.5	14.8	0.144*	14.9	0.157*	14.9	0.145*	8.1	0.064*
3.0	17.5	0.184*	18.0	0.210*	18.0	0.182*	8.6	0.072*
4.0	17.1	0.169*	17.4	0.180*	17.4	0.127*	5.4	0.048*
5.0	13.0	0.200*	13.7	0.222*	13.7	0.127*	2.2	0.046*

[a] Error-free calculated data were used throughout. The minimum separation of atoms in the structures used was 1.3 Å. The asterisk denotes rescaling prior to calculating the R factor. In the last column, $\mathbf{h}$, $\mathbf{k}$, and $\mathbf{h} - \mathbf{k}$ were restricted to the 50% strongest E values.

where $K = 2N^{-1/2} |E_{\mathbf{h}}| |E_{\mathbf{k}}| |E_{\mathbf{h-k}}|$. Here the condition on equality is reinserted, and the accuracy of the relation is also increased.

Table 8.1 gives an indication of the accuracies achieved by the above systems. A comparison of the forms of (8.1) and (8.7) shows that (8.1) must always hold at least as well as (8.7), with free selection of f. Thus (8.1) matches (8.7) generally and surpasses it when the condition r (atomic resolution) holds, as indicated by inspection of the upper left-hand corner of Table 8.1. Equation (8.10) also matches (8.7) generally, but its real value is seen in the lower right-hand corner of Table 8.1, where it achieves excellent accuracy on the basis of strong, low-resolution structure factors only.

8.3. Outlines of the Phasing Methods Based on the Convolutional Equation Systems

The phasing methods to be described in this section consist of a solution technique applied either to one of the equation systems given in the preceding section or to a related system. In the first two methods the equation system is first converted into a fixed-point problem, to which a simple and inexpensive, but possibly unstable, solution procedure is applied. In the third method the system first undergoes a smaller change to a minimization problem, and the minimization is carried out by a comparatively complex and expensive but stable procedure. In the first method, and probably in the second, a loss of information occurs in the initial conversion to fixed-point form.

A fixed-point problem has the form $h(y) = y$, a solution y being called a fixed point of h. The problem may be posed in any number of dimensions. The simple iterative procedure

$$h(y^i) = y^{i+1} \tag{8.11}$$

in which h itself is the operator which produces the $(i + 1)$st approximation from the ith, frequently but not always discovers a fixed point of h: iteration (8.11) is called the method of successive approximations. Consider the one-dimensional example shown in Fig. 8.1. Then (8.11) converges to the fixed point if $h(y)$ is like curve 1 in Fig. 8.1, connecting regions a and c, but diverges, no matter how close to the solution it is started, if $h(y)$ is like curve 2, connecting regions b and d. Thus as simple a problem as $2y = y$ will diverge from its solution, $y = 0$, under (8.11). More generally, (8.11) will converge if h is a contraction mapping (Luenberger, 1969). The iteration

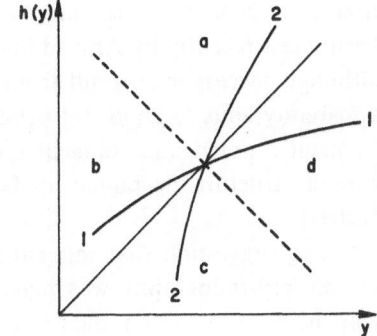

FIGURE 8.1. The fixed-point problem $h(y)$ = y in one dimension. It is safe to use the iteration (8.11) as a solution technique if $h(y)$ is known to resemble curve 1 in the neighborhood of the solution, but not if it may resemble curve 2.

(8.11) requires special properties of h because it asks that h serve as a corrector function for y as well as the function defining y.

8.3.1. Tangent-Formula Refinement

Since the coefficients of E_h on the right-hand sides of (8.7)–(8.10) are in each case real, the angular portions of these equations may be summarized thus:

$$\text{angle}(\langle E_k E_{h-k} \rangle_k) \stackrel{(r'',e)}{\approx} \phi_h \qquad (8.12)$$

or

$$\text{angle}(\langle E_k E_{h-k} \rangle_k) \stackrel{(r'')}{\sim} \phi_h \qquad (8.13)$$

These formulas, which are in fixed-point form, are usually not distinguished and are called the tangent formula. It should be noted that (8.12) or (8.13) present only half as many phase-determining relations as (8.7)–(8.10), because of the loss of the magnitude portions of the latter equations. Thus information has been lost at this point. It was suggested by Karle and Hauptman (1956) that phase extension and refinement might be carried out by solution of the tangent formula. The suggestion was taken up in the mid-1960's and is now applied very frequently in the analysis of small and medium-size structures. The solution technique is based upon the following iteration, which is of the type of (8.11):

$$\text{angle}(\langle E_k^i E_{h-k}^i \rangle_k) = \phi_h^{i+1} \qquad (8.14)$$

where the E^i are the structure factors phased with the output of the ith cycle, and the ϕ^{i+1} are the output of the $(i+1)$st cycle. Usually (8.14) is supplemented by additional rules which govern the acceptance or rejection of its results. A summary of some of the typical procedures used has been given recently by Ahmed and Hall (1975). Results are generally good, although success may result from one starting set of phases while another, apparently equally good, set produces failure (Lessinger, 1976). It is fairly frequently noted that tangent-formula refinement refines away from the correct structure or builds up false detail [see discussion following Hall (1969)].

The suggestion that tangent-formula refinement might be of value in protein crystallography was made by Coulter (1965). Among the authors who have reported on their experiences with the method are Weinzierl, Eisenberg, and Dickerson (1969) on protein models and cytochrome c; Reeke and Lipscomb (1969) on carboxypeptidase; Coulter and Dewar (1971) on myoglobin; Destro (1972) on insulin; Hendrickson, Love, and Karle (1973) on lamprey hemoglobin; Hendrickson (1973) on Glycera hemoglobin, carp muscle Ca-binding protein, and concanavalin A; and Hendrickson and Karle (1973) and Hendrickson (1975) on carp muscle Ca-binding protein. In evaluating the results of tangent-formula refinement on proteins, Hendrickson has drawn a distinction between improvement in map interpretability and improvements in average phase error. There has usually been a small improvement in map interpretability (Hendrickson, 1973), but improvement in the actual value of the average phase error was small or nonexistent in the one case in which it could be checked (Hendrickson, 1975). The tendency to develop false detail when using this method in protein structure determination has been noted by Hendrickson (1975).

8.3.2. Density Modification Methods

These methods (Hoppe and Gassmann, 1968; Hoppe, Gassmann and Zechmeister, 1969) consist in applying the following iteration, which is of the type of (8.11):

$$\text{angle } [G^{-1}(m(\varrho^i))] = \phi^{i+1} \tag{8.15}$$

where ϱ^i is the density function computed with magnitudes $|F_o|$ and phases ϕ^i, m is a modification function, and G is the Fourier transformation. In words, $|F_o|$ and ϕ^i are Fourier transformed to give ϱ^i, ϱ^i is modified to $m(\varrho^i)$, and $m(\varrho^i)$ is back-transformed to give ϕ^{i+1}. The iteration (8.15) may be regarded as a means of solving the fixed-point problem $m(\varrho) = \varrho$, with the incidental condition that $|G^{-1}(\varrho)| = |F|$, although this interpretation is optional. The inclusion of the incidental condition is reflected in the dropping of $|G^{-1}(m(\varrho))|$ in favor of $|F_o|$ in each cycle. The choice of the modification function m has varied somewhat in different versions of the method but at present is most commonly chosen to be of the form $a\varrho + b\varrho^2 + c\varrho^3$.

Historically, the method originated as a noniterative technique, $i = 0$ only, with $m(\varrho)$ defined as ϱ^2 (Hoppe, 1962). The technique was originally envisaged as a phase extension method for protein crystallography; given that m is nonlinear, (8.15) will generate phases outside the starting set. The method was tried on sperm whale myoglobin by Hoppe and Gassmann (1964), the phase extension being from 2 to 1.5 Å. Since that time the Munich group have utilized the method only very slightly on proteins or in that resolution range (see Gassmann, 1975, 1976), but they use the method in its iterative form very successfully in the resolution range 1.1 Å or better. Several applications to structures containing 17 to 108 nonhydrogen atoms are described by Hoppe *et al.* (1969), Gassmann and Zechmeister (1972), and Gassmann (1975). Gassmann and Zechmeister (1972) state that the method in that range is superior to tangent-formula refinement. The use of a related technique has been described by Simonov (1975, 1976). Zwick *et al.* (1976) describe *ab initio* attempts at structure solution with the method.

In protein crystallography, Kartha (1969) discussed the possible use of density modification methods. Barrett and Zwick (1971) tested the ability of the method to extend phases in myoglobin from 3 Å to 2 Å, with somewhat inconclusive results. Collins, Brice *et al.* (1975) and Collins, Cotton *et al.* (1975) have reported successes in producing easily interpretable 1.5-Å maps of rubredoxin, starting with 1.5-, 2.0-, and 2.5-Å phase sets, and also an improved 2-Å map of staphylococcus nuclease–Ca^{2+}–inhibitor complex, starting with a 2-Å phase set. Podjarny (1976) has found that density modification to eliminate or reduce regions of negative electron density is a useful preliminary step to phase extension by determinant maximization (see Chapter 9).

Gassmann (1976) has argued that with the forms of m currently used there is no loss of information corresponding to the dropping of the magnitude portions of the relations in Section 8.2, but his argument is not convincing. It has, for example, been emphasized by Gassmann and Zechmeister (1972) that with the earlier choice of $m(\varrho)$ as ϱ^2, (8.15) reduces exactly, except for a change from E to F, to (8.14), so that density modification essentially reduces in that case to tangent-formula refinement. For that choice of m, therefore, the loss of information certainly exists, and there is nothing in the argument given on this point that depends upon the form of m. Gassmann argues also that with the current forms of m the iteration (8.15) is more stable than (8.14), and this may be correct.

A consequence of the observation that density modification with one choice of m reduces to tangent-formula refinement is that density modification generally, except possibly for the change from E to F, must be at least as powerful as tangent-formula refinement. It should be noted, however, that this does not mean that density modification is as powerful as the solution of one or another of the full convolutional systems.

8.3.3. Least-Squares Phase Refinement

Methods for the solution of a full convolutional system have been described by Krabbendam and Kroon (1971) and Sayre (1972). Sayre chose (8.1) or (8.2) as the system to be solved and constructed:

$$S(\phi) = \tfrac{1}{2} \sum \mid C_{\mathbf{h}} F_{\mathbf{h}} + \sum F_{\mathbf{k}} F_{\mathbf{h-k}} \mid^2 \qquad (8.16)$$

where[†] $C_{\mathbf{h}} = -V f_{\mathbf{h}}^{sa}/f_{\mathbf{h}}$ as a measure of the degree to which a phase-set ϕ fails to satisfy (8.1). $S(\phi)$ was then minimized with respect to ϕ. The method is entirely analogous to ordinary least-squares structure refinement, with phases ϕ instead of structure parameters x and B and with (8.1) instead of the structure-factor equation. The minimization process was found to be smooth and reliable in tests on 100-atom model structures. The technique by which the minimization was carried out will be discussed in Section 8.5.

The method has subsequently been tested further on one small protein, rubredoxin, by Sayre (1974), and used on another, insulin, by Cutfield et al. (1975). Input data consisted of 1.5-Å experimental $\mid F \mid$ magnitudes in both cases, with a starting phase set consisting of 2.5-Å heavy-atom phases in the rubredoxin case and 1.9-Å heavy-atom phases in the insulin case. In both cases the phase set was first extended to 1.5 Å by use of the angular portion of (8.1), a squaring-method analog of (8.12) and (8.13). It was followed by the minimization process, which was again observed to be smooth and steady. In both cases the 1.5-Å map phased with the output of the refinement was easy to interpret in most regions. In the insulin case the map was used to provide starting coordinates for structure refinement. By comparison with the phases resulting from extensive structure refinement, Agarwal and Isaacs (private communication) have found that the original 1.9-Å heavy-atom phases had a mean error of 60.0°. The phase extension and refinement technique reduced the mean error of this group to 52.8° and supplied new phases to 1.5 Å with a mean error of 60.5°.

[†] See note on C's following (8.6).

The method of Krabbendam and Kroon is identical in concept but differs in technique, with the minimization carried out by trying alternative phase values one reflection at a time. With small structures the method is satisfactory, as was shown by the authors, but it is probably not efficient enough for large structures. Hoppe (1963) also suggested solving a convolutional system by least squares, but included only the magnitude portion of the system in the function to be minimized.

8.3.4. Davies–Rollett Technique

Davies and Rollett (1976) have described a technique that resembles somewhat that of Sayre (1972) except that the requirement that the F that appears on the left-hand side of (8.1) be the same as that on the right-hand side is dropped. This technique corresponds to requiring that ϱ have a real square root, that is, that ϱ be positive; the requirement that ϱ in addition resemble its square root is, however, dropped. In its potential scope the method should therefore resemble somewhat the determinant maximization methods. The authors express this as follows:

...there is no reason why it should not yield an accurate fit to a density function with peaks of varying heights. This makes our method attractive for application to protein structures with heavy atoms, variations of vibrational amplitude from one atom to the next, and regions of disordered solvent. So far as our present work has gone, the less restrictive condition of non-negativity (rather than the resemblance to the square) has been adequate to define the electron density.

The successful application of the method to a model 16-atom structure in $P2_12_12_1$ is described; the program is being extended to cope with protein structure applications.

8.4. Comments on the Phasing Methods Based on the Convolutional Equation Systems

To date the methods of this chapter have achieved only a moderate importance in the actual practice of protein crystallography. Least-squares phase refinement has produced excellent results, but only at the extreme high-resolution end of the field, where examples are few and the number of phases to be determined, relative to the size of structure, large. On the other hand, tangent-formula refinement and density modification have usually not, in proteins, produced results of sufficient quality to make them widely interesting.

Examination of the material presented in the two preceding sections will show, however, that the problem may be removable by better matching between equation system and solution method. Equation (8.10), which may be able to function at both high and low resolutions while allowing a large number of the weaker reflections to be ignored, has never been treated in its full form by a solution method. At the same time, the least-squares solution method develops stably a solution to a full convolutional system but has only been applied to (8.1), which because of its strong resolution assumption, r, is limited to high-resolution cases.

8.5. Computational and Practical Aspects

The methods available for the numerical treatment of convolutional systems are currently in a state of rapid change and improvement. This fact, coupled with the situation that computational costs in this field can be affected profoundly by technique, makes the topic highly important at the present time. In this section we attempt to summarize briefly some of the current developments.

The reader will be familiar already with the fact that with the advent of FFT it became markedly faster, whenever F is a large set, to compute $F * F$ by $G^{-1}([G(F)]^2)$ than by $\sum F_k F_{h-k}$. Thus, for some time, (8.15) with $m(\varrho) = \varrho^2$ has been the efficient computational form for (8.14) in large problems. (See, e.g., Barrett and Zwick, 1971.) In practice computing times of the order of 20 ms/phase per iteration of (8.15) are achieved on large machines.

Still more recent developments are carrying the matter one stage further. Faster convolution techniques and faster FFT techniques are now available, both based on new fast methods of performing circular convolution (Silverman, 1977; Kolba and Parks, 1977). It is unclear as yet whether the estimate $n \log_2 n$ for the computing time of Cooley–Tukey FFT is an overestimate for the newer algorithms.

In another development, a much-improved numerical minimization technique for the expression (8.16) has been proposed by Toupin in collaboration with the author (Sayre and Toupin, 1975; Sayre, 1975). Generally speaking, efficient techniques for the numerical minimization of a functional S involve the computation of the gradient of S. In a case like that of (8.16), where S is of the form $S = (1/2) \sum |\mathbf{r}_i|^2$, grad $S = \mathbf{J}^T\mathbf{r}$, where $J_{jk} = \partial r_j/\partial \phi_k$ and $\mathbf{J}^T$ stands for the transpose of $\mathbf{J}$. Thanks to the special form of

r_i in (8.16), it was not difficult for Toupin to show that in this case

$$\text{grad } S = \mathbf{J}^T \mathbf{r} = -2 \text{ Im}\{\mathbf{F} \times [\mathbf{C} \times \mathbf{r}^* + 2(\mathbf{F} * \mathbf{r})^*]\} \qquad (8.17)$$

where Im is the imaginary part, $\times$ stands for element-by-element multiplication of two vectors (Hadamard product), and superscript * is the complex conjugate. (8.17) is valid in space-group $P1$. Thus the computation of grad S is reduced effectively to the carrying out of two convolutions, $\mathbf{F} * \mathbf{F}$ to get $\mathbf{r}$ and $\mathbf{F} * \mathbf{r}$ to get grad S. In the computations on rubredoxin (Sayre, 1974) and insulin (Cutfield et al., 1975), grad S was calculated according to the formula obtained by straightforward differentiation of (8.16), a very much slower procedure. When grad S has been obtained, in some methods the direction of $-$grad S is adopted as the direction of the step to be taken in solution space, and the length of the step is determined by calculating S at one or more points along that direction and estimating from this information the step size that will minimize S. For each such point S costs one convolution. Alternatively, in other methods which are most commonly used when $\mathbf{J}^T\mathbf{J}$ is far from the identity matrix, the step $\mathbf{s}$ is determined by solving the system of normal equations $\mathbf{J}^T\mathbf{J}\mathbf{s} = -\mathbf{J}^T\mathbf{r}$. One efficient way of solving the normal equations in large cases is by conjugate gradients, a technique in which it is necessary to be able to compute $\mathbf{J}^T\mathbf{J}\mathbf{p}$ given an arbitrary vector $\mathbf{p}$. In the rubredoxin and insulin work, this was done by the straightforward technique of storing $\mathbf{J}$ during the computation of $\mathbf{J}^T\mathbf{r}$; $\mathbf{J}$ is an exceedingly large matrix, the storage and repeated recovery of which were costly and slow. Here Toupin was able to show that in space group $P1$

$$\mathbf{J}^T\mathbf{J}\mathbf{p} = 2\mathbf{F}^* \times [\mathbf{C} \times \mathbf{Y} + 2(\mathbf{F} * \mathbf{Y})] \qquad (8.18)$$

where $\mathbf{Y} = \mathbf{C} \times \mathbf{Z} + 2(\mathbf{F} * \mathbf{Z})$, $\mathbf{Z} = 2(\mathbf{F} \times \mathbf{p})$. Thus the computation of $\mathbf{J}^T\mathbf{J}\mathbf{p}$ for arbitrary $\mathbf{p}$ is also reduced to two convolutions, $\mathbf{F} * \mathbf{Z}$ to get $\mathbf{Y}$ and $\mathbf{F} * \mathbf{Y}$ to get $\mathbf{J}^T\mathbf{J}\mathbf{p}$, and no storage of $\mathbf{J}$ is required. Using these techniques it is estimated that computing times of the order of 100 ms/phase per iteration of the minimization will be realizable on large machines. In rubredoxin and insulin, computing times were of the order of 1000 ms/phase per iteration. An experimental least-squares refinement program which uses (8.17) and (8.18) has been run successfully. It appears that (8.17) and (8.18) can be generalized to all space groups, although for space groups that have structure factors with restricted phases, some special treatment of those discrete phase variables will be necessary. In the rubredoxin and insulin refinements, each of those variables was replaced by a continuous variable which was interpreted as a multiplier of the structure factor with its phase set to a standard value. According as the variable refined to $+1$ or -1 the phase was regarded as having refined to the standard value plus the value $0°$ or π, respectively.

8.6. Summary

Solution of (8.10) by a least-squares technique is suggested as a promising method of phase refinement. To date (8.10) has been converted to (8.12) and (8.13), resulting in serious loss of information, and solved by an iterative procedure of type (8.11), resulting sometimes in loss of stability (tangent-formula refinement); or iteration (8.15) has been employed, with a similar loss of information but perhaps greater stability (density modification); or (8.10) has been replaced by (8.1), restricting the technique to high-resolution cases and increasing the cost (the author's least-squares phase refinement technique). A computationally efficient least-squares technique is suggested (8.17, 8.18), which effectively reduces the computation to the carrying out of convolutions, and attention is called to the new superfast algorithms for convolution.

References[†]

Agarwal, R. C., and Isaacs, N. W. (1976). Private communication.

Ahmed, F. R., and Hall, S. R. (1975). In *Crystallographic Computing Techniques*, Ed. F. R. Ahmed, Munksgaard, Copenhagen, pp. 71–84.

Barrett, A. N., and Zwick, M. (1971). *Acta Crystallogr. Sect. A* **27**, 6.

Carpenter, G. B. (1969). *Principles of Crystal Structure Determination*, W. A. Benjamin, New York.

Collins, D. M., Brice, M. D., la Cour, T. F. M., and Legg, M. J. (1975). In *Crystallographic Computing Techniques*, Ed. F. R. Ahmed, Munksgaard, Copenhagen, pp. 330–335.

Collins, D. M., Cotton, F. A., Hazen, E. E., Jr., Meyer, E. F., Jr., and Morimoto, C. N. (1975). *Science* **190**, 1047.

Coulter, C. L. (1965). *J. Mol. Biol.* **12**, 292.

Coulter, C. L., and Dewar, R. B. K. (1971). *Acta Crystallogr. Sect. B* **27**, 1730.

Cutfield, J. F., Dodson, E. J., Dodson, G. G., Hodgkin, D. C., Isaacs, N. W., Sakabe, K., and Sakabe, N. (1975). *Acta Crystallogr. Sect. A* **31**, S21.

Davies, A. R., and Rollett, J. S. (1976). *Acta Crystallogr. Sect. A* **32**, 17.

Destro, R. (1972). Report of C.E.C.A.M. workshop, Orsay, France.

[†] Compiled October, 1976.

Gassmann, J. (1975). In *Crystallographic Computing Techniques*, Ed. F. R. Ahmed, Munksgaard, Copenhagen, pp. 144–154.

Gassmann, J. (1976). *Acta Crystallogr. Sect. A* **32**, 274.

Gassmann, J., and Zechmeister, K. (1972). *Acta Crystallogr. Sect. A* **28**, 270.

Hall, S. R. (1969). In *Crystallographic Computing*, Ed. F. R. Ahmed, Munksgaard, Copenhagen, pp. 67–70.

Hauptman, H. (1971). Abstract H3, American Crystallographic Association Winter Meeting, Columbia, South Carolina.

Hendrickson, W. A. (1973). *Trans. Am. Crystallogr. Assoc.* **9**, 61.

Hendrickson, W. A. (1975), *J. Mol. Biol.* **91**, 226.

Hendrickson, W. A., and Karle, J. (1973). *J. Biol. Chem.* **248**, 3327.

Hendrickson, W. A., Love, W. E., and Karle, J. (1973). *J. Mol. Biol.* **74**, 331.

Hoppe, W. (1962). *Acta Crystallogr.* **15**, 13.

Hoppe, W. (1963), *Z. Kristallogr.* **118**, 121.

Hoppe, W., and Gassmann, J. (1964). *Ber. Bunsenges. Phys. Chem.* **68**, 808.

Hoppe, W., and Gassmann, J. (1968). *Acta Crystallogr. Sect. B* **24**, 97.

Hoppe, W., Gassmann, J., and Zechmeister, K. (1969). In *Crystallographic Computing*, Ed. F. R. Ahmed, Munksgaard, Copenhagen, pp. 26–36.

Hughes, E. W. (1953). *Acta Crystallogr.* **6**, 871.

Karle, J. (1964). *Advances in Structure Research by Diffraction Methods*, Interscience, New York, pp. 55–89.

Karle, J. (1969). *Advances in Chemical Physics*, Interscience, New York, Vol. XVI, pp. 131–222.

Karle, J., and Hauptman, H. (1950). *Acta Crystallogr.* **3**, 181.

Karle, J., and Hauptman, H. (1956). *Acta Crystallogr.* **9**, 635.

Karle, J., and Karle, I. L. (1966). *Acta Crystallogr.* **21**, 849.

Kartha, G. (1969). *Acta Crystallogr. Sect. A* **25**, S87.

Kolba, D. P., and Parks, T. W. (1977). *IEEE Trans. Acoust. Speech Signal Process.* **25**, 281.

Krabbendam, H., and Kroon, J. (1971). *Acta Crystallogr. Sect. A* **27**, 48.

Lessinger, L. (1976). *Acta Crystallogr. Sect. A* **32**, 538.

Luenberger, D. G. (1969). *Optimization by Vector Space Methods*, Wiley, New York.

Podjarny, A. (1976). Talk given at International School for Crystallography, Erice, Sicily.

Reeke, G. N., Jr., and Lipscomb, W. N. (1969). *Acta Crystallogr. Sect. B* **25**, 2614.

Rothbauer, R. (1976). *Acta Crystallogr. Sect. A* **32**, 169.

Sayre, D. (1952). *Acta Crystallogr.* **5**, 60.

Sayre, D. (1972). *Acta Crystallogr. Sect. A* **28**, 210.

Sayre, D. (1974). *Acta Crystallogr. Sect. A* **30**, 180.

Sayre, D. (1975). In *Crystallographic Computing Techniques*, Ed. F. R. Ahmed, Munksgaard, Copenhagen, pp. 322–327.

Sayre, D., and Toupin, R. (1975). *Acta Crystallogr. Sect. A* **31**, S20.

Simonov, V. I. (1975). In *Crystallographic Computing Techniques*, Ed. F. R. Ahmed, Munksgaard, Copenhagen, pp. 138–143.

Simonov, V. I. (1976). Abstract SC1, American Crystallographic Association Summer Symposium, Buffalo, New York.

Silverman, H. F. (1977). *IEEE Trans. Acoust. Speech Signal Process.* **25**, 152.

Slepian, D., and Pollack, H. O. (1961). *Bell System Tech. J.* **40**, 43.

Stout, G. H., and Jensen, L. H. (1968). *X-Ray Structure Determination*, MacMillan, New York.

Tsoucaris, G. (1976). Private communication.

Weinzierl, J. E., Eisenberg, D., and Dickerson, R. E. (1969). *Acta Crystallogr. Sect. B* **25**, 380.

White, P. S., and Woolfson, M. M. (1975). *Acta Crystallogr. Sect. A* **31**, 53.

Zwick, M., Bantz, D., and Hughes, J. (1976). *Ultramicroscopy* **1**, 275.

Maximum Determinant Method

G. TSOUCARIS

9.1. Introduction

Inequalities (Harker and Kasper, 1948) and nonnegative determinants (Karle and Hauptman, 1950) are historically the first "direct methods" for phase determination from the moduli and the *a priori* known information on the electron-density function $\varrho(\mathbf{r})$. Although Harker and Kasper have convincingly shown that inequalities are efficient for the determination of small structures ($N \sim 10$), the simple low-order inequalities are not useful for more complex structures. In order to make practical use of the determinants (or the associated matrices), new work has been performed recently in the following directions: More "refined" properties of the Karle–Hauptman determinants and matrices (Sections 9.2 and 9.3)—in particular, the introduction of statistical considerations provides more restrictive relations than inequalities; construction of very-high-order determinants (for example, one of order 400 has been used for insulin; Section 9.4); new determinants, that make better use of the *a priori* information, or the introduction of new information (for example stereochemistry, Section 9.5).

Practical applications are described in Sections 9.4 and 9.5. Three problems are examined: phase invariant calculations (small-order determinants), *ab initio* determination of phases (medium-order), and phase refinement and extension in proteins (high-order).

G. TSOUCARIS • Laboratoire de Physique, Université de Paris—Sud, Centre Pharmaceutique, 92290 Chatenay-Malabry, France.

9.2. Inequalities and Algebraic Properties of Determinants

9.2.1. Unitary and Normalized Structure Factors

The power of inequalities, and other direct methods, is strongly enhanced by using not merely the "observed" structure factors, but rather the unitary or the normalized structure factors, which correspond to an idealized point-atom structure. Thus, before giving the essential results of the theory, we point out the hypothesis underlying the mathematical expressions involving these point-atom structure factors.

9.2.1.1. Positivity of the Diffracting Density

The electron density is everywhere positive, that is, $\varrho(\mathbf{r}) \geq 0$. Thus, classical theory does not apply to neutron diffraction data, for which the atomic scattering factor can be negative. However, it will be seen in Section 9.5 that it is possible to obtain mathematical expressions involving, for the same structure, both X-ray and neutron scattering factors.

9.2.1.2. Atomic Hypothesis

The general expression of the structure factor for any diffracting matter $\varrho(\mathbf{r})$ is

$$F_{\mathbf{H}} = \int_{V_{\mathbf{r}}} \varrho(\mathbf{r}) \exp(2\pi i \mathbf{H} \cdot \mathbf{r}) \, dV_{\mathbf{r}}$$

This expression can be further simplified if we introduce the following atomic hypothesis:

The electronic clouds belonging to different atoms may be distinguished and described completely by a separate function for each of them; we neglect the particular form of each electron orbital. The electron distribution of each atom does not depend on the thermal motion of other atoms: in particular, we assume that there are no phase relations between the vibrations of different atoms.

By introducing this hypothesis, we thus obtain the familiar form of the structure factor:

$$F_{\mathbf{H}} = \sum_{j=1}^{N} f_j \exp(2\pi i \mathbf{H} \cdot \mathbf{r}_j) \tag{9.1}$$

9.2.1.3. Point-Atom Model

Equation (9.1) can be simplified further under two additional assumptions.

(a) We consider that all atoms at rest have the same shape factor, i.e., the electron density distribution may be expressed as the product of the atomic number Z_j by a unitary atomic distribution that is common to all atoms.

(b) The thermal motion is considered to be isotropic and identical for all atoms in the unit cell. Hence,

$$f_j(\mathbf{H}) = Z_j f$$

where f is the common form factor. Therefore,

$$F_{\mathbf{H}} = f \sum_{j=1}^{N} Z_j \exp(2\pi i \mathbf{H} \cdot \mathbf{r}_j)$$

Introducing the total number of electrons in the unit cell,

$$Z = \sum_{j=1}^{N} Z_j$$

we obtain

$$\frac{F_{\mathbf{H}}}{f} = Z \sum_{j=1}^{N} \frac{Z_j}{Z} \exp(2\pi i \mathbf{H} \cdot \mathbf{r}_j)$$

By setting $Z_j/Z = n_j$, we obtain the unitary structure factor $U_{\mathbf{H}}$:

$$U_{\mathbf{H}} = \frac{F_{\mathbf{H}}}{Zf} = \sum_{j=1}^{N} n_j \exp(2\pi i \mathbf{H} \cdot \mathbf{r}_j) \tag{9.2}$$

where

$$\sum_{j=1}^{N} n_j = 1 \tag{9.3}$$

We now define the normalized structure factor:

$$E_{\mathbf{H}} = F_{\mathbf{H}} / (\sum Z_j^2 f^2)^{1/2} = \sum_{j=1}^{N} g_j \exp(2\pi i \mathbf{H} \cdot \mathbf{r}_j) \tag{9.4}$$

where

$$g_j = f_j / (\sum Z_j^2 f^2)^{1/2} \tag{9.5}$$

and

$$\langle | E_{\mathbf{H}} |^2 \rangle_{\mathbf{H}} = 1$$

For equal atom structures, we have

$$E_\mathbf{H} = N^{1/2}U_\mathbf{H}$$

It is clear that $U_\mathbf{H}$ is the structure factor where the electrons of each atom are concentrated at its center. Such a hypothesis is absurd for an electron density distribution. The important point to emphasize is that the mathematical relations used in direct methods are more powerful if we use U rather than F. The above treatment shows that we can obtain U from F by a reasonable, yet approximate, calculation, but we need to know the number and nature of all atoms in the unit cell.

9.2.2. Karle–Hauptman Determinants (1950)

9.2.2.1. General Scope

Let us choose arbitrarily m reciprocal-lattice vectors,

$$\mathbf{H}_1 \cdots \mathbf{H}_p \cdots \mathbf{H}_m$$

To the differences

$$\mathbf{H}_p - \mathbf{H}_q, \qquad p, q = 1, \ldots, m$$

there correspond m^2 unitary factors

$$U_{\mathbf{H}_p - \mathbf{H}_q} = U_{pq}$$

forming a square matrix of order m denoted by U_m. The determinant D_m of this matrix has the property of being nonnegative (Table 9.1, Fig. 9.1):

$$D_m = \det(U_m) = \det(U_{\mathbf{H}_p - \mathbf{H}_q}) \geq 0 \qquad (9.6)$$

Moreover, it has been shown that the positivity of the whole set of determinants (9.6) is a necessary and sufficient condition to ensure that $\varrho(\mathbf{r}) \geq 0$.

In practice, in order to use the inequalities (9.6), we calculate the value of a D_m determinant for all combinations of phases.

According to (9.6) all combinations leading to a negative value of D_m are forbidden and must be rejected as false. In other words, the correct combination is one of those that lead to a positive value of D_m.

TABLE 9.1. Karle–Hauptman Determinant

$$D_m = \begin{vmatrix} 1 & U_{12} & \cdots & U_{1p} & U_{1q} & \cdots & U_{1m} \\ U_{21} & 1 & \cdots & U_{2p} & U_{2q} & \cdots & U_{2m} \\ \cdot & \cdot & & \cdot & \cdot & & \cdot \\ \cdot & \cdot & & \cdot & \cdot & & \cdot \\ \cdot & \cdot & & \cdot & \cdot & & \cdot \\ U_{p1} & U_{p2} & \cdots & 1 & U_{pq} & \cdots & U_{pm} \\ U_{q1} & U_{q2} & \cdots & U_{qp} & 1 & \cdots & U_{qm} \\ \cdot & \cdot & & \cdot & \cdot & & \cdot \\ \cdot & \cdot & & \cdot & \cdot & & \cdot \\ \cdot & \cdot & & \cdot & \cdot & & \cdot \\ U_{m1} & U_{m2} & \cdots & U_{mp} & U_{mq} & \cdots & 1 \end{vmatrix}$$

9.2.2.2. The D_3 Example

First, we change the above general notation into a familiar one: among the vectors $\mathbf{H}_p$, we can arbitrarily set one vector to zero:

$$\mathbf{H}_1 = 0$$

$$\mathbf{H}_2 - \mathbf{H}_1 = \mathbf{H}_2 = \mathbf{H}$$

$$\mathbf{H}_3 - \mathbf{H}_1 = \mathbf{H}_3 = \mathbf{K}$$

Then,

$$0 \le D_3 = \begin{vmatrix} 1 & U_{\mathbf{H}_1-\mathbf{H}_2} & U_{\mathbf{H}_1-\mathbf{H}_3} \\ U_{\mathbf{H}_2-\mathbf{H}_1} & 1 & U_{\mathbf{H}_2-\mathbf{H}_3} \\ U_{\mathbf{H}_3-\mathbf{H}_1} & U_{\mathbf{H}_3-\mathbf{H}_2} & 1 \end{vmatrix} = \begin{vmatrix} 1 & U_{-\mathbf{H}} & U_{-\mathbf{K}} \\ U_{\mathbf{H}} & 1 & U_{\mathbf{H}-\mathbf{K}} \\ U_{\mathbf{K}} & U_{\mathbf{K}-\mathbf{H}} & 1 \end{vmatrix} \quad (9.7)$$

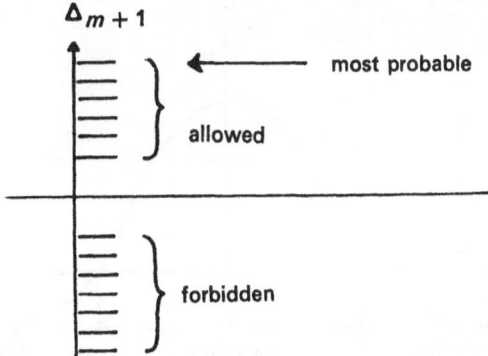

FIGURE 9.1. Diagram showing the possible and most probable solutions as a function of Δ_{m+1}.

The familiar calculation scheme

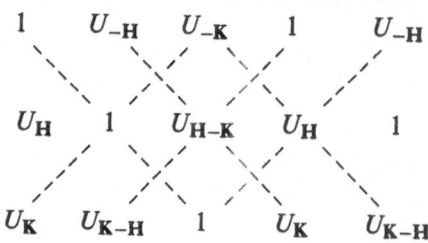

leads to

$$D_3 = 1 - |U_K|^2 - |U_{H-K}|^2 - |U_H|^2 + U_{-H}U_{H-K}U_K + U_{-K}U_HU_{K-H}$$

or

$$D_3 = 1 - |U_H|^2 - |U_K|^2 - |U_{K-H}|^2 + 2|U_HU_{-K}U_{K-H}| \cos \alpha \geq 0 \tag{9.8}$$

The phase invariant α is given by

$$\alpha = \Phi_H + \Phi_{-K} + \Phi_{K-H}$$

where Φ_H is the phase associated with U_H. For centrosymmetric structures, it becomes

$$D_3 = 1 - U_H^2 - U_K^2 - U_{K-H}^2 + 2U_HU_KU_{K-H} \tag{9.9}$$

If the moduli are larger than a certain limit, inequality (9.7) puts a restriction on the allowed values of α, as shown in Fig. 9.2.

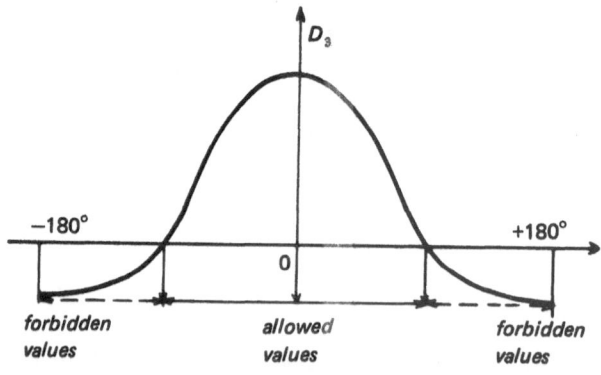

FIGURE 9.2. Variation of a D_3 determinant as a function of the phase invariant α.

The fact that D_3 depends solely upon a structure invariant and not upon the individual phases of the structure factor is a general property of these determinants, as stated in Section 9.2.3.

9.2.2.3. Need for Probability Theory and the Maximum Determinant Rule

In practice, the criterion of positivity of Δ_{m+1} is not sufficiently restrictive: generally, too large a number of allowed sets remains to make such a method alone sufficient to solve the structure directly, except for a simple structure.

However, the probability theories lead to a new result which further restricts the allowed range of phases: *Among all sets of phases that are compatible with inequalities, the most probable set is that which leads to a maximum value of the determinant.* Among the attempts to extract phase information from inequalities alone we mention those of Kitaigorodski (1961) and Diamond (1963).

9.2.3. Determinants as a Function of Structure Invariants

A fundamental property of determinants is that *the value of D_m depends solely upon structure invariants.* This means that D_m does not depend on the choice of the origin in direct space, since this choice affects only the phase of individual structure factors. Thus, the determinant D_m itself is also a structure invariant.

Let us consider a structure factor of the first row, say $U_{\mathbf{H}_1-\mathbf{H}_p} = U_{1p}$, and another one of the first column, say $U_{\mathbf{H}_q-\mathbf{H}_1} = U_{q1}$. The triplet structure invariant, or structure invariant, corresponding to these two structure factors is

$$U_{1p}U_{pq}U_{q1} = U_{p1}^{*}U_{pq}U_{1q}^{*} \tag{9.10}$$

where

$$(\mathbf{H}_1 - \mathbf{H}_p) + (\mathbf{H}_p - \mathbf{H}_q) + (\mathbf{H}_q - \mathbf{H}_1) = 0$$

It is easy to locate the third structure factor U_{pq} within the determinant: it lies on the intersection of the pth row (that of $U_{p1} = U_{1p}^{*}$) and the qth column (that of $U_{1q} = U_{q1}^{*}$), as seen in Table 9.2a.

Let us now prove the main statement: If all elements of the pth row are multiplied by $C_p = \exp(i\Phi_{1p})$, and all elements of the pth column by $1/C_p = \exp(i\Phi_{p1})$, then the value of D_m remains unchanged. By repeating

TABLE 9.2. Phase Invariants [a]

(a)

$$\begin{pmatrix}
1 & \cdots & \boxed{U_{1p}} & \cdots & \underline{U_{1q}} & \cdots & U_{1m} \\
\vdots & & & & & & \vdots \\
\underline{U_{p1}} & \cdots & 1 & \cdots & \boxed{U_{pq}} & \cdots & U_{pm} \\
\boxed{U_{q1}} & \cdots & U_{qp} & \cdots & 1 & \cdots & U_{qm} \\
\vdots & & & & & & \vdots \\
U_{m1} & \cdots & U_{mp} & \cdots & U_{mq} & \cdots & 1
\end{pmatrix}$$

(b)

$$\begin{pmatrix}
1 & \cdots & |U_{1p}| & \cdots & |U_{1q}| & \cdots & |U_{1m}| \\
|U_{p1}| & \cdots & 1 & \cdots & |U_{pq}|\exp(i\alpha_{pq}) & \cdots & \\
|U_{q1}| & \cdots & |U_{qp}|\exp(-i\alpha_{pq}) & \cdots & 1 & & \\
\vdots & & & & & & \\
|U_{m1}| & \cdots & & & & & 1
\end{pmatrix}$$

(c)

$$\begin{pmatrix}
1 & \cdots & U_{1r} & \cdots & U_{1p} & \cdots & U_{1q} & \cdots & U_{1m} \\
U_{r1} & \cdots & 1 & \cdots & \boxed{U_{rp}} & \cdots & \underline{U_{rq}} & \cdots & U_{rm} \\
U_{p1} & \cdots & \underline{U_{pr}} & \cdots & 1 & \cdots & \boxed{U_{pq}} & \cdots & U_{pm} \\
U_{q1} & \cdots & \boxed{U_{qr}} & \cdots & U_{qp} & \cdots & 1 & \cdots & U_{qm} \\
U_{m1} & \cdots & U_{mr} & \cdots & U_{mp} & \cdots & U_{mq} & \cdots & 1
\end{pmatrix}$$

(d)

$$\begin{pmatrix}
1 & \cdots & U_{1r} & \cdots & U_{1s} & \cdots & U_{1p} & \cdots & U_{1q} & \cdots & U_{1m} \\
U_{r1} & \cdots & 1 & \cdots & \boxed{U_{rs}} & \cdots & U_{rp} & \cdots & \underline{U_{rq}} & \cdots & U_{rm} \\
U_{s1} & \cdots & & & 1 & \cdots & U_{sp} & & & \cdots & U_{sm} \\
U_{p1} & \cdots & & & \underline{U_{ps}} & \cdots & 1 & \cdots & \boxed{U_{pq}} & \cdots & U_{pm} \\
U_{q1} & \cdots & & & & & 1 & \cdots & 1 & \cdots & U_{qm} \\
U_{m1} & \cdots & U_{mr} & \cdots & U_{ms} & \cdots & U_{mp} & \cdots & U_{mq} & \cdots & 1
\end{pmatrix}$$

[a] The initially chosen structure factors [and in parts (a) and (c) the third structure factor, as given in eqs. (9.10) and (9.12)] are enclosed in solid boxes; conjugates of the initially chosen structure factors are in dashed boxes.

the same operation for all rows and columns ($p = 1, \ldots, m$) the value of D_m is still unchanged, but the general element, which was U_{pq}, becomes that shown in Table 9.2b, since

$$U_{pq} \exp(i\Phi_{1p}) \exp(i\Phi_{q1}) = |U_{pq}| \exp[i(\Phi_{pq} + \Phi_{q1} + \Phi_{1p})]$$
$$= |U_{pq}| \exp(i\alpha_{pq})$$

This element depends solely on $|U_{pq}|$ and on the phase invariant:

$$\alpha_{pq} = \Phi_{1p} + \Phi_{pq} + \Phi_{q1} = \Phi_{H_1-H_p} + \Phi_{H_p-H_q} + \Phi_{H_q-H_1} \qquad (9.11)$$

Therefore, the value of D_m depends upon $(m-1)(m-2)/2$ independent structure invariants.[†] However, D_m involves a much larger number of structure invariants, which, of course, are not independent of each other, as illustrated by Table 9.2c. Indeed, in the above definition of the structure invariants one can replace $\mathbf{H}_1$ by $\mathbf{H}_r$, with $r \neq p \neq q \neq 1$, thus obtaining derived invariants of general form

$$U_{rp}U_{pq}U_{qr} = U_{pr}^* U_{pq} U_{rq}^* \qquad (9.12)$$

In other words, to each pair (U_{rp}, U_{qr}) belonging, respectively, to the rth row and rth column, one can associate through (9.12) the structure factor U_{pq} which lies on the intersection of the pth row (that of $U_{pr} = U_{rp}^*$) and qth column (that of $U_{rq} = U_{qr}^*$), as shown in Table 9.2c. The total number of triplet structure invariants is

$$\underbrace{\frac{(m-1)(m-2)}{2}}_{\text{primary}} + \underbrace{\frac{(m-1)(m-2)}{2} \cdot \frac{(m-3)}{3}}_{\text{derived}} = \underbrace{\frac{m(m-1)(m-2)}{6}}_{\text{total}}$$

For example, a D_{32} contains 465 independent structure invariants for a total number of 4960.

In the same way, one can define quartets by associating four structure factors (or their conjugates, in dashed boxes) located at the corners of a "rectangle," as shown in Table 9.2d:

$$T_{pqrs} = U_{pq}U_{qr}U_{rs}U_{sp} = U_{pq}U_{rq}^*U_{rs}U_{ps}^*$$

Anticipating the results of probability theory (Section 9.2), we emphasize that the elements of the first row can be chosen from among the largest

---

† These invariants are called "primary" or "fundamental" (Kitaigorodski, 1961).

TABLE 9.3. Characteristics of Karle–Hauptman Determinants for Actual Structures

Structures	Centrosymmetric case				Noncentrosymmetric case	
	Theoretical distribution	Hydrate of trigenelline	Jamine	Allenic phosphine oxide	Theoretical distribution	Isoniazide
Space group		$P\bar{1}$	$P\bar{1}$	$P\bar{1}$		$P2_12_12_1$
N		22	48	32		40
m		18	18	17		30
$\lvert E_H \rvert > 3$	0.3%	32%	29%	13%	0.01%	3%
$\lvert E_H \rvert > 2$	5%	60%	69%	37%	1.8%	20%
$\lvert E_H \rvert > 1$	33%	97%	93%	79%	37%	65%

$\lvert E \rvert$ values; then the average value of all elements in D_m is larger than the general average of the structure factors. An indication of the distribution of E's for such determinants (i.e., chosen according to the criterion of "high E's in the first row") is given in Table 9.3.

We conclude by stating that the matrix U_m of the determinant D_m can be used as a very compact table for looking up large structure invariants. This result is independent of the detailed theory to be developed in this chapter. This theory can be stated either in the initial "structure-factor form" of Table 9.1, or in the "structure-invariant form" of Table 9.2b. We will adopt the first or the second form according to the specific goal in each section.

9.2.4. The Ellipsoid Representation of Inequalities, and Efficiency in Phase Determination

9.2.4.1. Ellipsoid Representation of Inequalities

In the course of scanning through all the phases of a determinant, the use of the geometrical picture described here will be both convenient and helpful. Moreover, this geometrical representation provides a link with the study of the eigenvalue spectrum and the probability theory (Section 9.3).

We explore first the D_3 determinant. Simple manipulations lead to an alternative form of the inequality $D_3 \geq 0$, as follows[†]:

$$1 - \frac{D_3}{D_2} = \frac{Q_2}{N} = \frac{|U_K|^2 + |U_{K-H}|^2 - 2|U_H U_{-K} U_{K-H}|\cos\alpha}{1 - |U_H|^2} \leq 1 \tag{9.13}$$

where α is the invariant (9.11) for $m = 3$ with $p = 3$, and $q = 2$; in the familiar notation, $\alpha = \Phi_{-K} + \Phi_{K-H} + \Phi_H$. For centrosymmetric structures, it becomes:

$$\frac{Q_2}{N} = \frac{|U_K|^2 + |U_{K-H}|^2 - 2U_H U_K U_{K-H}}{1 - U_H^2} \leq 1 \tag{9.14}$$

Assume now that, in the centrosymmetric case, U_H is given and fixed (Fig. 9.3), and we are looking for the limitations on the values of U_K and U_{K-H} considered as variables according to the scheme developed below. Before examining these limitations, we point out, in view of generalization to any determinant D_m with $m \geq 3$, that U_H (fixed) is in fact the only structure factor of the principal minor of D_2 (9.7):

$$D_2 = \begin{vmatrix} 1 & U_H^* \\ U_H & 1 \end{vmatrix} = 1 - |U_H|^2$$

Similarly we note that the variables U_K and U_{K-H} are the elements of the last row of D_3. The idea of considering the elements of the D_2 minor as fixed, in contrast with those of the last row of D_3, which are variable, is not only useful in the studies of inequalities, but also is essential in the probability theory of the maximum determinant rule.

By plotting the contour $Q_2(U_K, U_{K-H}) = 1$ on orthogonal axes, it becomes clear that the allowed region is an ellipse (Fig. 9.3, solid line). If we assume that U_H is positive, the major axis is directed at 45° from the coordinate axis U_K; this is not generally true for hyperellipsoids corresponding to $m \geq 3$. Indeed, a 45° rotation of the coordinate axes transforms (9.14) into the familiar equation of an ellipse, where the new coordinate axes x and y coincide with the two axes of the ellipse:

$$\frac{(U_K + U_{K-H})^2}{2(1 + U_H)} + \frac{(U_K - U_{K-H})^2}{2(1 - U_H)} = 1 \tag{9.15}$$

or

$$(x/\lambda_1^{1/2})^2 + (y/\lambda_2^{1/2})^2 = 1 \tag{9.16}$$

[†] N is the number of atoms in the unit cell. The notation will be explained in Section 9.2.

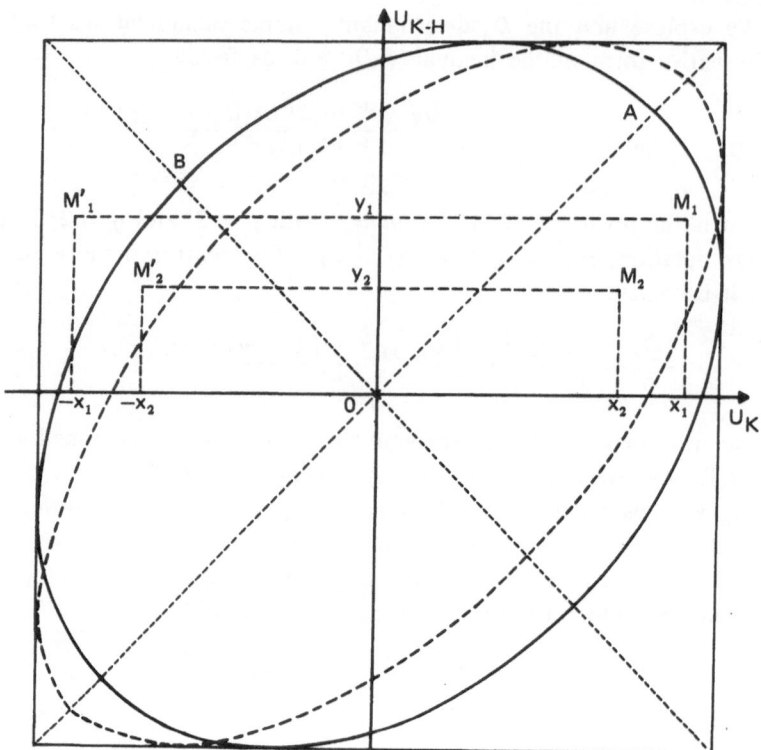

FIGURE 9.3. Limiting ellipses defining the allowed region inside the ellipse and the forbidden region in (U_K, U_{K-H}) space for a D_3 determinant. $U_H = 0.3$ (full line), and $U_H = 0.6$ (dotted line).

with

$$x = (U_K + U_{K-H})/2^{1/2}, \qquad y = (U_K - U_{K-H})/2^{1/2}$$

$$\lambda_1 = 1 + U_H, \qquad \lambda_2 = 1 - U_H, \qquad \text{and} \qquad U_H > 0 \qquad (9.17)$$

The half-lengths of the axes are, respectively, $OA = \lambda_1^{1/2}$, $OB = \lambda_1^{1/2}$, λ_1 and λ_2 being the eigenvalues of the matrix

$$U_2 = \begin{vmatrix} 1 & U_{-H} \\ U_H & 1 \end{vmatrix}$$

For given moduli $|U_K| = x_1$ and $|U_{K-H}| = y_1$, it is shown in Fig. 9.3 that the combination abbreviated $(++)$, corresponding to $U_K = +x_1$, $U_{K-H} = +y_1$, is allowed (point M_1 corresponding to $s = \text{sign}(U_H U_K U_{K-H}) = +1$]; the combination $(-+)$ (point M_1', $s = -1$) is forbidden. But for the absolute values x_2, y_2 (points M_2 and M_2'), both combinations are allowed and $D_3 \geq 0$ is useless in determining signs.

However, for a positive value of $|U_H|$ larger than that corresponding to the ellipse in the full line, we obtain another ellipse represented by a dotted line in Fig. 9.3. We see that for this second ellipse the combination $(-+)$ is now forbidden. This means that for a larger value of $|U_H|$, inequality $D_3 \geq 0$ or, equivalently, inequality (9.14) is more powerful. A last step is useful before generalizing for any m. We write Q_2/N in a form involving the inverse matrix U_2^{-1}, starting from (9.13) in the noncentrosymmetric case:

$$\frac{Q_2}{N} = \frac{|U_K|^2 + |U_{K-H}|^2 - U_H^* U_K U_{K-H}^* - U_H U_K^* U_{K-H}}{1 - |U_H|^2} \leq 1$$

where we recognize the Hermitian form

$$\frac{Q_2}{N} = D_{11}|U_1|^2 + D_{22}|U_2|^2 + D_{12}U_1 U_2^* + D_{21}U_2 U_1^* \qquad (9.18)$$

with

$$D_{11} = D_{22} = \frac{1}{1 - |U_H|^2}, \qquad D_{12}^* = D_{21} = \frac{-U_H}{1 - |U_H|^2}$$

$$U_1 = U_K, \qquad U_2 = U_{K-H}$$

We recognize in the D_{11}, D_{12}, $\ldots$, coefficients the elements of the inverse matrix (Appendix 9A.1):

$$U_2^{-1} = \begin{vmatrix} 1 & U_H^* \\ U_H & 1 \end{vmatrix}^{-1} = \begin{vmatrix} D_{11} & D_{12} \\ D_{21} & D_{22} \end{vmatrix} = \frac{1}{1 - |U_H|^2} \begin{vmatrix} 1 & -U_H^* \\ -U_H & 1 \end{vmatrix}$$

and we can write Q_2/N finally in the form

$$\frac{Q_2}{N} = UU_2^{-1}U^+ = |U_1 U_2| \begin{vmatrix} \dfrac{1}{1-|U_H|^2} & -\dfrac{U_H^*}{1-|U_H|^2} \\[2mm] -\dfrac{U_H}{1-|U_H|^2} & \dfrac{1}{1-|U_H|^2} \end{vmatrix} \begin{vmatrix} U_1^* \\ U_2^* \end{vmatrix}$$

$$(9.19)$$

where U is a row vector $U(U_1, U_2)$. Indeed, after performing the elementary matrix operation on (9.19), we find exactly (9.13).

We come now to the general case of any m. For reasons that will appear in section 9.2, it is desirable to consider order D_{m+1} rather than D_m.

In Appendix 9A.1, the following fundamental formula is shown in connection with the maximum determinant rule [generalization of (9.13)]:

$$\frac{Q_m}{N} = 1 - \frac{D_{m+1}}{D_m} \le 1 \tag{9.20}$$

which involves the Hermitian form

$$\frac{Q_m}{N} = \sum_{p=1}^{m} \sum_{q=1}^{m} D_{pq} U_{m+1,p} U^*_{m+1,q} \tag{9.21}$$

$D_{pq} = (U_m^{-1})_{pq}$ is an element of the inverse matrix U_m^{-1} and $D_m = \text{Det}(U_m)$.

Assume now that $D_{m+1} > 0$ is strictly positive; then, anticipating inequality (9.27), $D_{m+1}/D_m \le 1$, and combining with (9.20), we find

$$D_{m+1} > 0 \Leftrightarrow \frac{Q_m}{N} < 1 \tag{9.22}$$

In other words, the Karle–Hauptman inequality $D_{m+1} > 0$ is strictly equivalent to the inequality $Q_m < N$.

We consider again, as in the D_3 example, that all elements of the principal minor D_m are fixed and known, in modulus and phase. Then Q_m/N is a function of the only *variables* $U_{m+1,q}$, the m elements of the $(m + 1)$th row (or column of D_{m+1}); for simplicity, we set [cf. (9.18)]

$$U_q = U_{m+1,q}, \qquad q = 1, \ldots, m$$

For centrosymmetric structures, Q/N involves only real structure factors and is of quadratic form for the m variables:

$$\frac{Q_m}{N} = \sum_{p=1}^{m} \sum_{q=1}^{m} D_{pq} U_p U_q \tag{9.23}$$

Next, we "plot" the surface $Q_m/N = 1$ in m-dimensional space, where each coordinate axis represents $U_q = U_{m+1,q}$; $q = 1, \ldots, m$. This surface is a hyperellipsoid, having m principal axes. The whole set of the m U_q's defines a point M in this space (compare to points M_1, M_2 in Fig. 9.3 for $m = 2$). The use of this general hyperellipsoid is exactly the same as in Fig. 9.3: if M is located outside the hypersurface $Q/N = 1$, the corresponding combination of signs of the U_q's is forbidden.

9.2.4.2. Link with Probability Theory. Multidimensional Gaussian Probabilities

Inequality (9.14) tells us that all points inside the ellipse of Fig. 9.3 represent allowed combinations. But nothing has yet been said about the probability associated with each point within this allowed ellipse. The theory in Section 9.3 establishes that the probability density associated with each pair of values (U_K, U_{K-H}) for a given U_H is simply $p(U_K, U_{K-H} \mid U_H)$ $= \text{const} \times e^{-Q_2}$. The equiprobability contours are concentric ellipses, similar to the "inequality ellipse" (9.15) defining the allowed region. Indeed, we recognize already that Q_2 (9.14) and Q_m (9.23) are the exponents of multidimensional Gaussian functions. Crystallographers are familiar with Gaussian functions in three dimensions: the thermal ellipsoids of atoms represent surfaces of equal probability of the presence of an atom (defined by x_1, x_2, x_3) at the surface of the ellipsoid. This probability is given by $p(x_1, x_2, x_3) = \text{const} \times e^{-Q_3}$, where Q_3 is given by

$$Q_3 = \sum_{i=1}^{3} \sum_{j=1}^{3} A_{ij} x_i x_j$$

The proof of the approximate Gaussian character of the structure-factor probabilities is the main goal of Section 9.3. It is to be noted that a Gaussian function falls off smoothly to infinity, although the inequality theory states that the probability is strictly zero outside the limiting ellipse of Fig. 9.3. In fact a strict probability theory should yield the value $p = 0$ in the forbidden region. However, the value of the Gaussian function at the limiting ellipse (9.15), although not zero, is very small.

9.2.4.3. Efficiency in Phase Determination

In the D_3 example, it has been seen from Fig. 9.3 that the smaller the allowed region, the greater is the power of inequality (9.14) for discriminating between false and correct combinations of signs. The area (S) of the allowed ellipse is given by

$$S = \pi(\lambda_1 \lambda_2)^{1/2} = \pi D_2^{1/2}$$

Clearly, the maximum efficiency is achieved when $S \rightarrow 0$, or $U_H \rightarrow \pm 1$.

These results are immediately generalized for any m. The smaller the volume V_m of the allowed hyperellipsoid $Q_m/N \leq 1$, the smaller is the region of allowed combinations. The highest efficiency occurs when $D_m \rightarrow 0$.

Conversely, the smallest efficiency occurs when all $|U|$ values in D_m are zero: the allowed hyperellipsoid degenerates to an m-dimensional sphere and has the maximum possible volume; no phase information is obtained. A similar conclusion is reached for phase determination in the noncentro-symmetric case, where the hyperellipsoid is defined in $2m$-dimensional space (m variables for, respectively, the real and imaginary part of $U_1 \cdots U_m$).

On the other hand, mathematical theory shows that the volume V_m is proportional to $D_m^{1/2}$ [equation (23) in de Rango, Tsoucaris, and Zelwer, 1974]:

$$V_m = \frac{2}{m} \, D_m^{1/2} \, \frac{\pi^{m/2}}{\Gamma(m/2)} \tag{9.24}$$

where Γ is the gamma function. Therefore, we state a simple criterion for the efficiency of D_{m+1}, under the condition that all elements of D_m are known:

The smaller the value of D_m, the higher the power of D_{m+1} for phase restriction of the elements of the $(m+1)$th row. This statement, established for inequalities, is valid also for phase determination by the maximum determinant rule.

Next, we ask if it is possible to choose a determinant D_m, through the defining vectors $\mathbf{H}_1 \cdots \mathbf{H}_m$, such that it is small. In Section 9.4.2, we discuss how to achieve this goal.

9.2.4.4. Eigenvalues

The square half-lengths of the m axis in the m-dimensional hyper-ellipsoid (9.23) are the eigenvalues of the matrix U_m.

Mathematical theory establishes that, for the correct set of phases, all eigenvalues must be nonnegative[†]:

$$\lambda_j \geq 0, \qquad j = 1, \ldots, m \tag{9.25}$$

This requirement is stronger than the inequality $D_m \geq 0$. Indeed, the relation

$$D_m = \prod_{j=1}^{m} \lambda_j \tag{9.26}$$

[†] It is to be noted that Karle and Hauptman (1950) have formulated the inequalities in the following form, which is strictly equivalent to (9.25). The necessary and sufficient condition that $\varrho(\mathbf{r})$ be nonnegative is that every determinant D_m and all its principal minors be nonnegative.

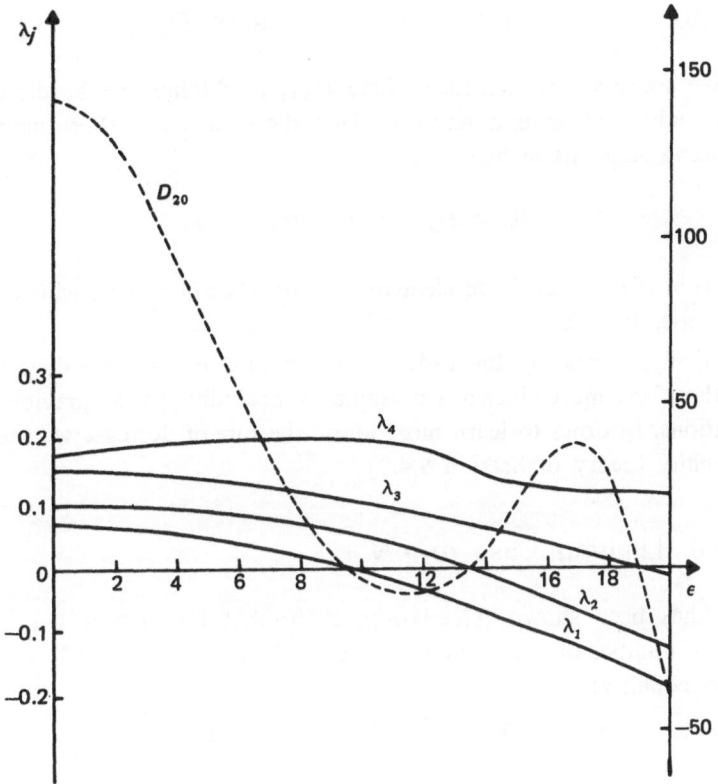

FIGURE 9.4. Variation of a D_{20} determinant ($N = 40$) and of the four lowest eigenvalues $\lambda_1, \lambda_2, \lambda_3, \lambda_4$, as a function of an increasing error, with random sign, introduced into the phases. The abscissa represents the absolute average value of the error ε in degrees.

shows that D_m can be positive if an even number of λ_j's is negative; however, a set of phases leading to 2, 4, ... negative eigenvalues is certainly false according to (9.25). Figure 9.4 shows the variation of the four smallest eigenvalues as a function of an increasing error ε (in degrees), with a random sign, introduced into the phases of isoniazide ($P2_12_12_1$, $N = 40$). The value of D_m is alternatively positive and negative as the number of negative eigenvalues is, respectively, even or odd. This feature emphasizes that a positive determinant does not ensure a correct set of phases, and that examination of the eigenvalues is also necessary.

Intuitively, since an ellipse with imaginary axes (square roots of negative eigenvalues) is inconceivable, as we are dealing with a hyperellipsoid region in structure-factor space, we can understand the conditions (9.25).

9.2.4.5. The Nonincreasing Sequence of D_m

We consider the sequence where D_{m+1} is obtained by bordering D_m with an additional column and row. Then the sequence of D_m forms a non-decreasing series as m increases:

$$D_m \geq D_{m+1} \geq 0 \quad \text{for any } m \tag{9.27}$$

The proof is obtained from elementary properties in m-dimensional space (Tsoucaris, 1970a).

Finally, increasing the order of D_m results in smaller values of D_m and, therefore, more efficient determinants, according to the previous considerations. In order to learn more about the rate of decrease we need the probability theory of Section 9.4.2).

9.2.4.6. Limiting Case $m = N + 1$

It has been shown (Goedkoop, 1950) that for $m \geq N + 1$, where N is the number of atoms in the unit cell, the inequality (9.6) becomes a strict equality:

$$D_m = 0 \quad \text{for } m \geq N + 1 \tag{9.28}$$

9.3. Maximum Determinant Rule

9.3.1. Definition

First, we state more precisely the maximum determinant rule.

We assume that all structure factors involved in a D_m determinant are known in both modulus and phase. Then, we consider the D_{m+1} determinant, obtained from D_m by adding a new column and row. The phases of the $(m + 1)$th column and row are unknown. We shall prove that the most probable phases of the $(m + 1)$th column and row of D_{m+1} are those that lead to the maximum value of D_{m+1} and for which all eigenvalues λ_i are positive:

$$(D_{m+1})_{\text{most probable phases}} = \max \lambda_i \geq 0, \tag{9.29}$$

$$i = 1, \ldots, m + 1$$

9.3.2. Proof

The proof of the maximum determinant rule involves three steps.

(a) First, we give a probabilistic interpretation of the Sayre equation (correlation coefficient) and the Karle–Hauptman determinant (matrix of correlation coefficients).

(b) This interpretation allows the application of the central limit theorem which immediately leads to the Gaussian expression of the probability law.

(c) This probability law is then very simply expressed in terms of determinants.

We will now develop these three steps in detail.

9.3.2.1. Connection between Sayre's Equation, Determinants, and Probabilities

It will be shown that Sayre's equation provides an expression for the correlation coefficient between two structure factors. Let us write this equation in a form given first by Hughes (1953):

$$\langle E_K E_{H-K} \rangle_K = U_H \qquad (9.30)$$

The symbol $\langle \ \rangle_K$ means the mean value of the expression within the brackets as K sweeps out uniformly the whole of reciprocal space. It is important to emphasize that the vector H is kept fixed, in contrast to the random (i.e., variable) K. Therefore, $H - K$, the difference between a random vector and a fixed vector, is also a random vector, and the normalized structure factors E_K and E_{K-H} are both random variables. This idea is used to interpret the left-hand member of Sayre's equation as the correlation coefficient between two structure factors. Indeed, we have

$$E_{H-K} = E_{K-H}^*$$

and

$$\langle E_K E_{K-H}^* \rangle_K = U_H \qquad (9.31)$$

However, in probability theory, the correlation coefficient r between two random variables x and y is defined as the mean value of the product of one random variable and the complex conjugate of the other, $\langle xy^* \rangle = r$.

Thus, according to (9.31), the correlation coefficient between the random variables E_K and E_{K-H} is equal to the unitary structure factor U_H. Recalling that both E_K and E_{H-K} are random variables displaying the same role in (9.31), it seems interesting to rewrite Sayre's equation in a more symmetrical form by the following change of notation:

$$K = L - H_q \qquad \text{random vector}$$
$$H = H_p - H_q \qquad \text{fixed vector} \tag{9.32}$$

with H_p, H_q fixed vectors and L a random vector. Let us check that the random and constant characters of the vectors are respected by this change of notation.

Vector H is the difference between two fixed vectors and is also a fixed vector; K is the difference between a random vector L and a fixed vector H_q and is a random vector.

By substituting (9.32) in (9.31), we have

$$H - K = H_p - L$$

and

$$E_{H-K} = E_{H_p-L} = E_{L-H_p}^{*}$$

Finally

$$\langle E_{L-H_q} E_{L-H_p}^{*} \rangle_L = U_{H_p-H_q} = U_{pq} \tag{9.33}$$

Next, we consider that

$$p, q = 1, \ldots, m$$

which means that we choose arbitrarily m reciprocal-lattice vectors

$$H_1, \ldots, H_p, \ldots, H_q, \ldots, H_m$$

that will be kept fixed, while L behaves randomly. The families of m random variables $E_{L-H_1}, \ldots, E_{L-H_m}$ are generated by considering vector L as sweeping out all reciprocal space. As a consequence, the "figure" formed by the m points $L - H_p$ moves randomly in reciprocal space, in the sense that this figure sweeps out all reciprocal space, while keeping a constant orientation. By setting

$$E_p = E_{L-H_p}, \qquad p = 1, \ldots, m$$

we write the Sayre–Hughes equation as

$$\langle E_q E_p{}^* \rangle = U_{pq}$$

Clearly, this generates m^2 Sayre's equations of the form (9.33). We recognize in the second member of (9.33) the general element of the determinant. Thus, we may state that *the right-hand members of m^2 Sayre's equations* (9.33) *form the matrix of a Karle–Hauptman determinant D_m*. The above probabilistic interpretation will allow us to obtain a probability law by using a very simple mathematical apparatus.

9.3.2.2. Conditional Probability Law and Central Limit Theorem

We show first that the set of random variables E_p may be considered also as elements of a Karle–Hauptman determinant, of order $m + 1$. We recall that for the point-atom model, U's and E's differ only by a constant scale factor:

$$E_p = E_{\mathbf{L}-\mathbf{H}_p} = N^{1/2} U_{\mathbf{L}-\mathbf{H}_p} \tag{9.34}$$

For simplicity, we assume that all atoms are equal. For unequal atoms, the theory has to be slightly modified, but the results are essentially the same.

Clearly, the U's of (9.34) form the $(m + 1)$st column and row of a D_{m+1} determinant [Table 9.4(a)]. We recall that D_m is completely defined when we choose the m vectors

$$\mathbf{H}_1, \dots, \mathbf{H}_p, \dots, \mathbf{H}_m$$

The determinant D_{m+1} is generated from D_m by adding a final vector $\mathbf{H}_{m+1} = \mathbf{L}$ to the above sequence:

$$\mathbf{H}_1, \dots, \mathbf{H}_p, \dots, \mathbf{H}_m, \mathbf{L}$$

We obtain thus the D_{m+1} determinant given in Table 9.4(a). In other words, D_{m+1} is generated from D_m by adding to it an $(m + 1)$th row and column containing the $U_{\mathbf{L}-\mathbf{H}_p}$'s. To be consistent with the meaning of (9.32) and (9.33), we assume that (a) all structure factors $U_{\mathbf{H}_p-\mathbf{H}_q}$ involved in D_m are also fixed (constant); i.e., all U_{pq} are assumed known in both modulus and phase; (b) all structure factors $U_{\mathbf{L}-\mathbf{H}_p}$ of the last column and row of D_{m+1} are random variables, that is, they are assumed to be unknown.

TABLE 9.4. Notation for the Determinants Involved in Probability Theory [a]

		D_m (fixed values of U_{pq}'s)			Random variables
		$\mathbf{H_1}$ $\cdots$	$\mathbf{H_p}$	$\cdots$ $\mathbf{H_m}$	$\mathbf{L}$
(a) $\Delta_{m+1} =$		$\begin{array}{ccccc} 1 & \cdots & U_{1p} & \cdots & U_{1m} \\ \vdots & & \vdots & & \vdots \\ U_{p1} & \cdots & 1 & \cdots & U_{pm} \\ \vdots & & \vdots & & \vdots \\ U_{m1} & \cdots & U_{mp} & \cdots & 1 \end{array}$			$\begin{array}{c} U_{\mathbf{H_1}-\mathbf{L}} \\ \vdots \\ U_{\mathbf{H_p}-\mathbf{L}} \\ \vdots \\ U_{\mathbf{H_m}-\mathbf{L}} \end{array}$
		$U_{\mathbf{L}-\mathbf{H_1}}$ $\cdots$	$U_{\mathbf{L}-\mathbf{H_p}}$	$\cdots$ $U_{\mathbf{L}-\mathbf{H_m}}$	1
(b) $\Delta_{m+1} = \dfrac{1}{N}$		$\begin{array}{ccccc} 1 & \cdots & U_{1p} & \cdots & U_{1m} \\ \vdots & & \vdots & & \vdots \\ U_{p1} & \cdots & 1 & \cdots & U_{pm} \\ \vdots & & & \cdots & \vdots \\ U_{m1} & \cdots & U_{mp} & & 1 \end{array}$			$\begin{array}{c} E_{\mathbf{H_1}-\mathbf{L}} \\ \vdots \\ E_{\mathbf{H_p}-\mathbf{L}} \\ \vdots \\ E_{\mathbf{H_m}-\mathbf{L}} \end{array}$
		$E_{\mathbf{L}-\mathbf{H_1}}$ $\cdots$	$E_{\mathbf{L}-\mathbf{H_p}}$	$\cdots$ $E_{\mathbf{L}-\mathbf{H_m}}$	N

[a] $\Delta_{m+1} \equiv D_{m+1}$ (cf. Sec. 9.3.2.3).

We now summarize these hypotheses which are fundamental for the theory to be developed below:

$$
\boxed{\begin{aligned}
&\mathbf{L}: \text{random vector} \\
&U_{\mathbf{L}-\mathbf{H_p}} = \frac{1}{N^{1/2}} E_{\mathbf{L}-\mathbf{H_p}} = \frac{1}{N^{1/2}} E_p \quad \text{random variables} \\
&\mathbf{H_p}, \mathbf{H_q}: \text{fixed vectors } (p, q = 1, \ldots, m) \\
&\quad \langle E_{\mathbf{L}-\mathbf{H_q}} E^{*}_{\mathbf{L}-\mathbf{H_p}} \rangle_{\mathbf{L}} = U_{\mathbf{H_p}-\mathbf{H_q}} = U_{pq} = \text{const} \\
&\quad\quad D_m = \det(U_{pq}) \geq 0
\end{aligned}}
$$

(9.35)

9.3.2.3. Notation

In order to emphasize the random character of the structure factors of the last column and row of D_{m+1}, we shall denote them by $E/N^{1/2}$, instead of U. Also, D_{m+1}, involving random variables, will be denoted differently, by Δ_{m+1}, in contrast with D_m, which involves only structure factors whose values are assumed known.

The factor $1/N^{1/2}$, common to all elements of the last row and column, can be taken twice out of the determinant, the diagonal term being written as $N/N^{1/2}N^{1/2} = 1$, as indicated in Table 9.4(b). Moreover, this notation has the advantage of emphasizing that the correlation coefficient U_{pq} between the normalized structure factors

$$E_q = E_{L-H_q} \quad \text{and} \quad E_p = E_{L-H_p}$$

is the unitary structure factor U_{pq} which lies on the intersection of the pth row and qth column of D_m, according to (9.33). In terms of determinants it is required to find the probability law of the structure factors of the last column and row of a D_{m+1} determinant, assuming that all structure factors involved in D_m are fixed and known. In terms of probability theory it is required to find the conditional joint probability law of m random variables, under the condition that the $m \times m$ matrix of the correlation coefficients is known. This probability law is denoted by

$$P(E_1 \cdots E_m \mid U_{pq} \cdots)$$

where the m random variables E are the m elements of the last column and row of Δ_{m+1}, as defined in (9.34). The vertical bar means "under the condition that all U_{pq}'s are known and fixed."

9.3.2.4. Central Limit Theorem

This theorem enables us to obtain immediately the expression of the probability law, in the case where each random variable is itself the sum of n independent random variables. Indeed, if there are n symmetry-independent atoms in the unit cell, the structure factor is a sum of n independent terms. The conditions for the validity of this theorem are, in crystallographic terms, as follows: (a) The number of atoms N in the unit cell is very large; (b) the positions of the atoms in the asymmetric unit are mutually independent; (c) there are no heavy atoms in the structure. The theorem was introduced into crystallography by Wilson (1949) for the case of one random variable. It leads to the familiar Gaussian distribution of structure factors:

$$P(E) = (2\pi)^{-1/2} \exp(-\tfrac{1}{2}E^2) \quad \text{(centrosymmetric case)} \tag{9.36}$$

$$P(E) = \pi \exp(-|E|^2) \qquad \text{(noncentrosymmetric case)} \tag{9.37}$$

We recall that (9.37) is a two-dimensional joint probability density which has the following meaning: if the primitive random variable **H** sweeps out reciprocal space, $E_{\mathbf{H}}$ is a random function of **H**; in its turn $E_{\mathbf{H}}$ is considered a random variable. The expression

$$P(E)\, dA\, dB = P(A, B)\, dA\, dB$$

is the chance of having observed jointly that (a) the real part $A_{\mathbf{H}}$ lies between A and $A + dA$; (b) the imaginary part $B_{\mathbf{H}}$ lies between B and $B + dB$.

Often, in order to avoid a double notation, we use indifferently the "name" of the random variable $E_{\mathbf{H}}$ or the "value" E associated with $E_{\mathbf{H}}$ in the probability density expression. Also, instead of the real number notation, we use the abbreviated complex notation $P(E)$, without forgetting that this is in fact a two-dimensional function of two real variables. Similarly, in the general joint probability density expression, we choose the abreviated complex notation.

The generalization of the familiar Gaussian in the case of m random variables leads to the m-dimensional Gauss–Laplace law (Tsoucaris, 1969, 1970a):

$$P\!\left(\frac{E_1 \cdots E_q \cdots E_m}{U_{pq} \cdots}\right) = \frac{1}{(2\pi)^{m/2}D_m^{1/2}} \exp\!\left(-\frac{1}{2}Q_m\right)$$

$$\text{(centrosymmetric case)} \tag{9.38}$$

$$P\!\left(\frac{E_1 \cdots E_q \cdots E_m}{U_{pq} \cdots}\right) = \frac{1}{\pi^m D_m} \exp(-Q_m)$$

$$\text{(noncentrosymmetric case)} \tag{9.39}$$

where Q_m is a Hermitian form [(9.20) and (9.21)]:

$$Q_m = N \frac{D_m - \Delta_{m+1}}{D_m} = \sum_{p,q} E_p E_q^* D_{pq} = EU^{-1}E^+ \tag{9.40}$$

where $E(E_1, \ldots, E_p, \ldots, E_m)$ is a row vector, similar to $U(U_1, U_2)$ of (9.19) and D_{pq} is the element of the inverse matrix U^{-1}. The exponent Q_m is still always positive, as in the Gaussian function, but it involves now not only the m E_p values, but also the $m(m-1)/2$ U_{pq} values through the

D_{pq}'s. The equiprobability surfaces are concentric hyperellipsoids (see Section 9.2). The special cases $m = 1, 2, 3$ are considered in Appendix 9A.1.

We mention here that Q_2 is the expression of ellipses similar to inequality ellipse of Fig. 9.3. Indeed, in Appendix 9A.1, we obtain the expression of (9.14) from the general Hermitian form:

$$Q_2 = \frac{|E_{\mathbf{K}}|^2 + |E_{\mathbf{K-H}}|^2 - 2U_{\mathbf{H}}E_{-\mathbf{K}}E_{\mathbf{K-H}}}{1 - U_{\mathbf{H}}^2}$$

$$= N \frac{|U_{\mathbf{K}}|^2 + |U_{\mathbf{K-H}}|^2 - 2U_{\mathbf{H}}U_{-\mathbf{K}}U_{\mathbf{K-H}}}{1 - U_{\mathbf{H}}^2} \qquad (9.41)$$

and (9.38) yields

$$P(E_{\mathbf{K}}, E_{\mathbf{K-H}} \mid U_{\mathbf{H}}) = \frac{1}{2\pi D_2^{1/2}} \exp\left(-\frac{1}{2} Q_2\right) \qquad (9.42)$$

9.3.2.5. Maximum Determinant Rule

The probability law can be expressed very simply in terms of two Karle–Hauptman determinants, D_m and Δ_{m+1}. The proof, involving only elementary operations, is given in Appendix 9A.1. The final result is

$$P(E_1 \cdots E_m \mid U_{pq} \cdots) = \begin{cases} [(2\pi)^{-m/2}D_m^{-1/2} \exp(-N)] \exp\left(N\frac{\Delta_{m+1}}{2D_m}\right) \\ \text{(centrosymmetric case)} \\ [\pi^{-m}D_m^{-1} \exp(-N)] \exp\left(N\frac{\Delta_{m+1}}{D_m}\right) \\ \text{(noncentrosymmetric case)} \end{cases} \qquad (9.43)$$

Clearly, the *largest* probability occurs when Δ_{m+1} is maximum. As we assume that the moduli of the E's are known, we now have the maximum determinant rule, (9.29). The remarkable feature of (9.43) lies in the fact that the joint probability of m structure factors depends solely on one parameter, the value of the ratio Δ_{m+1}/D_m, whatever the number m of the random variables. Moreover, the presence of D_m in the denominator, both in the exponent and in the multiplicative constant, ensures that the sensitivity of the probability with respect to the variables (E) included in Δ_{m+1} is higher if D_m is smaller.

Summarizing, the procedure for phase determination, or, more properly, for phase extension, is the following: first we choose m vectors $\mathbf{H}_1, \ldots,$ $\mathbf{H}_m$ so that the corresponding determinant D_m has a small value and all its elements are already known. Then, we choose arbitrarily an additional $(m + 1)$th vector **L** and we form the corresponding Δ_{m+1} determinant. The phases of the additional $(m + 1)$th row are unknown; they will be "efficiently" determined by maximizing Δ_{m+1}.

We recall that in (9.43) all elements of D_m are assumed fixed and known and the maximization process of Δ_{m+1} involves only the structure factors of the last row and column. One could then pose the questions: what is the joint probability with respect to all structure factors involved in Δ_{m+1}? Is the determinant a maximum with respect to all included structure factors? The following reasoning provides a good approach to answering the first question.

By applying the compound probability law (see Appendix 9A.2), we obtain the joint probability for all structure factors U_{pq} $(p, q = 1, \ldots, m)$ of D_{m+1} (noncentrosymmetric structure)

$$P(\cdots U_{pq} \cdots) = \text{const} \times \sum_{r=1}^{m} D_r^{-1} \exp\left(N \sum_{r=1}^{m} \frac{D_{r+1}}{D_r}\right) \qquad (9.44)$$

The formula is strictly valid, from the probability theory point of view, only if there is no redundancy, i.e., if the same structure factor does not appear several times in Δ_{m+1}. However, it can be hoped that this is only a minor problem.

It appears quite difficult to develop the mathematics necessary to answer the second question. Here we give only an intuitive argument. Concentrating our attention on the exponent in (9.44), we recall that the ratios D_{r+1}/D_r steadily decrease as r increases (Section 9.2.3). We now anticipate a result to be stated in Section 9.4 [(9.80); Fig. 9.8]: this ratio is quite close to the value $1 - m/N$ for all structures. Therefore, it is likely that maximizing the value of D_{m+1} will result in maximizing, on the average, all the other ratios, and, finally, in maximizing the probability expression of (9.44).

Equation (9.43) corresponds to the approximation of the central limit theorem. A higher degree of approximation is reached by using the methods developed by Bertaut (1956) and Klug (1958); Heinerman and Kroon (1976) have studied the meaning of higher-order terms for the expressions of (9.43).

9.3.3. Eigenvectors and the Ellipsoid Representation

The aim of this section is to provide a convenient way of searching for the set of phases that maximize Δ_{m+1} (given the U_{pq}'s of D_m). The set of m structure factors of the last row of Δ_{m+1} defines a point M in the m-dimensional complex space, a generalization of two-dimensional real space $(U_K, U_{K-H}$ of Fig. 9.3). In other words, these structure factors are the coordinates of vector $\overrightarrow{OM} = \mathbf{E}(E_1, \ldots, E_m)$. The m eigenvectors of the matrix U_m provide another orthonormal basis for this same space; they will be denoted by $\mathbf{Y}^{(i)}$. The directions of these eigenvectors coincide with the principal axes of the hyperellipsoid Q_m (9.40); in Fig. 9.3 ($m = 2$), they coincide, respectively, with the major and minor axes of the ellipse.

We state that the correct vector $\mathbf{E}(E_1 \cdots E_m)$ is a linear combination of the orthonormal eigenvectors of U_m:

$$\mathbf{E} = \sum_{i=1}^{m} C_i \mathbf{Y}^{(i)}, \qquad \mathbf{Y}^{(i)} \cdot \mathbf{Y}^{(j)} = \delta_{ij} \tag{9.45}$$

Or, for each coordinate of $\mathbf{E}$,

$$E_p = \sum_{i=1}^{m} C_i Y_p^{(i)} \tag{9.46}$$

where $Y_p^{(i)}$ is the pth coordinate of the ith eigenvector $\mathbf{Y}^{(i)}(Y_1^{(i)} \cdots Y_m^{(i)})$. The square of the length of $\mathbf{E}$ is

$$|E|^2 = \sum_{p=1}^{m} |E_p|^2 = \sum_{i=1}^{m} |C_i|^2 \tag{9.47}$$

The values of $Y_p^{(i)}$ are obtained from the known matrix U_m of D_m by standard techniques. But the C_i values are unknown, and (9.45) does not seem helpful as we have replaced the m unknown E_p by the m unknown C_i. However, empirical calculations with known structures have shown (Fig. 9.6) an interesting fact: the coefficient C_1 corresponding to the largest eigenvalue λ_1 of the matrix U_m is considerably larger than all other C_i. Therefore, we can write (9.46) in a very simplified form:

$$E_p \approx C_1 Y_p^{(1)}, \qquad p = 1, \ldots, m \tag{9.48}$$

This equation shows that all phases of the unknown E_p can be approximated by the corresponding known $Y_p^{(1)}$, except for an additive constant, the phase of C_1. Therefore, (9.48) provides a *starting set* of phases.

This starting set can be refined by a steepest-descent technique: Q_m is expressed in terms of phases, the moduli $|E_p|$ being fixed:

$$Q_m = \sum_{p=1}^{m} D_{pp} |E_p|^2 + \sum_{\substack{p,q=1 \\ p>q}}^{m} |D_{pq} E_p E_q| \cos(\gamma_{pq} + \phi_p - \phi_q) \tag{9.49}$$

where γ_{pq} is the phase of the complex element D_{pq}. The argument of the cosine is a phase invariant, which for $m = 2$ amounts to the familiar triplet invariants.

The set of phases $\phi_1 \cdots \phi_m$ minimizing Q_m can be obtained by standard least-squares or steepest-descent techniques, the starting set of phases being given by (9.48).

The validity of the approximation (9.48) to the correct formula (9.45) can be understood by using again the hyperellipsoid representation of Q_m and the special two-dimensional case $m = 2$ (Fig. 9.3).

The fundamental formulas (9.38) and (9.39) express the fact that the *equiprobability contours corresponding to constant values of Q_m are concentric hyperellipsoids*, similar to the "limiting" ellipse [(9.14) and (9.50)] defining the region allowed by inequalities. This argument establishes a pronounced link between inequalities and probabilities for any order m, through the formulas[†]

$$Q_m/N \leq 1 \tag{9.50}$$

$$p \propto \exp(-Q_m) \tag{9.51}$$

For $m = 2$ (centrosymmetrical case), we obtain a set of concentric ellipses, similar to and included in the "inequality ellipse" $Q_2 \leq N$, as illustrated in Fig. 9.5 with the following data: $N = 100$, $U_H = 0.3$; for $U_K = U_1 = E_1/N^{1/2} = 4/10 = 0.4$; $U_2 = U_{K-H} = E_{K-H}/N^{1/2} = 2/10 = 0.2$, the value of Q_2^+ (9.41) is 16.7, the invariant $U_H U_{-K} U_{K-H}$ is positive. Recalling the proportionality $E = N^{1/2}U$, we plot E rather than U to emphasize that the probability depends explicitly on Q_2, whereas the inequalities depend explicitly on Q_2/N. For $E_1 = -4$, $E_2 = +2$, we find $Q_2^- = 27.2$; the invariant is negative. The ratio of the corresponding probabilities is, from (9.42),

$$\frac{P_+}{P_-} = \frac{\exp(-\tfrac{1}{2}Q_2^+)}{\exp(-\tfrac{1}{2}Q_2^-)} = \exp(5.25) = 191$$

The value of this ratio is close to that predicted from the theory of Woolfson (1954) and Cochran (1955).

For $m \geq 3$ one can imagine, in m dimensions, a set of concentric hyperellipsoids. The half-lengths of their principal axes are equal, respectively, to the m square roots of the eigenvalues of the matrix U_m. It has been found

---

[†] The probability expression (9.51) depends solely on Q_m, which gives the reason for using the notation Q_m/N in (9.13) and (9.21).

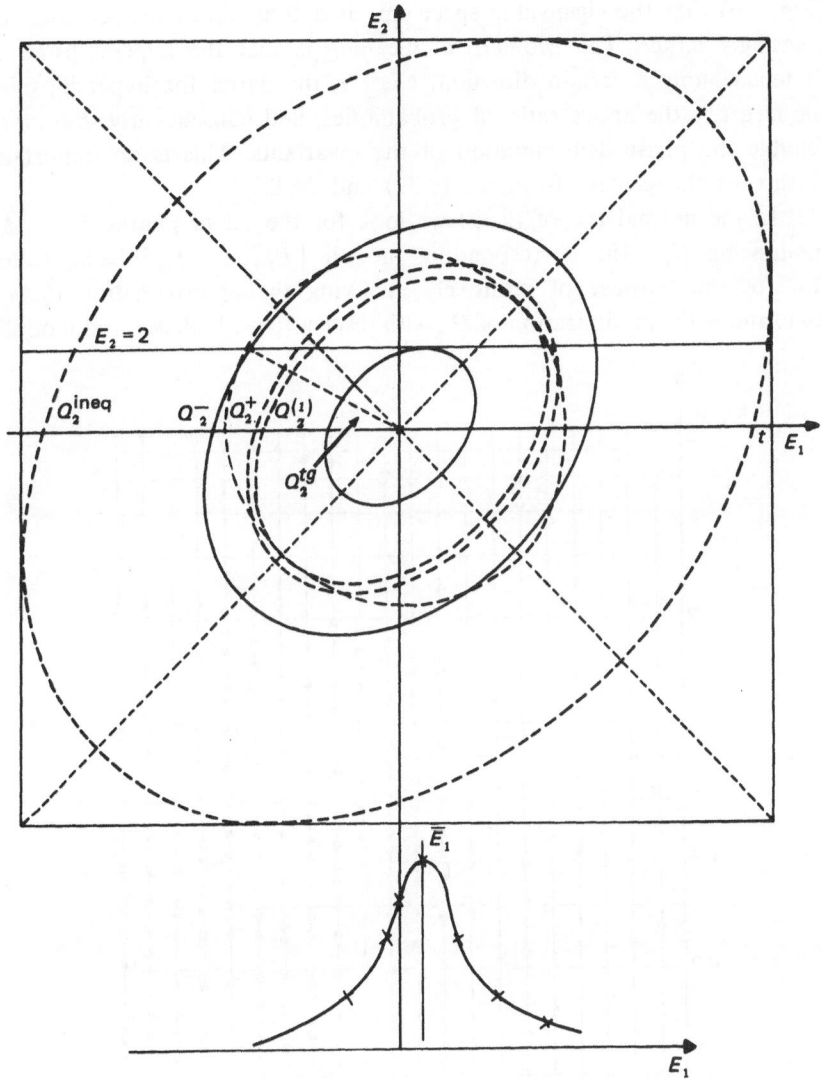

FIGURE 9.5. Equiprobability ellipses in $(E_{\mathbf{K}}, E_{\mathbf{K-H}})$ space. $Q_2^{\text{ineq}} = N = 100$: inequality ellipse for $E_{\mathbf{H}} = 3$; $Q_2^- = 27.2$: negative invariant for $E_{\mathbf{K}} = -4$, $E_{\mathbf{K-H}} = 2$; $Q_2^+ = 16.7$: positive invariant for $E_{\mathbf{K}} = 4$, $E_{\mathbf{K-H}} = 2$; $Q_2^{(1)} = 15.4$: smallest value of Q_2 for $E_{\mathbf{H}} = 3$ and $E_{\mathbf{K}}^2 + E_{\mathbf{K-H}}^2 = 20$ (circle drawn); $Q_2^{tg} = 4$: smallest value of Q_2 for $E_{\mathbf{H}} = 3$ and $E_{\mathbf{K-H}} = 2$. The Gaussian function corresponds to (9.55).

(Fig. 9.6) that the eigenvalue spectrum, as a function of m, becomes increasingly larger. The geometrical meaning is that the hyperellipsoid is flattened along a certain direction. Clearly, the flatter the hyperellipsoid, the larger is the above ratio of probabilities, and consequently, the more reliable the phase determination of the invariants. This is an important feature of the general formulas (9.39) and (9.43).

In the normal use of (9.39) we look for the set of phases $\phi_1 \cdots \phi_m$ minimizing Q_m, the corresponding moduli $|E_1| \cdots |E_m|$ being fixed. But for the purpose of intuitively justifying the approximation (9.48), we examine the minimization of Q_m with respect to both phases and moduli,

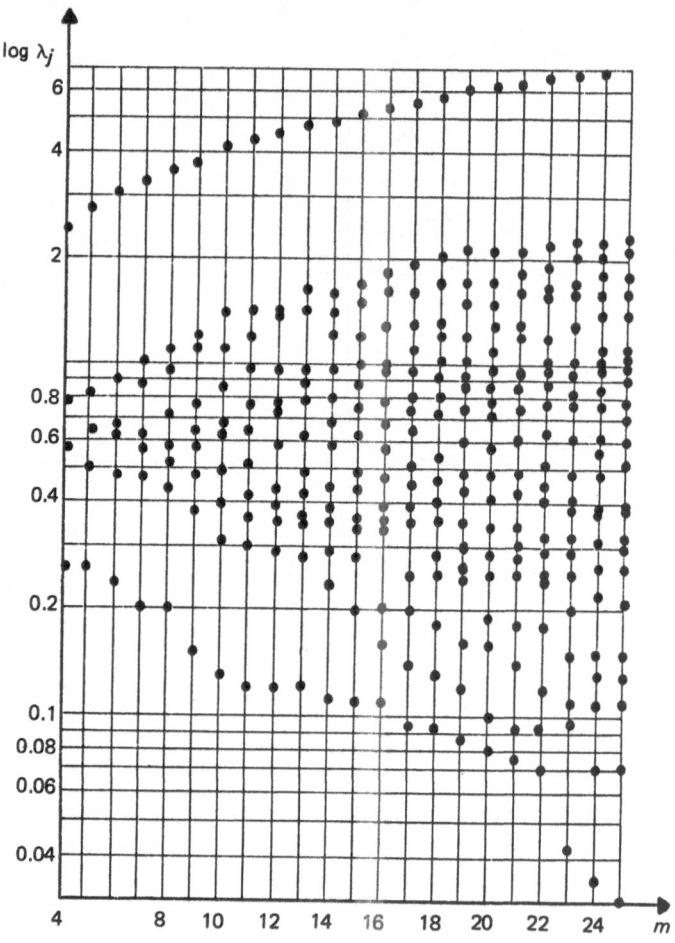

FIGURE 9.6. Plot of the eigenvalues for a series of Karle–Hauptman determinants of increasing order.

under the only restriction that the sum of the m moduli be constant:

$$|\mathbf{E}|^2 = \sum_{p=1}^{m} |E_p|^2 = M = \text{const} \tag{9.52}$$

This requirement is readily achieved with the help of Fig. 9.5; the end of vector $\overrightarrow{OM} = \mathbf{E}(E_1, E_2)$ must lie on the circumference of radius

$$E_1^2 + E_2^2 = M = 20, \qquad E_1 = E_{\mathbf{K}}, \qquad E_2 = E_{\mathbf{K-H}}$$

Clearly, the point P belonging to this circumference and leading to a minimum value of Q_2 is the point where the circle is tangential to an ellipsoid. The associated direction is simply the largest axis of the ellipse, and vector $\overrightarrow{OP}$ is the eigenvector of the matrix U_2 corresponding to the largest eigenvalue λ_1. Clearly[†] this value is the smallest that Q_2 can ever take with the given data ($U_{\mathbf{H}} = 0.3$, $E_1^2 + E_2^2 = M = 20$, $N = 100$): it will be denoted by $Q_2^{(1)} = 15.4$.

Returning to the m-dimensional case, we state from a direct generalization of Fig. 9.5 that if the moduli, as well as the phases of $\mathbf{E}(E_1 \cdots E_m)$, are allowed to vary under condition (9.52), then the smallest possible value for Q_m would be reached for the eigenvector $\mathbf{E}^{(1)}(E_1^{(1)} \cdots E_m^{(1)})$ corresponding to the largest eigenvalue λ_1 of the matrix U_m.

Of course, we cannot expect that all vectors $\mathbf{E}(E_1 \cdots E_m)$, i.e., all sets $E_{\mathbf{L-H_1}}, \ldots, E_{\mathbf{L-H_m}}$ have the same moduli, irrespectively of $\mathbf{L}$. However, for an actual set $|E_1|, \ldots, |E_m|$, we can expect that the vector $\mathbf{E}(E_1 \cdots E_m)$ minimizing Q_m is not too far from the eigenvector $\mathbf{E}_1^{(1)}(E_1^{(1)} \cdots E_m^{(1)})$; such is the geometrical meaning of the approximation (9.48). A better approximation would be to include in the sum of (9.45) not only the $C_1 \mathbf{Y}^{(1)}$ term but also the eigenvectors corresponding to a few of the largest eigenvalues.

We note also another empirical fact: the largest eigenvalue λ_1, for all orders m, is considerably larger than the next one (Fig. 9.6; Sarrazin, 1970). A further study on eigenvalues and eigenvectors has been initiated by Gifkins (1972), Main (1974), and Taylor and Woolfson (1975). It has been shown that each eigenvalue can be considered as a measure of the structural information contained in the corresponding eigenvector. Thus, we obtain an independent indication of the correctness of the assumption (9.48).

[†] The eigenvector corresponding to the maximum eigenvalue λ_1 of U is identical to that corresponding to the minimum eigenvalue $1/\lambda_1$ of U^{-1} involved in Q_m.

9.3.4. Regression Equation

The practical use of the above rule requires the maximization of a function depending on m variables. It is often useful to consider a special case of the problem, where only one variable is involved at a time.

9.3.4.1. Most Probable Value of E_q

We assume next not only that the U_{pq} of D_m are known, but also that all structure factors of the last row of D_{m+1} are known, *except one*, say E_q.

The most probable value of E_q, denoted by $\bar{E}_q$, is obtained by maximizing the probability density (9.39), or by minimizing Q_m (de Rango, 1969):

$$\left| \delta Q_m / \delta E_q \right|_{\bar{E}_q} = 0$$

$$\bar{E}_q = -\frac{1}{D_{qq}} \sum_{\substack{p=1 \\ p \neq q}}^{m} D_{pq} E_p, \qquad m \leq N + 1 \qquad (9.53)$$

The expressions D_{qq} are elements of the matrix U^{-1}. This remarkable probabilistic equation, often called a regression equation, expresses the most probable phase of one structure factor as a function of all other structure factors involved in D_{m+1}. It can be written in a way that recalls Sayre's equation, after a convenient change of notation:

$$E_H = \sum_{K} C_{HK} E_K$$

where C_{HK} is the ratio of two minors in (9.53). The denominator is always positive and the numerator can be developed to show that the first term is precisely U_{H-K}:

$$C_{HK} \approx (\delta_{m-1}^{qq})^{-1} \left(U_{H-K} - \sum_{L} U_{H-L} U_{L-K} + \cdots \right) \qquad (9.54)$$

where δ_{m-1}^{qq} denotes the $(m-1)$th-order minor obtained from D_{m+1}, by suppressing the qth row and column. The first term corresponds to Sayre's equation. The higher terms involve quartets, quintets, and so on.

For $m = 2$, the regression equation reduces to a single term of the tangent formula or to Sayre's equation. But for a sufficiently high-order m, C_{HK} can be very different from U_{H-K}. We note also that (9.53) is closely connected with the theory of inequalities (Karle and Hauptman, 1950; Karle, 1972).

9.3.4.2. Assessment of the Probability

Let us consider the m-dimensional vector $E(E_1 \cdots E_m)$ and the corresponding m-dimensional Gaussian function, i.e., the set of concentric ellipsoids of Section 9.3. We now consider a one-dimensional "section," the straight line defined by constant values of $E_1, \ldots, E_m$ except for E_q, which is the only variable. Mathematical theory shows that this section is still a one-dimensional Gaussian function which, in the centrosymmetrical case, is

$$P(E) = \frac{1}{(2\pi)^{1/2}\sigma} \exp\left[-\frac{(E_q - \bar{E}_q)^2}{2\sigma_q^2} \right] \qquad (9.55)$$

with a mean value $E_q = \bar{E}_q$ given by (9.53) and a variance

$$\sigma_q^2 = 1/D_{qq} = D_m/\delta_{m-1}^{qq} \qquad (9.56)$$

where δ_{m-1}^{qq} represents an $(m-1)$-order principal minor of D_m; (9.55) is illustrated by Fig. 9.5 ($m = 2$), where E_1 is the variable ($q = 1$) and E_2 ($p = 2$) is fixed.

We now consider a straight line taken at a constant value of E_2, and the values taken by the two-dimensional Gaussian function (9.42) along the line. These values correspond to a one-dimensional Gaussian function, illustrated at the bottom, which is centered at

$$\bar{E}_1 = -\frac{1}{D_{11}} (D_{21}E_2) = \frac{U_H E_2}{1 - U_H^2}$$

with $\sigma_1^2 = D_{11} = (1 - U_H^2)$, $D_{21} = -U_H$, $E_1 = E_K$, $E_2 = E_{K-H}$. The abscissa $\bar{E}_1$ corresponds to the tangent point of the straight line ($E_2 =$ const) to the ellipse which is precisely that associated with the minimum value of Q_2, among all the ellipses encountered by this straight line; it will be denoted by Q_2^{tq}; with the numerical values of Section 9.3, we obtain $\bar{E}_1 = 0.60$; $Q_2^{tq} = 4$. Similarly, the Gaussian function in the general complex case is

$$P(E_q) = \frac{1}{\pi\sigma_q} \exp\left(-\frac{|E_q - \bar{E}_q|^2}{\sigma_q} \right) \qquad (9.57)$$

Then the probability law for the phase error ε_q is deduced:

$$P(\varepsilon_q) = \frac{\exp(A_q \cos \varepsilon_q)}{2\pi I_0(A_q)} \qquad (9.58)$$

where

$$\varepsilon_q = \phi(E_q) - \phi(\bar{E}_q), \qquad A_q = 2\,|\,\bar{E}_q E_q\,|/\sigma_q{}^2, \qquad \sigma_q{}^2 = 1/D_{qq}$$

It is worthwhile noting that for $m = N + 1$ the above probabilistic equation becomes strict a algebraic equation, $E_q = \bar{E}_q$, whether there is redundancy or not.

9.3.4.3. Average over Several Determinants

The principle resides in the fact that, in the above theory, $\mathbf{L}$ is a random vector. When $\mathbf{L}$ sweeps out reciprocal space (the $\mathbf{H}_p$'s being fixed), a family of $\Delta_{m+1}(\mathbf{L})$ determinants is generated having the same principle minor D_m. Also, one can construct several determinants D_m, and generate in the same way several families of $\Delta_{m+1}(\mathbf{L})$.

In all cases, the same structure factor (or a symmetry related to it) say $E_{\mathbf{L}-\mathbf{H}_q}$, may appear several times in the last row of different determinants, labeled by $1, \ldots, i, j, \ldots$. For instance, the reflection $\mathbf{H} = \mathbf{L}_j - \mathbf{H}_q$ may appear in the jth determinant $\Delta_{m+1}^{(j)}$ *and* in the ith determinant $\Delta_{m+1}^{(i)}$, provided that

$$\mathbf{H} = \mathbf{L}_i - \mathbf{H}_p = \mathbf{L}_j - \mathbf{H}_q = \cdots$$

For each of these determinants, a different expected value is determined by equation (9.53) for the same reflection $E_{\mathbf{H}} = |\,E_{\mathbf{H}}\,|\exp(i\phi_{\mathbf{H}})$.

By taking into account the above definition of $\mathbf{H}$, these determinations are denoted

$$\bar{E}_{\mathbf{H}}{}^{(j)} = \bar{E}_{\mathbf{L}_j - \mathbf{H}_q} \qquad \text{from } \Delta_{m+1}^{(j)}$$

$$\bar{E}_{\mathbf{H}}{}^{(i)} = \bar{E}_{\mathbf{L}_i - \mathbf{H}_p} \qquad \text{from } \Delta_{m+1}^{(i)}$$

A better approximation to $E_{\mathbf{H}}$ will be obtained by the average over j determinations:

$$E_{\mathbf{H}} \simeq \langle \bar{E}_{\mathbf{H}}{}^{(j)} \rangle_j$$

The last equation is strictly valid, of course, only if all determinations are independent. After introducing a weighting factor

$$1/\sigma_q{}^{2(j)} = D_{qq}^{(j)}$$

we have

$$E_{\mathbf{H}} \simeq \langle D_{qq}^{(j)} \bar{E}_{\mathbf{H}}{}^{(j)} \rangle_j \tag{9.59}$$

Finally, the phase Φ_H of the right-hand side of (9.59) is

$$\tan \Phi_H = \frac{\sum_j D_{qq}^{(j)} |\bar{E}_H^{(j)}| \sin \bar{\phi}_H^{(j)}}{\sum_j D_{qq}^{(j)} |\bar{E}_H^{(j)}| \cos \bar{\phi}_H^{(j)}} \qquad (9.60)$$

with

$$\bar{E}_H^{(j)} = |\bar{E}_H|^{(j)} \exp(i\bar{\phi}_H^{(j)})$$

and can be associated with the probability law

$$P(\phi_H - \Phi_H) = \frac{1}{\pi I_0(A_H)} \exp[A_H \cos(\phi_H - \Phi_H)] \qquad (9.61)$$

where ϕ_H is the exact phase, I_0 denotes a Bessel function and A_H is given by

$$A_H^2 = \left[\sum_j A_H^{(j)} \cos \bar{\phi}_H^{(j)}\right]^2 + \left[\sum_j A_H^{(j)} \sin \bar{\phi}_H^{(j)}\right]^2 \qquad (9.62)$$

In this last equation $A_H^{(j)}$ indicates the value of A_q, as given by equation (9.58) associated with the occurrence of reflection $H = L_j - H_q$ at the qth column of the jth determinant $\Delta_{m+1}^{(j)}$.

9.3.4.4. Sensitivity of the Determinant to Phase Variations

The regression equation offers a simple way to evaluate the sensitivity of D_m to the phase variation of one structure factor, say, $U_H = U_s = U_{L-H_s}$, whatever is its redundancy.

We assume for the moment that there is no redundancy for U_s. For simplicity, let us assume that the phase of U_s is shifted by π, and we are looking for the corresponding variation of Q_m, and, consequently, D_{m+1}. We consider first all terms in the Q_m expression such as

$$\sum_{p \neq s} D_{ps} U_p U_s^* = -D_{ss} U_s^* \bar{U}_s$$

recalling the definition (9.53) or $\bar{U}_s = \bar{E}_s/N^{1/2}$.

Clearly, the variation of this set of terms is

$$D_{ss}(U_s^* e^{i\pi} \bar{U}_s - U_s^* \bar{U}_s) = -2D_{ss} U_s^* \bar{U}_s$$

We consider now the set of terms

$$\sum_{q \neq s} D_{sq} U_s U_q^* = -D_{ss} U_s \bar{U}_s^*$$

and the corresponding variation, which is $-2D_{ss} U_s \bar{U}_s^*$. The total variation of Q_m is

$$-\delta(Q_m \mid N) = \delta(D_{m+1}) = 2D_{ss}(U_s^* \bar{U}_s + U_s \bar{U}_s^*)$$
$$= 4D_{ss} \operatorname{Re}(U_s^* \bar{U}_s) = 4D_{ss} \mid U_s^* \bar{U}_s \mid \cos(-\phi_s + \bar{\phi}_s) \quad (9.63)$$

The upper limit of $\mid \delta(D_{m+1}) \mid$ is $4D_{ss} \mid U_s^* \bar{U}_s \mid$.

Next, if the same reflection $U_H = U_s$ appears several times in D_{m+1}, the exact calculation would be very complicated. However, as the redundancy is generally small as compared to the total number of elements, we can accept the approximation that the variation of D_{ps} as a result of the variation of U_H included in D_{ps} is small.

Thus, for each time the reflection U_H appears in D_m, a term like (9.63) is to be added to the variation of Q_m. To simplify the notation, we call $(U_s)_r$ the reflection which appears for the rth time, regardless of the index of its column and row. Then, the upper limit is

$$\mid \delta(Q_m \mid N)_{\max} \mid = \mid -\delta(D_{m+1})_{\max} \mid = 4 \sum_{r=1}^{R} (D_{ss})_r \mid (U_s^*)_r (\bar{U}_s)_r \mid \quad (9.64)$$

where R is the redundancy.

9.3.4.5. Goedkoop Determinants

Most of the properties of the Karle–Hauptman determinants are still valid for the Goedkoop G_m determinants. Their elements are linear combinations of U_{pq}:

$$G_m^i = \det(V_{pq}^i) \qquad V_{pq}^i = \sum_s \chi_s^i \exp(iH_p \cdot t_s) U_{\phi_s H_p - H_q} \quad (9.65)$$

where i denotes the ith irreducible representation of the group, χ_s^i the character of the symmetry element $(\hat{\phi}_s, t_s)$, $\hat{\phi}_s$ being the rotation operator, and t_s the translation associated with $\hat{\phi}_s$. Their interest resides in more powerful relations for a given order m. The maximization properties and the corresponding regression equations have been worked out by Mauguen, de Rango and Tsoucaris (1973) and Castellano, Podjarny, and Navaza (1973). Here we quote only two outstanding results (Goedkoop, 1950):

$$G_m = 0 \qquad \text{for } m \geq N/S + 1$$

where S is the symmetry multiplicity. For any m, D_m can be factorized in a product involving G_m:

$$D_m = \prod_i G_{m/S}^i$$

9.4. Practical Applications

Three radically different ways to exploit the power of determinants have been devised, as described below.

9.4.1. Low-Order Determinants, and Determination of Structure Invariants

For order $m < 10$, even if all phases could be determined, the number of reflections is not enough to determine even a moderately complex structure. However, one can still use the invariant formulation of D_m (Section 9.2.3) in order to determine a priori phase invariants from the moduli. As an example, the Harker–Kasper (1948) inequalities provide directly phase invariants, for instance, the sign of U_{2H} from $|U_{2H}|$ and $|U_H|$.

The philosophy of the calculation is simple in principle: starting from the joint probability expression (9.43), one can integrate over all phase invariants, except one (or a few). We are left then with a joint probability density of only one (or a few) phases. It is not difficult to extract the phase-invariant information from the very-low-order determinants; for D_3 (9.8) the maximum D_3 determinant rule leads immediately to the familiar relation

$$D_3 = \text{Max} \Rightarrow \alpha_{\text{most probable}} = 0$$

But it is very useful to detect phase invariants that are different from zero. The D_4 is the lowest-order determinant offering such possibilities, as shown by Messager and Tsoucaris (1972) and Sarrazin (1971, personal communication). We illustrate below the usefulness of this approach. The expression of D_4 is

$$D_4 = \begin{vmatrix} 1 & U_{-H} & U_{-K} & U_{-L} \\ U_H & 1 & U_{H-K} & U_{H-L} \\ U_K & U_{K-H} & 1 & U_{K-L} \\ U_L & U_{L-H} & U_{L-K} & 1 \end{vmatrix} \tag{9.66}$$

$$
\begin{aligned}
D_4 = {} & 1 - (1 - |U_{K-L}|^2)|U_H|^2 - (1 - |U_{H-L}|^2)|U_K|^2 \\
& - (1 - |U_{H-K}|^2)|U_L|^2 - |U_{H-K}|^2 - |U_{H-L}|^2 - |U_{K-L}|^2 \\
& + 2\,\text{Re}[(U_{-H} - U_{-K}U_{K-H})U_L U_{H-L} + (U_{-K} - U_{-H}U_{H-K})U_L U_{K-L} \\
& + (U_{H-K} - U_H U_{-K})U_{L-H}U_{K-L} + (U_H U_{-K}U_{K-H})] \tag{9.67}
\end{aligned}
$$

9.4.1.1. Qualitative Approach

In the special case where $E_{\mathbf{K}} = E_{\mathbf{L-K}} = 0$, the maximum determinant rule leads simply to

$$D_4 = \text{moduli} - 2 \mid U_{-\mathbf{H}}U_{\mathbf{H-K}}U_{\mathbf{L}}U_{\mathbf{K-L}} \mid \cos \beta = \text{Max} \qquad (9.68)$$

$$\beta = \Phi_{-\mathbf{H}} + \Phi_{\mathbf{H-K}} + \Phi_{\mathbf{K-L}} + \Phi_{\mathbf{L}}, \qquad D_4 = \text{Max} \Rightarrow \beta_{\text{most probable}} = \pi$$

Also, in the special case where $\mathbf{L} = \mathbf{H} + \mathbf{K}$ and $U_{\mathbf{K}} = 0$, we have

$$\beta = 2\Phi_{-\mathbf{H}} + \Phi_{\mathbf{H-K}} + \Phi_{\mathbf{H+K}} \Rightarrow \beta_{\text{most probable}} = \pi$$

9.4.1.2. Quantitative Approach

Next, we return to the general D_4 and we write the joint probability law for $m = 4$ from the general law (9.39) or (9.43):

$$P(E_{\mathbf{L}}, E_{\mathbf{L-H}}, E_{\mathbf{L-K}} \mid U_{\mathbf{H}}, U_{\mathbf{K}}, U_{\mathbf{H-K}}) = \pi^{-3}D_3{}^{-1} \exp\left[N\left(\frac{D_3 - D_4}{D_4}\right)\right]$$

The expansion of the numerator is given in Appendix 9A.1. This expression involves four triple phase invariants:

$$\alpha = \Phi_{\mathbf{H}} + \Phi_{-\mathbf{K}} + \Phi_{\mathbf{K-H}}, \qquad \alpha_1 = \Phi_{\mathbf{H}} + \Phi_{-\mathbf{L}} + \Phi_{\mathbf{L-H}}$$

$$\alpha_2 = \Phi_{\mathbf{K}} + \Phi_{-\mathbf{L}} + \Phi_{\mathbf{L-K}}, \qquad \alpha_3 = \Phi_{\mathbf{H-K}} + \Phi_{\mathbf{K-L}} + \Phi_{\mathbf{L-H}}$$

Clearly, only three of these invariants are independent. In order to obtain information about one of them, say α, one should first integrate over the other two, for instance, α_1 and α_2. The result of the integration is the probability density (Appendix 9A.3):

$$P(R_1, R_2, R_3 \mid \alpha)$$

with

$$R_1 = \mid E_1 \mid = \mid E_{\mathbf{L}} \mid, \qquad R_2 = \mid E_2 \mid = \mid E_{\mathbf{L-H}} \mid, \qquad R_3 = \mid E_3 \mid = \mid E_{\mathbf{L-K}} \mid$$

The final result, given below, is identical with that obtained by Hauptman (1971) by using a more general, but lengthy, mathematical approach:

$$P(\mathbf{R_L} \mid \alpha) = P(R_1, R_2, R_3 \mid \alpha)$$

$$= R_1 R_2 R_3 D_3{}^{-1} \exp(-D_{11}R_1{}^2 - D_{22}R_2{}^2 - D_{33}R_3{}^2)I_0(x)I_0(y)I_0(z)$$

$$+ 2 \sum_{r=1}^{\infty} (-1)^r I_r(x)I_r(y)I_r(z) \cos r\gamma \qquad (9.69)$$

where, in a more concise notation, we use the three-dimensional vector $(R_1, R_2, R_3) = \mathbf{R}_L$; $I_r(x)$ is a modified Bessel function of order r, and

$$x = 2R_2R_3 \,|\, D_{32} \,|, \qquad y = 2R_1R_3 \,|\, D_{13} \,|, \qquad z = 2R_1R_2 \,|\, D_{12} \,|$$

$|\, D_{12} \,|$ etc. are the moduli of minors of the D_3 determinant indicated in dashed lines in (9.66):

$$|\, D_{23} \,| = D_3^{-1}(|\, U_\mathbf{H} \,|^2 \,|\, U_\mathbf{K} \,|^2 + |\, U_{\mathbf{K}-\mathbf{H}} \,|^2 - 2 \,|\, U_\mathbf{H} U_\mathbf{K} U_{\mathbf{K}-\mathbf{H}} \,|\cos \alpha)^{1/2}$$

$$\gamma = \alpha - \tan^{-1} a - \tan^{-1} b - \tan^{-1} c$$

$$a = \frac{|\, U_\mathbf{H} U_\mathbf{K} \,|\sin \alpha}{|\, U_\mathbf{H} U_\mathbf{K} \,|\cos \alpha - |\, U_{\mathbf{K}-\mathbf{H}} \,|}$$

b and c are obtained by sequential permutation of the subscripts $\mathbf{H}$, $\mathbf{K}$, $\mathbf{K} - \mathbf{H}$. The above formula expresses the joint probability law of the moduli R_1, R_2, R_3 under the condition that α is fixed.

Next, in order to introduce a larger number of moduli in the calculation of the invariant α, we can apply the axiom of compound probabilities for vectors $\mathbf{L}$ running over all reciprocal-lattice points:

$$P(\mathbf{R}_{L_1} \cdots \mathbf{R}_{L_j} \cdots |\, \alpha) = \prod_j P(\mathbf{R}_{L_j} |\, \alpha) \tag{9.70}$$

where $\prod$ denotes the product over j, comprising all vectors $\mathbf{L}_j$ for which the three corresponding moduli are located within the observable Ewald sphere. Therefore, equation (9.70) has the general form

$$P(\text{all moduli} \,|\, \alpha)$$

In the above formula, we assume that the $\mathbf{R}_L$ are mutually independent, which is, of course, only an approximation.

Conversely, one can obtain probabilistic information on α under the condition that all moduli are fixed, by using a fundamental theorem of conditional joint probabilities, also called Bayes theorem. We will denote now by $\mathbf{R}$ the whole set of moduli involved in (9.70).

Then, this theorem states that for two random variables $\mathbf{R}$ and α, we have

$$P(\mathbf{R} \,|\, \alpha) \cdot P(\alpha) = P(\mathbf{R}, \alpha) = P(\alpha \,|\, \mathbf{R}) \cdot P(\mathbf{R}) \tag{9.71}$$

$$P(\alpha \,|\, \mathbf{R}) = P(\mathbf{R} \,|\, \alpha) \cdot P(\alpha)/P(\mathbf{R}) \tag{9.72}$$

where the denominator depends solely on $\mathbf{R}$ and can be considered as a

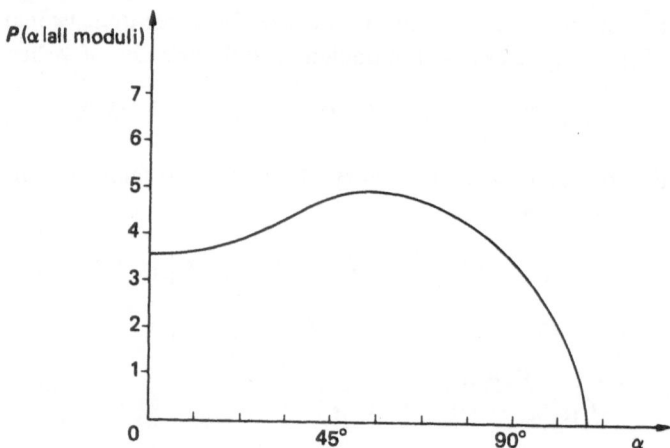

FIGURE 9.7. Conditional probability density given by (9.74); the scale is arbitrary.

constant with respect to α:

$$K = 1/P(\mathbf{R})$$

The expression $P(\alpha)$ is the *a priori* probability density of α, i.e., assuming only the values of the moduli $|U_\mathbf{H}|$, $|U_\mathbf{K}|$, and $|U_{\mathbf{K}-\mathbf{H}}|$, and no information about the moduli R_1, R_2, R_3. The expression, already known, is

$$P(\alpha) = \frac{\exp(A \cos \alpha)}{2\pi I_0(A)} \tag{9.73}$$

$$A = 2N^{-1/2} |E_\mathbf{H} E_\mathbf{K} E_{\mathbf{H}-\mathbf{K}}|$$

We obtain finally from (9.70) and (9.72)

$$P(\alpha \,|\, \text{all moduli}) = KP(\alpha) \prod_j P(\mathbf{R}_{\mathbf{L}_j} \,|\, \alpha) \tag{9.74}$$

The same final result is obtained by starting from (9.44) instead of (9.43) combined with Bayes' theorem. The important point is that the probability expression (9.69), as well as (9.74) obtained from D_4, is maximum for a value of α which is generally different from zero, in contrast with the corresponding formula (9.73) obtained from a D_3 determinant (Fig. 9.1). Figure 9.7 shows an example[†] of variation of the expression (9.74), where the maximum occurs for a value $\alpha_{\max} \approx 110°$, close to $\alpha_{\text{correct}} = 88°$.

[†] The data are taken from the structure of estriol (Cooper, Norton, and Hauptman, 1969). In the product (9.74), only terms with $|E_\mathbf{L}| = 1.32$ have been included.

Next, we can try to use determinants with $m \geq 5$ in order to include more moduli in the determination of one structure invariant. As an example,

$$D_5 = \begin{vmatrix} 1 & U_{-K} & U_{-K-L} & U_{-L} & U_H \\ U_K & 1 & U_{-L} & U_{K-L} & U_{H+K} \\ U_{K+L} & U_L & 1 & U_K & U_{-m} \\ U_L & U_{L-K} & U_{-K} & 1 & U_{H+L} \\ U_{-H} & U_{-H-K} & U_m & U_{-H-L} & 1 \end{vmatrix} \qquad (9.75)$$

with $H + K + L + m = 0$ and $U_{K-L} = 0$, leads through (9.44) to the probability density

$$P(\beta \mid |U_H|, |U_K|, |U_L|, |U_m|, |U_{H+K}|, |U_{H+L}|, |U_{K+L}|)$$

Again, in the simplified case where $U_{H+K} = U_{K+L} = U_{H+L} = U_{L-K} = 0$, we obtain immediately

$$D_5 = \text{moduli} - 4 \, |U_H U_K U_L U_m| \cos \beta = \text{Max} \Rightarrow \beta_{\text{most probable}} = \pi \qquad (9.76)$$

Even higher determinants can be constructed and used along these lines, for example

$$D_8 = \begin{vmatrix} 1 \\ U_H & 1 \\ U_K & U_{K-H} & 1 \\ U_L & U_{L-H} & U_{L-K} & 1 \\ U_{H+K} & U_K & U_H & U_{H+K-L} & 1 \\ U_{H+L} & U_L & U_{H+L-K} & U_H & U_{L-K} & 1 \\ U_{K+L} & U_{K+L-H} & U_L & U_K & U_{L-H} & U_{K-H} & 1 \\ U_{-m} & U_{K+L} & U_{H+L} & U_{K+H} & U_L & U_K & U_H & 1 \end{vmatrix}$$
$$(9.77)$$

with $H + K + L + m = 0$.

Again, as an extreme simplification, if nine structure factors out of 13 are zero,

$$U_{H+K} = U_{H+L} = U_{L+H} = U_{L-H} = U_m = U_K = U_{H+L-K} = U_{K+L-H}$$
$$= U_{H+K-L} = 0$$

then there is a higher probability that the "quartet" β is close to π:

$$D_8 = \text{moduli} - 6 \, |U_{-H} U_{H-K} U_{K-L} U_L| \cos \beta = \text{Max} \Rightarrow \beta_{\text{most probable}} = \pi$$
$$(9.78)$$

We note also that the probability law for the moduli of the last row of D_{m+1}, given all structure factors of D_m, has been derived by Mauguen (1972) for any order m.

Let us summarize the main ideas and results of this section.

(a) The global maximum determinant rule is an excellent mathematical apparatus to extract phase-invariant information from a limited number of moduli (say from 3 to 30). So far, the results coincide with those obtained by Hauptman *et al.* by using the "general characteristic function" technique, where the calculations are much more involved. Conversely, this coincidence is an indication for the correctness of the global maximum rule [(9.44)].

(b) The reliability of the results depends on both the number of moduli involved and the degree of approximation accepted in the formulas. The results obtained so far are encouraging, but we cannot say today that they are decisive. More work is necessary along the lines of high-order determinants (more moduli) and higher approximation in the mathematical treatment.

9.4.2. Medium-Order Determinants, and Their Use in *ab initio* Structure Determination

The optimum order depends primarily on the number of atoms N. Approximately, we can consider m of the order of $N/4$ to $N/2$. For centrosymmetric structures, an algorithm has been described and worked out by de Rango (1969). In noncentrosymmetric structures, the essential problem is the determination of an initial set of phases which can be subsequently extended by the regression formula or other methods. Generally, three steps are involved: (a) choice of a "good determinant"; (b) phase determination for a "nucleus" of phases; (c) phase extension and refinement. Steps (a) and (c) are very similar for centrosymmetric and noncentrosymmetric structures. Step (b) is quite different in the two cases.

9.4.2.1. Choice of a Good Determinant

The following criteria, or their combinations, are useful in practice.

(a) *The value of D_m must be minimum. The order m must be as high as possible*. This point has been developed in Section 9.2.4. The decrease of D_m as function of m is illustrated in Fig. 9.8. The experimental points are to be compared to a theoretical relation (full line) derived as follows.

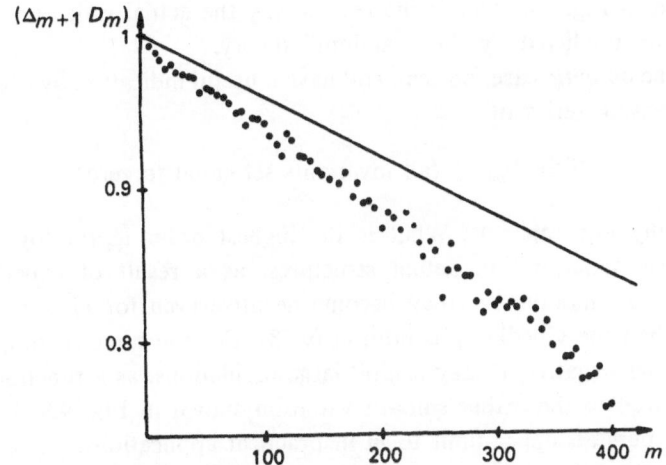

FIGURE 9.8. Variation of the value of the ratio (Δ_{m+1}/D_m) as a function of the determinant order m; the straight line indicates the theoretical mean value of $(1 - m/N)$.

Assume, the conditions (9.35), i.e., $H_1 \ldots H_m$ are fixed, L is variable; it has been shown (Tsoucaris, 1970a, p. 495) that

$$\langle Q_m \rangle_L = m \qquad (9.79)$$

recalling that Q_m depends solely on *one* random vector L through the set $(E_{L-H_1} \cdots E_{L-H_m})$. Therefore, from (9.20)

$$\left\langle \frac{\Delta_{m+1}}{D_m} \right\rangle_L = \left\langle 1 - \frac{Q_m}{N} \right\rangle_L = 1 - \frac{m}{N} \qquad (9.80)$$

Now, the above reasoning can be applied by recurrence from $m = 1$; it leads to the straight line of Fig. 9.8. From (9.80) one can compute immediately the average value of D_m:

$$(D_m)_{av} = \prod_{r=1}^{m} \left(1 - \frac{r}{N}\right) \qquad (9.81)$$

Clearly the validity of (9.80) is connected with the notion of a determinant chosen randomly. Indeed, the statistical average over L in (9.79) combined with the recurrence character of (9.80), is equivalent to a random choice of all defining vectors $H_1 \cdots H_m$. But in the actual determinant of insulin, the choice of the $H_1 \cdots H_m$ was not random; criteria (b) and (c) as well as condition (a) led to a special set $H_1 \cdots H_m$ whose determinant is plotted in Fig. 9.8. This is the reason why the actual ratios are smaller than those predicted by the "random" theory.

In the *ab initio* case, one can still have a useful indication by examining the decreasing series of

$$D_m^{(0)} = D_m \qquad \text{(all invariants set equal to zero)} \qquad (9.82)$$

Finally one can ask: What is the highest order useful for efficient phase determination? In actual structures, as a result of experimental errors on the moduli, D_m may become negative even for m considerably smaller than the Goedkoop condition (9.28). For the same reason, before D_m becomes negative, it may exhibit large oscillations, as a function of m, as compared to the rather smooth variation shown in Fig. 9.8. It is this fact that puts an upper limit to m in practical applications.

(*b*) *The average value of the moduli of the elements must be large.* This has been pointed out by Kitaigorodski (1961, p. 161), who has studied, with a simplified model, the dependence of D_m on the moduli of its elements. However, there are indications that the efficiency of determining phase invariants far from zero is enhanced if a row contains both large E's and very small ones; this suggests a criterion of the type $\langle \mid \mid E \mid^2 - 1 \mid \rangle$.

(*c*) *The moduli of the first row must be large.* Indeed, in this case there are theoretical and empirical indications that (i) the moduli of the other moduli are also large, (ii) the phase invariants are closer to zero than required by Cochran's distribution (de Rango, 1969).

(*d*) *The redundancy may be high.* Here also, we have a quantitative indication in a special case. In the paper of Knossow *et al.* (1977) it is shown that the value of D_m in the case of the highest possible redundancy (cyclic determinant) is considerably smaller than the average value (9.81). We recall also the theoretical difficulties discussed in connection with (9.44).

The construction of D_m can be performed either by starting from D_2 and "growing" the determinant to the desired order, or by a "convergence" process. Sarrazin (1973) has devised a program starting from D_{400} and deleting rows and columns according to the above criteria; similar results are being achieved by the York school.

9.4.2.2. Phase Determination of the "Nucleus"

The "nucleus" is the set of reflections that are to be determined by a single algorithm. The phases of structure factors included in a single determinant (except the very small ones) can be this nucleus. Perhaps two or a few more or less "independent" determinants are necessary for more complex structures.

Centrosymmetric Case. A method has been devised by de Rango (1969) and successfully applied to actual structures with about 30 independent atoms. Although this method has to be replaced by a more modern procedure, we give, for pedagogical reasons, the essential steps for the structure of jamine (Karle, 1966; $P\bar{1}$, $N = 48$), summarized as follows.

(1) A D_{18} determinant was constructed by the "growing" technique. Then, it was written in the "phase-invariant" form (Section 9.2.3), or in the "sign-invariant" form; each element U_{pq} is replaced by the invariant expression

$$u_{pq} = t_{pq} \mid U_{pq} \mid \qquad (9.83)$$

with

$$t_{pq} = \exp(i\alpha_{pq}) = s_{1p} s_{pq} s_{q1}$$

where α_{pq} is the phase invariant (9.11) and t_{pq}, for centrosymmetrical structures, involves only signs [$\text{sign}(U_{pq}) = s_{pq}$].

(2) A general algorithm was devised to determine the sign invariants of the last row of Δ_{m+1}, given the sign invariants of D_m (Appendix 9A.4). This algorithm was applied by recurrence from $m = 2$ to $m = 18$; we obtained a D_{18} with known sign invariants. An essential trend is the determination of negative invariants, through a single formula, the regression equation (9.53) (shown in a phase-invariant form in Appendix 9A.4).

(3) Symbolic signs were introduced for the structure factors of the first row (s_{1p}) and subsequently determined by a method similar to the symbolic addition (Karle and Karle, 1966). Then all signs of D_m were immediately obtained (9.83) from the symbols s_{1p} and t_{pq} determined in (2) above:

$$s_{pq} = s_{1p} s_{1q} t_{pq} \qquad (9.84)$$

It is possible that the "nucleus D_m determinant" may not have enough redundancies to determine the signs of all its structure factors. But this is not necessary at this stage; the signs of the invariants of D_m determined in (2) are used directly to extend the sign invariants in stage (c). Then, having obtained a very large number of sign invariants, the determination of the structure-factor sign is almost trivial.

Noncentrosymmetric Case. The procedure is basically the same as in classical direct methods: (a) choice of origin- and enantiomorph-defining reflections; (b) determination of the order in which reflections are treated [an appropriate criterion could be (9.81), where the D_{pq} are obtained *ab initio* through the approximation "all invariants set to zero," as in (9.82)]; (c) use of a criterion to accept or reject a calculated phase [again, (9.64) may be a useful one].

Two schemes are considered in practice.

(a) Simultaneous determination of all phases in D_m through the maximum determinant rule, with respect to all phases. We can apply either the global rule or (9.43). Of course, the main problem is the computing time necessary to scan through all combinations of phases. Several devices can be considered including the magic integer approach developed in York.

(b) Stepwise determination by an integration formula, recalling the algorithm developed in Section 9.3:

$$P(\phi_{rs}) = \int_{\text{except } \phi_{rs}} \cdots \int P(\cdots U_{pq} \cdots)\, d\phi_{12} \cdots d\phi_{pq} \cdots \qquad (9.85)$$

where the integration of (9.44) is performed numerically over all reflections except one at a time, say, ϕ_{rs}.

9.4.2.3. Phase Refinement

For a given D_m it is possible to generate a very large number of Δ_{m+1} determinants, each associated with a vector $\mathbf{L}$ ($\mathbf{L} \neq \mathbf{H}_p$, $p = 1, \ldots, m$). A particular $\Delta_{m+1}(\mathbf{L})$ is selected for the refinement if the following hold: (a) the phase of at least one of the structure factor $E_p = E_{\mathbf{L}-\mathbf{H}_p}$ is already known; and (b) the number of $E_{\mathbf{L}-\mathbf{H}_p}$ outside the "observed" sphere is smaller than a limit fixed in advance.

For each $\Delta_{m+1}(\mathbf{L})$ selected, the phases of the last row are determined by maximization of Δ_{m+1}. This can be achieved either by a steepest-descent technique (see Section 9.2), or by the regression formula (9.53) (Appendix 9A.4). Then, for the same reflection, several determinations are obtained; they are recombined according to (9.60).

9.4.3. High-Order Determinants, and Their Use in Protein Structure Determination

Phase refinement and phase extension in protein structure determination is a problem particularly suitable for the determinantal method. Indeed, the method of multiple isomorphous replacement (MIR) provides approximate phases for low-resolution reflections. Ideally, known reflections will be the elements of D_m; the reflections outside the "phased sphere" will appear in the last row of D_{m+1} (denoted Δ_{m+1}), and will be determined by maximization. In the actual computer program, maximization is achieved through the regression (9.53). As in the refinement (see Section 9.4.2.3) for small molecules, for a given D_m, a set of $\Delta_{m+1}(\mathbf{L})$ determinants is selected, and the phases are determined for each of them. Finally, for each symmetry-independent reflection, the phase determinations from different $\Delta_{m+1}(\mathbf{L})$'s are recombined according to (9.60).

However, before applying this procedure to protein structures, we must make sure that these formulas are still valid for very-high-order determinants. Such orders are necessary according to the discussion on the efficiency and accuracy of phase determinants: the value of D_m must be small (see Section 9.2.4.3) as well as the value of the ratio Δ_{m+1}/D_m, which has the meaning of a variance [(9.56)]. As a consequence of the large value of N (generally over 2000), a correspondingly large-order m is necessary, according to (9.80), to ensure a value of Δ_{m+1}/D_m significantly smaller than 1.

9.4.3.1. Choice of Order m

In theory, one could use determinants up to order N $[D_{N+1} = 0$, (9.28)], the ratio D_{m+1}/D_m reaching zero for $m = N$. But, owing to experimental errors, this ratio tends to zero, or oscillates for a much smaller value of m (about 500, instead of $N \approx 2400$, as for insulin). A convenient order is the range just below this oscillation range, say, for insulin between 300 and 450. This limiting value depends on the choice of the defining reflections (Fig. 9.8).

It is worth noticing that the optimum value of m is better defined in terms of the ratio m/N. For small structures, i.e., $N = 20\text{--}200$, $(m)_{\text{optimum}} \approx N/2$ or $N/4$. For $N \approx 2500$, $(m)_{\text{optimum}} \approx N/6$. For more details, one should consult the original paper (de Rango, Mauguen, and Tsoucaris, 1975).

9.4.3.2. Validity of the Regression Formula

The theoretical distribution of errors associated with a single regression equation is given by (9.58). The corresponding histograms have been established from a set of normalized structure factors calculated from the atomic models of myoglobin and insulin. Figure 9.9 shows the histogram for a fixed value of A_q in (9.58). For $m = 400$ and for high-resolution data (1.9 Å) the agreement is satisfactory.

The final value of the calculated phase is obtained from the "phase recombination" formula (9.60), which is a sort of generalized tangent formula.

A measure of the overall efficiency of the method, for a given set of reflections, may be obtained from the average absolute value of the differences between the exact value ϕ_H and the computed value Φ_H (9.60):

$$\Delta\Phi = \langle |\,\phi_H - \Phi_H\,|\rangle_H \tag{9.86}$$

where H belongs to a given set. A measure of the precision associated with the determination of each contributor to $\tan\Phi_H$ in (9.60) is evaluated by recalling that $\phi_H^{(j)}$ is, in fact, the value of $\bar\phi_q = \bar\phi_{L-H_q}$ obtained from the jth determinant, (9.58); we define

$$\delta\phi = \langle |\,\varepsilon_q\,|\rangle_{q,L} = \langle |\,\phi_{L-H_q} - \bar\phi_{L-H_q}\,|\rangle_{q,L} \tag{9.87}$$

for fixed A_q where $\phi_q = \phi_{L-H_q}$ is the exact value, and $\bar\phi_q = \bar\phi_{L-H_q}$ is a single determination obtained by (9.53). Here, also, the variable indices q and L may belong to a given set which will be indicated at each use of

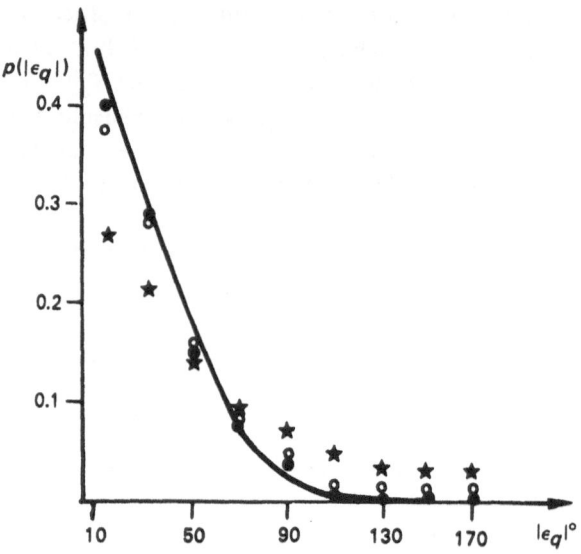

FIGURE 9.9. Distribution of the differences $|\,\varepsilon_q\,|$ for $A_q = 3$. The curve indicates the theoretical distribution given by (9.58). ★, Myoglobin ($m = 150$, $R > 4$ Å); ○, insulin ($m = 150$, $R > 1.9$ Å); ●, insulin ($m = 400$, $R > 1.9$ Å).

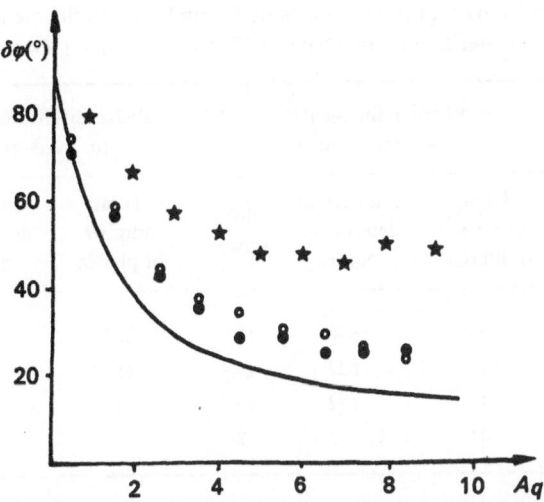

FIGURE 9.10. Variation of $\delta\phi$ as a function of A_q for myoglobin and insulin. The curve indicates the theoretical distribution given by equation (9.88). ★, Myoglobin ($m = 150$, $R > 4$ Å); ○, insulin ($m = 150$, $R > 1.9$ Å); ●, insulin ($m = 400$, $R > 1.9$ Å).

(9.87). Figure 9.10 shows the experimentally found dependence of $\delta\phi$ on A_q, as compared to the theoretical one obtained from (9.58):

$$(\delta\phi)_{\text{th}} = \frac{1}{\pi} \int_0^{\pi} \varepsilon P(\varepsilon)\, d\varepsilon \qquad (9.88)$$

9.4.3.3. Phase Refinement and Extension

Table 9.5 shows the final values of $\Delta\Phi$. In the refinement process (9.60), D_m includes all structure factors. In the extension process, it is not practical to introduce into D_m only reflections with known phase, the unknown phases appearing exclusively in the last row of D_{m+1}. Fortunately, the presence of unknown phases in D_m can be treated very simply: the structure factors of unknown phase are set equal to zero (actually a third of the structure factors involved in D_{300} of Table 9.5). Despite this approximation, the value of $\Delta\phi$ is reasonably small. We must keep in mind, however, that in the initial data used to determine the phases indicated in Table 9.5, no error is assumed, neither in the moduli (in the 2.2-Å resolution sphere), nor in the phases (in the 2.8-Å sphere).

Practical applications have been performed by using observed data of insulin (de Rango, Mauguen, and Tsoucaris, 1975b), triclinic lysozyme (Podjarny, Yonath, and Traub, 1976), and tRNA (Podjarny and Yonath, 1977).

TABLE 9.5. Summary of Results Obtained from Phase Refinement and Extension
with a D_m Determinant of Order 300, for Calculated Data of Insulin

$\|E\|$	Phase refinement (1.9-Å resolution)			Phase extension from 2.8-Å to 2.2-Å resolution		
	Total number of phases	Number of determined phases	$\Delta\Phi$ (deg)	Total number of phases	Number of determined phases	$\Delta\Phi$ (deg)
>1	2653	1640	17	893	621	23
>1.5	836	627	14	215	203	19
>2	214	172	13	41	39	19
>2.5	41	37	9	4	4	10

9.5. New Gram Determinants

The Karle–Hauptman determinants have been devised to translate
into reciprocal space the positivity of the electron density. Mathematically,
they are a special case of Gram determinants. One could then consider
another type of information, or another type of crystallographic problem,
and study a Gram determinant adapted to this problem. Three such types
of determinants are described in this section. The general mathematical
theory will be given in Appendix 9A.5 (Knossow *et al.*, 1977; Knossow,
1975).

9.5.1. Use of Stereochemical Information—Known Fragments

9.5.1.1. The "Moduli-Model" Determinant

Here, it is assumed that a fragment of the structure has a known
configuration. The problem of orienting and determining the translational
parameters in the unit cell can be solved by the Patterson search or rotation
function technique (Tollin and Cochran, 1964; Braun, Hornstra, and
Leenhouts, 1967). The Gram determinants are an alternative, potentially
more powerful.

We consider first a D_m determinant whose elements are $|U_{pq}|^2 = |U_{H_p - H_q}|^2$. We know already that $D_m \geq 0$, as the Patterson function is
positive, and, therefore, its Fourier coefficients must fulfill the Karle–

Hauptman fundamental inequalities. Next, as in Δ_{m+1} of Section 9.3, we add to D_m a last column and row, whose elements, denoted by

$$| C_{H_{m+1}-H} |^2 = | C_{m+1,p} |^2 = | C_p |^2$$

are the squared moduli of the contributions of the known fragment to the corresponding structure factor $U_{H_{m+1}-H_p} = U_{m+1,p} = U_p$. It is shown in Appendix 9A.5 that Δ_{m+1} is a Gram determinant, and therefore it is non-negative. Moreover, there are indications—not a proof—that Δ_{m+1} is maximum for the correct orientation and position of the fragment. It is worthwhile noting that the Δ_{m+1} expansion is proportional to a modified rotation function.

We now foresee practical possibilities by expressing the C_p values as a function of three orientation and three position parameters, and writing $\Delta_{m+1}(\theta_1, \theta_2, \theta_3, x, y, z)$. Furthermore, the orientation and position parameters can be explored separately. Indeed, if we consider the interatomic vectors within only *one* molecule in the unit cell, they all ought to end at points lying inside a sphere centered at the origin of Patterson space of radius equal to the "diameter" of the molecule. Their number, for all S symmetry-related molecules, is $SN_f^2 - 1$ as compared to the total number of Patterson peaks involving all symmetry-related fragments, which is $S^2 N_f^2 - 1$, where N_f is the number of atoms of the fragment. Therefore, by setting the fragment at an arbitrary position, for instance at the origin, we can compute $| C_{L-H_p}(\theta_1, \theta_2, \theta_3) |^2$ as a function of the angular parameters alone, and hope that the corresponding Δ_{m+1} determinant has still a maximum at the correct values of $\theta_1, \theta_2, \theta_3$. Practical applications, as shown below, justify this expectation.

After having determined the orientation $(\theta_1^0, \theta_2^0, \theta_3^0)$ of the fragment, the contributions of the fragment $| C_{L-H_p}(\theta_1^0, \theta_2^0, \theta_3^0, x, y, z) |^2$ are computed and plotted as a function of the positional parameters x, y, z alone.

9.5.1.2. Full Use of Crystal Symmetry

We have examined already the implications of crystal symmetry for Karle–Hauptman determinants (containing U's) and we have shown that their factorization as a product of Goedkoop determinants is very useful in practice. Here also, it is possible to factorize Δ_{m+1} and use each $G_{m+1}^{(i)}$ determinant separately (i denotes the ith irreducible representation). In practice, the totally symmetric (trivial) representation is the most convenient (Knossow *et al.*, 1977).

The elements of the Goedkoop determinant trivial representation are sums of $|U|^2$'s as follows:

$$V_{pq} = \sum_{s=1}^{S'} |U_{\hat{\phi}_s \mathbf{H}_p - \mathbf{H}_q}|^2 \tag{9.89}$$

where $\hat{\phi}_s$ is the rotation operator associated with the sth symmetry element of the point group; S' is the multiplicity in the Patterson function. Let us compare the above equation with (9.65): as the Patterson function does not involve helicoidal axes, the corresponding factor $\exp(i\mathbf{H} \cdot \mathbf{t}_s)$ does not appear here. Also, the characters being $\chi_s = +1$ for all symmetry elements ($s = 1, \ldots, S$), we arrive at the above expression.

We will examine now the decrease of $G_m = \det(V_{pq})$ for m increasing. According to the general theory, G_m is zero for m larger than the number n of distinct atoms or pseudoatoms in the asymmetric unit:

$$G_m = 0 \qquad \text{if } m \geq n+1$$

If there is no superposition in the Patterson function, then

$$n = \frac{N(N-1)}{S'} + 1$$

But, generally, for most structures there exists a number of strict superpositions (benzene rings, for instance), so that n is smaller, and sometimes considerably smaller, than the above limit. No strict theory has been developed for the exact form of the curve G_m as a function of m. For the Karle–Hauptman determinants we know that $D_{m+1}/D_m = 1 - m/N$. There are indications that a similar relation occurs also for Goedkoop determinants involving $|U|^2$:

$$\frac{G_{m+1}}{G_m} \approx 1 - \frac{m}{n}$$

9.5.1.3. Example: The Structure of Tri-ortho-thymotide (TOT)

For the structure of tri-ortho-thymotide (TOT; Brunie and Tsoucaris, 1974; $Pna2_1$, $Z = 4$, $N = 160$), Fig. 9.11 shows the decrease of G_m. The straight line represents the limiting equation

$$\frac{G_{m+1}}{G_m} = 1 - \frac{mS'}{N(N-1)} \approx 1 - \frac{m}{3150}$$

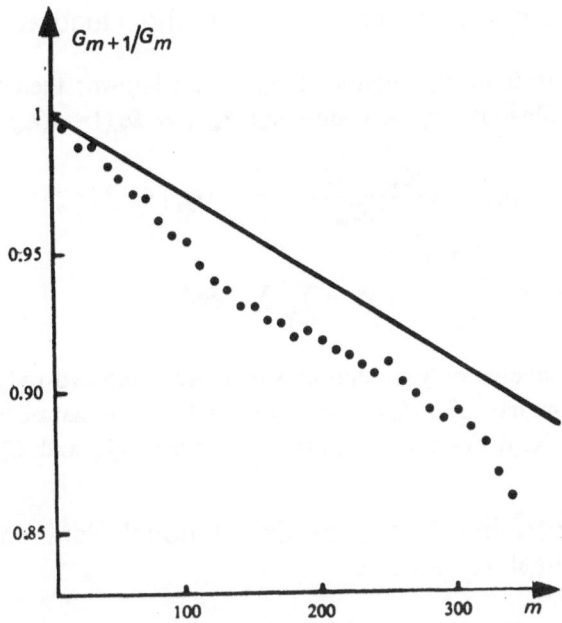

FIGURE 9.11. Variations of the value of the ratio (G_{m+1}/G_m) as a function of the determinant's order m; the straight line indicates the theoretical mean value $(1 - m/n)$.

The actual curve decreases more quickly for several reasons. First, there are certainly superpositions, and n is smaller than the limiting expression. Next, the determinants involved in this calculation are chosen, in view of practical applications, such that the average value of $|U|^2$ is higher than the general average for all structure factors (which is $1/N$). Finally, in the calculation, the observed intensities have been used, which are affected by experimental errors.

From the global maximum rule, we know that any error in the elements of the determinant, in modulus or in phase, results in a decrease of the determinant.

We notice that a similar curve has been found for the determinant of insulin (Fig. 8; de Rango, Mauguen, and Tsoucaris, 1975a) which involves U's. However, it happens that the number of atoms in the unit cell of insulin (≈ 2400) is comparable to the number of symmetry-independent pseudoatoms in the Patterson unit cell of TOT. This, in turn, indicates that the simple linear formula $(1 - m/n)$ is of wide application to any collection of point atoms or pseudoatoms.

9.5.1.4. Computation of G_{m+1} from the Quadratic Form Q_m

All elements of D_m, minor of Δ_{m+1}, are known; then the quadratic form is the obvious way of computing $\Delta_{m+1} = D_m(1 - Q_m/A)$:

$$Q_m = A \frac{G_m - G_{m+1}}{G_m} = \sum_{pq} G_{pq} \, |C_p|^2 \, |C_q|^2$$

where

$$A = \sum_{i=1} \sum_{j=1} g_i^2 g_j^2$$

Indeed, G_{pq} involve only moduli and they are computed only once by inverting the matrix of $|U_p|^2$; then, for each set of parameters only the $C_p(\theta_1, \theta_2, \theta_3, x, y, z)$'s are computed to evaluate Q_m and G_{m+1}.

9.5.1.5. Scanning Range of Orientational Parameters: Cheshire Groups

In the generation of trial structures by systematic variation of $(\theta_1, \theta_2, \theta_3, x, y, z)$ of a molecule, the ranges to be scanned by the parameters depend on the symmetry of the molecule and on the space group. The problem has been treated by Hirshfeld (1968) who tabulated the thirty "Cheshire" groups: they describe the symmetry the structure could acquire if all its atoms were removed and only its symmetry elements left behind, like the grin of the vanishing Cheshire cat. The direct product of this group with the point group of the molecule specifies the symmetry in six-dimensional space of the trial parameters. The asymmetric unit in this space is the region to be scanned by the several parameters.

9.5.1.6. Application to TOT

This example deals with the determination of the orientational parameters of the TOT molecule, which consists of three identical fragments, of known stereochemistry, arranged around a pseudoternary axis. The orientational parameters are the three Euler angles $\theta_1, \theta_2, \theta_3$. We recall that the efficiency of G_{m+1} for yielding information on $\theta_1, \theta_2, \theta_3$ is maximum for the highest order possible (by analogy with the efficiency for phase information discussed in Section 9.2.4.3). The order of G_{m+1} used in the calculations is 340. This order is the highest before the value of the ratio G_{m+1}/G_m starts to oscillate, becoming eventually negative. Such behavior is ascribed to the

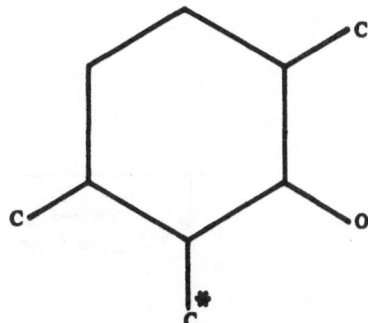

FIGURE 9.12. The stereochemically known fragment of TOT; the three identical fragments that build the molecule are bound through the oxygen and the carbon atom marked with an asterisk.

experimental errors in the moduli. It is felt that if no such error existed, the rounding errors during computing of G_m could display their effect for much higher orders. We recall that if no errors of any kind occurred, the ratio should be strictly equal to zero only for $m = n$, the number of distinct Patterson peaks, i.e., a number approximately between 2000 and 3000.

The scanning range of the parameters $(\theta_1, \theta_2, \theta_3)$ is, for $P2_12_12_1$,

$$0 \le \theta_1 < \tfrac{1}{2}\pi, \qquad 0 \le \theta_2 < \tfrac{1}{2}\pi, \qquad 0 \le \theta_3 \le \theta$$

Scanning this range, we should find, strictly, three maxima of the most probable region determinant, each maximum corresponding to one fragment of the pseudoternary molecule. But as can be seen in Fig. 9.12, the pseudoatom set of a fragment is left nearly invariant by a $2k\pi/6$ $(k = 1, 6)$ rotation about the axis normal to the benzene ring. So we must expect 18 peaks in the whole range. Remembering that D_m is constant, the equidistant contour drawn in Fig. 9.13 indicates the level of 750 when 1000 corresponds to the absolute maximum which can be reached by $G_{m+1}(\theta_1, \theta_2, \theta_3)/G_m$.

The 20 highest zones are examined, recalling that the above stereochemical considerations led us to expect 18 significant peaks. Three out of the 20 highest zones do not correspond to an actual fragment orientation; however, the height of these three peaks is among the lowest. On the other hand, one of the true orientations is at 15° from the maximum. The remaining maxima represent fairly well the correct fragment orientation.

This method is to be compared with the Patterson search, or rotation function method. First, we quote an interesting mathematical result: the expansion of Δ_{m+1} or Q_m, limited to the first terms, is identical to the rotation function $P(\mathbf{r})P_{\mathrm{rot}}^2(\mathbf{r})$. The results of the determinant method seem encouraging, in the light of the contrast between peaks and background,

$\theta_1 = 30°$

θ_2 \ θ_3	0	15	30	45	60	75
0	700	640	648	404	412	617
15	764	776	635	369	285	273
30	655	292	31	188	465	538
45	564	91	391	663	700	658
60	545	607	723	646	565	662
75	405	635	620	673	620	686
90	303	749	692	697	716	599
105	762	781	723	766	734	435
120	730	690	738	751	600	535
135	684	655	699	447	576	600
150	719	597	139	264	693	673
165	763	782	489	680	651	597

$\theta_1 = 75°$

θ_2 \ θ_3	0	15	30	45	60	75
0	404	412	617	643	679	681
15	630	632	511	675	720	606
30	570	254	−117	596	720	649
45	384	38	342	785	759	652
60	−25	361	740	737	771	751
75	16	645	603	638	723	771
90	200	675	677	631	732	724
105	500	577	540	623	691	513
120	295	599	667	690	595	513
135	571	627	662	778	685	547
150	720	754	699	787	777	641
165	716	709	742	654	738	711

FIGURE 9.13. Rotation search for the orientation of the stereochemically known fragment of TOT. The variations of the determinant are scaled to the interval (−1000, 1000); dots indicate the true positions of the fragment.

and the number of significant peaks. It is also important to emphasize that in difficult cases one can always make a better choice of G_m to improve the contrast. This is a general feature of determinantal methods: often the efficiency depends critically on the choice of an adequate determinant among the huge number of combinations of vectors $H_1 \cdots H_m$. It is hoped that both theory and experience will help crystallographers to make the best choice of determinants, eventually through an almost automatic program.

9.5.2. The "Moduli-Model-Phase" Determinant

By an appropriate choice of the Hilbert space vector V_{m+1} as explained in Appendix 9A.5, a new determinant is obtained: as in the "moduli and fragment" determinant described above, the m-order minor contains the $|U_{pq}|^2$'s. The last column and row elements are

$$V_{m+1,q} = C_{m+1,q} U^*_{m+1,q}$$

where $V_{m+1,q}$ is the element of the last, $(m+1)$th, row and qth column; $C_{m+1,q}$ is the contribution of a known fragment to the corresponding structure factor $U_{m+1,q}$. The Gram determinant Δ_{m+1} must be nonnegative for the correct values of the phases and the correct orientation and position of the fragment. Again, there are indications that Δ_{m+1} is maximum for these correct values. Therefore, it can be used as a test for both phases and an identified fragment. As an example, the test can be applied to select the correct solution (and the correct fragment interpretation) in the multi-solution method.

9.5.3. The Equiprobability Function in Direct Space

Here we examine a fundamental problem; let us consider a function $\varrho(\mathbf{r})$ and the set of its Fourier coefficients F_H. We then pose the question: given a limited set of Fourier coefficients and certain a priori information about $\varrho(\mathbf{r})$, what is the best function $\tau(\mathbf{r})$ approximating $\varrho(\mathbf{r})$. We give a few simple examples: if there is no a priori information, $\tau(\mathbf{r})$ is simply the Fourier series with the available (limited) Fourier coefficients. If we know, in addition, that $N = 2$ in space group $P\bar{1}$, without thermal motion, obviously the coordinates are exactly obtained from a few structure factor equations ($U_H = \cos \mathbf{H} \cdot \mathbf{r}$). If we accept the usual atomic model with the Gaussian thermal ellipsoid, the $\tau(\mathbf{r})$ is usually obtained from the least-squares method, and not from a Fourier series.

However, in the last two examples, a given atomic model is necessary. But is it possible, at least in principle, to obtain directly $\tau(\mathbf{r})$, or a function close to it, without model and without Fourier series? The "forbidden domain" theory of von Eller (1962) provides a first answer. The main idea is quite simple: if one subtracts from a given structure the electron density of one "test atom," at its correct position $\mathbf{r}_j$ the electron density of the remaining $(N - 1)$-atom structure corresponds to a nonnegative Karle–Hauptman determinant:

$$^sD_m = \det\{U_{pq} - n_j \exp[2\pi i(\mathbf{H}_p - \mathbf{H}_q) \cdot \mathbf{r}_j]\} \geq 0$$

But if the subtracted test atom is located at an incorrect position, say $\mathbf{r} \neq \mathbf{r}_j$, this generally results in a $(N + 1)$-atom structure, i.e., N initial "positive" atoms and a $(N + 1)$th "negative" atom. The $\varrho(\mathbf{r})$ becomes negative around the point $\mathbf{r}$ and the inequalities may not hold. Consequently, if the "test atom" sweeps out the whole asymmetric unit, the positions for which the $^sD_m(\mathbf{r})$ is negative,

$$^sD_m(\mathbf{r}) = \det\{U_{pq} - n_j \exp[2\pi i(\mathbf{H}_p - \mathbf{H}_q) \cdot \mathbf{r}]\} \leq 0$$

are forbidden regions for the presence of an atom. Generally, the forbidden (by inequalities) regions are not large enough to locate atoms directly. However, we can call on probability theory in order to provide further restrictions inside the allowed regions. We obtain, thus, the "most probable regions" around the true atomic positions. In the paper of Knossow et al. (1977) it is shown that the maximum determinant theory may be used in connection with such investigations: the "most probable regions" are the regions where $^sD_m(\mathbf{r})$ is maximum, as a function of $\mathbf{r}$. In other words, the plotting of $^sD_m(\mathbf{r})$ in the asymmetric unit, as a function of $\mathbf{r}$, provides contours which are considered as equiprobability contour for the presence of an atom, given the phases of ^sD_m.

This theory involves a remarkable feature: the $^sD_m(\mathbf{r})$ determinant can be written in the form of a Fourier series whose coefficients are the elements of the inverse matrix of D_m (denoted by D_{pq}):

$$\tau(\mathbf{r}) = {}^sD_m(\mathbf{r}) = \sum_{p,q} D_{pq} \exp[2\pi i(\mathbf{H}_p - \mathbf{H}_q) \cdot \mathbf{r}]$$

The problem is then amenable to the standard techniques of matrix inversion and Fourier transform.

Summarizing, the knowledge of the phases in $^sD_m(\mathbf{r})$, combined with a priori known positivity of $\varrho(\mathbf{r})$ leads to a map indicating the equiprobability contours for the presence of an atom. Practical applications are under investigation in protein structure determination. The possibility of phase refinement in direct and reciprocal spaces has been thoroughly investigated by Gassmann (1976). It is remarkable that, starting from integrals involving the electron density in direct space, this author provides a new proof of the maximum determinant rule. Another important piece of information has been provided by Laszerowicz and Laszerowicz (1966): by forbidding two atoms to coincide, these authors obtained an improved probability density for structure factors. As a consequence, the special determinant D_N ($N =$ number of atoms in the unit cell) must be maximum.

Acknowledgment

The author wishes to thank Professor D. Harker for his numerous useful suggestions and for his careful reading of the manuscript throughout the preparation of this chapter.

Appendix 9A.1. Expression of the Hermitian Form Q_m for $m = 1, 2, \ldots$

We recall first how to compute the elements of an inverse matrix:

(a) Take the transpose U^t of U.
(b) Replace each element $(U^t)_{pq}$ by the corresponding minor δ_{pq}.
(c) Multiply δ_{pq} by $(-1)^{p+q}$.
(d) Divide by $D_m = \det(U_m)$.

Finally

$$D_{pq} = (U^{-1})_{pq} = (-1)^{p+q} \frac{\delta_{pq}}{D_m}$$

We illustrate the operation with a simple example:

$$U_2 = \begin{pmatrix} 1 & U_{-H} \\ U_H & 1 \end{pmatrix}, \qquad U_2^t = \begin{pmatrix} 1 & U_H \\ U_{-H} & 1 \end{pmatrix}, \qquad D_2 = 1 - |U_H|^2$$

$$\delta_{11} = \delta_{22} = 1, \qquad \delta_{12} = U_{-H}, \qquad \delta_{21} = U_H$$

$$D_{11} = D_{22} = \frac{1}{D_2}, \qquad D_{12} = -\frac{U_{-H}}{D_2}, \qquad D_{21} = -\frac{U_H}{D_2}$$

The Hermitian form Q_m has already been discussed and illustrated in Section 9.2; we will take from there the necessary expressions [(9.40) and Table 9.4]. We shall prove that

$$\mathbf{E}^+ U^{-1} \mathbf{E} = N \frac{D_m - \Delta_{m+1}}{D_m}$$

Let us develop the Δ_{m+1} determinant along the elements of the last column, then along the elements of the last line:

$$\Delta_{m+1} = \frac{1}{N} \left\{ \sum_{p=1}^{m} \sum_{q=1}^{m} (-1)^{p+q+1} \delta_{pq} E_p E_q^* + N D_m \right\}$$

where δ_{pq} denotes the minor order $m - 1$ obtained from D_m by suppressing the pth line and qth column. The above equation can be written as

$$N(D_m - \Delta_{m+1}) = \sum_{p=1}^{m} \sum_{q=1}^{m} (-1)^{p+q} \delta_{pq} E_p E_q^*$$

However, the elements of U^{-1} are

$$D_{pq} = (U^{-1})_{pq} = (-1)^{p+q} \delta_{pq} / D_m$$

Therefore

$$N \frac{D_m - \Delta_{m+1}}{D_m} = \sum_{p=1}^{m} \sum_{q=1}^{m} D_{pq} E_p E_q^*$$

which is (9.40).

For $m = 1$ we have

$$\Delta_2 = \frac{1}{N} \begin{vmatrix} 1 & E_{-H} \\ \hline E_H & N \end{vmatrix}$$

There is only one $E_1 = E_{-H}$; the principal minor reduces to one element $U_m = 1$, whose inverse is 1. Finally

$$Q_1 = D_{11} |E_1|^2 = |E_H|^2$$

and (9.39) yields immediately the Wilson law.

For $m = 2$ we have

$$\Delta_3 = \frac{1}{N} \begin{vmatrix} 1 & U_{-H} & E_{-K} \\ U_H & 1 & E_{H-K} \\ \hline E_K & E_{K-H} & N \end{vmatrix}$$

Taking the elements from above and recalling that $E_1 = E_K$, $E_2 = E_{K-H}$, we obtain [see also (9.19)]

$$Q_2 = D_{11} |E_1|^2 + D_{22} |E_2|^2 + D_{12}E_1E_2^* + D_{21}E_2E_1^*$$

$$= \frac{1}{D_2} [|E_K|^2 + |E_{K-H}|^2 - U_{-H}E_KE_{K-H}^* - U_HE_{K-H}E_K^*]$$

$$= \frac{1}{D_2} [|E_K|^2 + |E_{K-H}|^2 - 2\,\mathrm{Re}(U_{-H}E_KE_{H-K})]$$

$$= \frac{1}{D_2} [|E_K|^2 + |E_{K-H}|^2 - 2\,|U_{-H}E_KE_{H-K}|\cos\alpha]$$

where α is the familiar phase invariant. Replacing E_K by $E_K = N^{1/2}U_K$, and dividing both members by N, we arrive at (9.41).

For $m = 3$ we have:

The expression of Δ_4 is given in (9.66). Q_4 could be obtained already from $Q_3 = N(D_3 - D_4)/D_3$. Here, we apply (9.40):

$$D_{11} = \frac{1}{D_3} \begin{vmatrix} 1 & U_{K-H} \\ U_{H-K} & 1 \end{vmatrix} = \frac{1}{D_3} (1 - |U_{K-H}|^2)$$

$$D_{12} = -\frac{1}{D_3} \begin{vmatrix} U_{-H} & U_{K-H} \\ U_{-K} & 1 \end{vmatrix} = -\frac{1}{D_3} (U_{-H} - U_{-K}U_{K-H})$$

$$D_{13} = \frac{1}{D_3} \begin{vmatrix} U_{-H} & 1 \\ U_{-K} & U_{H-K} \end{vmatrix} = \frac{1}{D_3} (U_{-H}U_{H-K} - U_{-K})$$

and so on. We arrive at

$$D_3Q_3 = (1 - |U_{K-H}|^2)|E_L|^2 + (1 - |U_K|^2)|E_{L-H}|^2 + (1 - |U_H|^2)|E_{L-K}|^2$$
$$-2\,\mathrm{Re}[(U_{-H} - U_{-K}U_{K-H})E_LE_{H-L} + (U_{-K} - U_{-H}U_{H-K})E_LE_{K-L}$$
$$+(U_{H-K} - U_HU_{-K})E_{L-H}E_{K-L}]$$

Appendix 9A.2. Joint Probability for All Structure Factors: Compound Probability Law

The fundamental theorem of compound probabilities states that

$$P(A, B) = P(A)P(B \mid A) \tag{9.90}$$

where $P(B \mid A)$ is the conditional probability density of B, given A. Similarly, for a series of m random variables $A_1 \cdots A_m$, we obtain

$$P(A_1 \cdots A_m) = P(A_m \mid A_{m-1} \cdots A_1)P(A_{m-1} \cdots A_1)$$
$$P(A_1 \cdots A_{m-1}) = P(A_{m-1} \mid A_{m-2} \cdots A_1)P(A_{m-2} \cdots A_1)$$
$$P(A_1, A_2) = P(A_2 \mid A_1)P(A_1)$$

Finally

$$P(A_1 \cdots A_m) = \prod_{r=2}^{m} (A_r \mid A_1 \cdots A_{r-1})P(A_1)$$

In our case, the above equation means that we consider, successively, the probability law of the $(m + 1)$th-row elements, given all elements of D_m, then the law of the mth-row elements of D_m, given all elements of D_{m-1}, then, the law of the last-row elements of D_{m-1}, given all elements of D_{m-2}, and so on. We arrive finally at the desired expression for the probability of all elements in D_{m+1}:

$$P(\cdots U_{pq} \cdots) = P(\mathbf{U}^{(m)} \mid D_m)P(\mathbf{U}^{(m-1)} \mid D_{m-1}) \cdots P(\mathbf{U}^{(2)} \mid D_2)P(U_{12})$$

where in the abbreviated expressions $P(\mathbf{U}^{(m)} \mid D_m)$, which is exactly (9.43), D_m stands for all elements of D_m, $\mathbf{U}^{(m)}$ stands for the set of m elements of the last row of D_{m+1}, $\mathbf{U}^{(m-1)}$ for the set of $(m - 1)$ elements of the last row of D_m, and so on. At the end, D_2 contains only U_{12}. By introducing (9.43) for each order and dropping the constant factor, we arrive at (9.44).

In the case of redundancy, a theoretical difficulty arises, as the same structure factor may appear both in the last column of D_r and in the D_{r-1} minor. Then, it is not possible to consider that a given structure factor is, simultaneously, a random variable in the rth row, and a fixed correlation coefficient in the D_{r-1} minor.

Appendix 9A.3. Negative Triplet Determination

For the purpose of integration it is convenient and more general to use the structure-factor phase notation:

$$\Delta_4 = \frac{1}{N} \begin{vmatrix} 1 & U_{12} & U_{13} & E_1^* \\ U_{21} & 1 & U_{23} & E_2^* \\ U_{31} & U_{32} & 1 & E_3^* \\ \hline E_1 & E_2 & E_3 & N \end{vmatrix}$$

$$P(E_1, E_2, E_3 \mid U_{12}, U_{13}, U_{23}) = P(E_1, E_2, E_3 \mid \alpha) = \pi^{-3}D_3^{-1}\exp(-Q_3)$$

keeping in mind that the D_3 minor contains only one phase invariant α. The change of variable (polar coordinates instead of Cartesian coordinates) of the complex number $E_1 = |E_1| \exp(i\phi_1)$

$$(A_1, B_1) \rightarrow (R_1, \phi_1), \text{ etc.}$$

yields a Jacobian: $R_1 \cdot R_2 \cdot R_3$. Thus, in order to obtain the probability density of the moduli, we have to integrate over the three phases:

$$P(R_1, R_2, R_3 \mid \alpha) = \int_{\phi_1=0}^{2\pi} \int_{\phi_2=0}^{2\pi} \int_{\phi_3=0}^{2\pi} R_1 R_2 R_3 \cdot P(E_1, E_2, E_3 \mid \alpha) \, d\phi_1 \, d\phi_2 \, d\phi_3 \tag{9.91}$$

Next, we write Q_3 as

$$Q_3 = \sum_{p=1}^{3} D_{pp} |E_p|^2 + 2 \sum_{p<q=1}^{3} |D_{pq} E_p E_q| \cos(\phi_p - \phi_q + x_{pq})$$

with

$$|D_{pq} E_p E_q| = A_{pq}, \qquad D_{pq} = |D_{pq}| \exp(i x_{pq})$$

We obtain

$$P(R_1, R_2, R_3 \mid \alpha) = \pi^{-3} D_3^{-1} \exp(-D_{11} R_1^2 - D_{22} R_2^2 - D_{33} R_3^2)$$

$$\times 2 \sum_{p<q=1}^{3} \exp[-A_{pq} \cos(\Phi_p - \Phi_q + x_{pq})] \tag{9.92}$$

We now expand the three exponentials within the sign Σ, according to

$$\exp(A \cos \theta) = \sum_{r=-\infty}^{\infty} I_r(A) \exp(ir\theta) \tag{9.93}$$

where I_r denotes a Bessel function.

Introducing this expansion of (9.92) in the integral (9.91) we arrive at the general term

$$\iiint \exp[ir(\Phi_1 - \Phi_2 + x_{12})] + s(\Phi_2 - \Phi_3 + x_{23})$$

$$+ it(\Phi_3 - \Phi_1 + x_{31}) \, d\Phi_1 \, d\Phi_2 \, d\Phi_3$$

which is zero, unless $r = s = t$, in which case it is simply

$$\exp[ir(x_{12} + x_{23} + x_{31})]$$

We obtain thus immediately (9.69). The above procedure, due to Mauguen (1972) may be generalized to any order m.

Appendix 9A.4. Determination of the Sign Invariants of the Last Row of Δ_{m+1}

We assume that all the sign invariants of D_m are known.

9A.4.1. Notation

We write Δ_{m+1} (Table 9.4) in the invariant form given in Table 9.2b; each element U_{pq} or $E_{m+1,q} = E_q$ is replaced by the corresponding invariant. Thus,

$$u_{pq} = t_{pq} \, | \, U_{pq} \, |, \qquad e_q = t_q \, | \, E_q \, |$$

where t_q is a short notation for $t_{m+1,q}$ and E_q for $E_{m+1,q}$. After these transformations, the determinants D_m and Δ_{m+1}, whose values remain invariant, take on the following form:

$$\Delta_{m+1} = \frac{1}{N}
\begin{vmatrix}
1 & \cdots & u_{1p} & \cdots & u_{1q} & \cdots & \vdots & e_1{}^* \\
\vdots & & \vdots & & \vdots & & \vdots & \vdots \\
u_{p1} & \cdots & 1 & \cdots & u_{pq} & \cdots & \vdots & e_p{}^* \\
\vdots & & \vdots & & \vdots & & \vdots & \vdots \\
u_{q1} & \cdots & u_{qp} & \cdots & 1 & \cdots & \vdots & e_q{}^* \\
\vdots & & \vdots & & \vdots & & \vdots & \vdots \\
\hline
e_1 & \cdots & e_p & \cdots & e_q & \cdots & & N
\end{vmatrix}$$

Let us call U_m the matrix of elements u_{pq} and d_{pq} the elements of the inverse matrix U_m^{-1}. The regression equation (9.53) is now written as

$$e_q = -\frac{1}{d_{qq}} \sum_{\substack{p=1 \\ p \neq q}}^{m} d_{pq} e_p \qquad (9.94)$$

9A.4.2. Selection of Δ_{m+1}

The U_{18} matrix (the sign-invariant form) has been constructed in Table 9.6 and inverted (matrix U_{18}^{-1} in Table 9.7). A vector L is chosen and the corresponding Δ_{m+1} determinant is obtained from D_m by "edging" D_m with the set of m structure factors $(e_{L-H_p}, p = 1, \ldots, m)$. If the moduli of some E_{L-H_q} are unobtainable under the experimental conditions, the

TABLE 9.6. Determinant [a] Δ_{m+1}

$p=$	1	2	3	4	5	6	7	8	9	10	11	12	13	14	15	16	17	18	$|e_{L-H_p}|$
E_0U_{1p}	6.92	4.85	*4.70*	4.35	4.34	4.24	4.12	4.08	3.94	3.84	3.80	3.80	*3.64*	3.61	3.59	3.53	3.28	3.10	2.03
E_0U_{2p}	4.85	6.92	*2.52*	2.29	2.72	2.09	3.04	3.28	1.95	4.24	4.20	2.78	*3.16*	2.21	0.63	2.17	−0.24	3.46	2.89
⋮	*4.70*	*2.52*	*6.92*	*4.08*	*2.69*	*4.14*	*2.19*	*2.19*	*2.10*	*2.71*	*2.47*	*1.72*	*1.82*	*1.53*	*1.60*	*4.12*	*2.28*	*2.43*	*—*
	4.35	2.29	*4.08*	6.92	3.80	3.49	2.18	4.70	2.55	2.65	4.08	0.78	*1.30*	1.39	1.77	3.71	4.14	2.05	2.35
	4.34	2.72	*2.69*	3.80	6.92	1.61	2.60	3.25	3.89	1.31	3.68	4.35	*3.46*	2.61	2.65	1.89	2.78	2.79	1.85
	4.24	2.09	*4.14*	3.49	1.61	6.92	1.67	1.27	1.04	1.94	2.23	0.82	*2.18*	2.31	3.11	2.39	1.94	1.49	0.37
	4.12	3.04	*2.19*	2.18	2.60	1.67	6.92	3.71	1.55	1.73	2.26	3.64	*1.08*	1.78	2.44	4.70	2.44	1.37	2.22
	4.08	3.28	*2.19*	4.70	3.25	1.27	3.71	6.92	1.04	1.94	2.26	2.36	*1.44*	2.04	2.21	3.07	4.85	1.81	0.87
	3.94	1.95	*2.10*	2.55	3.89	1.04	1.55	1.04	6.92	2.22	3.25	3.27	*3.23*	3.89	2.66	1.69	1.55	2.65	1.88
	3.84	4.24	*2.71*	2.65	1.31	1.94	1.73	1.94	2.22	6.92	3.30	2.54	*2.54*	2.26	1.26	1.99	0.24	2.55	0.99
	3.80	4.20	*2.47*	4.08	3.68	2.23	2.26	2.26	3.25	3.30	6.92	4.12	*2.23*	1.05	−0.21	2.37	0.74	2.21	3.36
	3.80	2.78	*1.72*	0.78	4.35	0.82	3.64	2.36	3.27	2.54	4.12	6.92	*4.12*	3.30	2.54	1.82	2.04	3.24	0.49
	3.64	*3.16*	*1.82*	*1.30*	*3.46*	*2.18*	*1.08*	*1.44*	*3.23*	*2.54*	*2.23*	*4.12*	*6.92*	*2.13*	*1.82*	*0.20*	*1.10*	*2.61*	*—*
	3.61	2.21	*1.53*	1.39	2.61	2.31	1.78	2.04	3.89	2.26	1.05	3.30	*2.13*	6.92	3.28	1.46	2.41	2.60	0.72
	3.59	0.63	*1.60*	1.77	2.65	3.11	2.44	2.21	2.66	1.26	−0.21	2.54	*1.82*	3.28	6.92	0.92	3.61	0.76	0.63
	3.53	2.17	*4.12*	3.71	1.89	2.39	4.70	3.07	1.69	1.99	2.37	1.82	*0.20*	1.46	0.92	6.92	2.52	1.62	2.60
	3.28	−0.24	*2.28*	4.14	2.78	1.94	2.44	4.85	1.55	0.24	0.74	2.04	*1.10*	2.41	3.61	2.52	6.92	1.10	1.23
E_0U_{18p}	3.10	3.46	*2.43*	2.05	2.79	1.49	1.37	1.81	2.65	2.55	2.21	3.24	*2.61*	2.60	0.76	1.62	1.10	6.92	1.20
$|e_{L-H_p}|$	2.03	2.89	*—*	2.35	1.85	0.37	2.22	0.87	1.88	0.99	3.36	0.49	*—*	0.72	0.63	2.60	1.23	1.20	6.92

[a] Rows and columns set in italics are suppressed (see text).

TABLE 9.7. Determinant of Inverse Matrix U^{-1}

$p=$	1	2	3	4	5	6	7	8	9	10	11	12	13	14	15	16	e_p
d_{1p}	8.62	-4.60	-2.66	-0.33	-1.71	-1.06	0.97	-0.76	0.57	0.79	-1.31	-0.40	-1.16	-0.02	-0.91	0.56	2.03
d_{2p}	-4.60	5.79	2.21	-0.26	0.61	0.05	-2.06	0.76	-1.17	-1.31	0.78	-0.16	0.79	0.09	1.58	-1.10	2.89
	-2.66	2.21	7.89	-2.39	-1.54	1.41	-2.83	0.30	0.14	-2.09	2.87	0.52	0.72	-1.80	-1.09	-1.01	2.35
	-0.33	-0.26	-2.39	3.54	0.88	-0.08	0.50	-0.64	0.25	-0.14	-1.74	0.09	-0.50	0.61	0.26	0.11	1.85
	-1.71	0.61	-1.54	0.88	4.99	-0.12	1.94	-1.05	-2.94	0.32	0.19	0.47	-0.88	-0.05	-0.85	0.37	0.37
	-1.06	0.05	1.41	-0.08	-0.12	3.71	-1.51	0.19	0.17	-0.09	-0.81	0.49	-0.64	-2.17	0.57	0.32	2.22
	0.97	-2.06	-2.83	0.50	1.94	-1.51	6.11	0.71	-0.94	-0.62	-0.36	-0.44	-0.31	0.76	-3.18	0.49	0.87
	-0.76	0.76	0.30	-0.64	-1.05	0.19	0.71	3.13	0.64	-1.35	-0.18	-1.13	-0.01	-0.06	0.10	-0.36	1.88
	0.57	-1.17	0.14	0.25	-2.94	0.17	-0.94	0.64	3.94	-0.53	-0.06	-0.63	-0.13	0.01	1.02	-0.40	0.99
	0.79	-1.31	-2.09	-0.14	0.32	-0.09	-0.62	-1.35	-0.53	3.58	-0.82	0.62	0.29	0.20	1.01	0.57	3.36
	-1.31	0.78	2.87	-1.74	0.19	-0.81	-0.36	-0.18	-0.06	-0.82	3.81	-0.26	0.19	-0.14	-0.74	-0.95	0.49
	-0.40	-0.16	0.52	0.09	0.47	0.49	-0.44	-1.13	-0.63	0.62	-0.26	2.25	-0.42	-0.33	-0.29	-0.10	0.72
	-1.16	0.79	0.72	-0.50	-0.88	-0.64	-0.31	-0.01	-0.13	0.29	0.19	-0.42	2.55	0.72	-0.57	0.08	0.63
	-0.02	0.09	-1.80	0.61	-0.05	-2.17	0.76	-0.06	0.01	0.20	-0.14	-0.33	0.72	3.00	-0.22	-0.11	2.60
	-0.91	1.58	-1.09	0.26	-0.85	0.57	-3.18	0.10	1.02	1.01	-0.74	-0.29	-0.57	-0.22	4.75	-0.18	1.23
d_{16p}	0.56	-1.10	-1.01	0.11	0.37	0.32	0.49	-0.36	-0.40	0.57	-0.95	-0.10	0.08	-0.11	-0.18	1.91	1.20
e_p	2.03	2.89	2.35	1.85	0.37	2.22	0.87	1.88	0.99	3.36	0.49	0.72	0.63	2.60	1.23	1.20	6.92

corresponding rows and columns are suppressed in D_m. A determinant Δ_{m+1} is finally selected only if the order is higher than a fixed number r_0. In our example, two rows and two columns (those set in italics) are suppressed (Table 9.6).

9A.4.3. Determination of the Sign Invariants

Four steps are involved:

(1) A starting set can be any allowed set of signs for which $\Delta_{m+1} \geq 0$; for instance, the set where all signs are positive, as in row $e_p(0)$ for $n = 1$ in Table 9.8. Although this set is a convenient starting point, the method developed below does not assume that the invariants are positive; any other set could be tried equally well, for example, a set involving one negative sign.

(2) From this starting set of signs, an approximate set is calculated with the regression equation (9.53). The allowed sets are searched "around" this approximate set. Several cycles of this equation may be necessary until convergence is reached.

This step is important, as negative invariants are now determined from a single formula where the right-hand member of (9.53) includes only positive invariants. For each cycle, the following results are given in Table 9.8: (a) the new value of the ratio $k = \Delta_{m+1}/D_m$ and the experimental value of the variance σ_{exp}^2, defined by

$$\sigma_{\text{exp}}^2 = \langle (e_p - \bar{e}_p)^2 \rangle$$

(b) the "expected value" $\bar{e}_p$ given by (9.94); (c) the associated probability $P_0(e_p) = \frac{1}{2} + \frac{1}{2} \tanh(e_p \bar{e}_p d_{pp})$; (d) the finally selected value of e_p. We notice that the value of k increases as the value of σ_{exp}^2 decreases; the value of σ_{exp}^2 decreases steadily with each cycle.

In each cycle the changes of signs t_p are taken into account only if the corresponding probabilities P_0 are higher than a certain value fixed in advance. In this structure, we have chosen a value $P_0 = 0.94$. In the first cycle, only one sign ($p = 15$) of e_p switches from positive to negative with a probability P_0 higher than 0.94, but in the second cycle the change of another structure factor sign is also accepted. In the third cycle, convergence was reached: the signs of two negative t_p are determined and the positive sign of the other t_p's are confirmed except for $p = 12$ ($e_p = 0.72$, $P_0 = 0.56$). Therefore, the corresponding row and column are suppressed in the

TABLE 9.8. Iterative Calculations of $\bar{e}_p$ and $P_0(e_p)$ [a],[b]

$p =$	1	2	3	4	5	6	7	8	9	10	11	12	13	14	15	16
$\sigma_p =$	0.34	0.42	0.36	0.53	0.45	0.52	0.40	0.56	0.50	0.53	0.51	0.67	0.65	0.58	0.46	0.72
$n = 1$, $k(1) = 0.07$, $\sigma^2_{\exp} = 0.81$																
$e_p(0)$	2.03	2.89	2.35	1.86	0.37	2.22	0.87	1.88	0.99	3.36	0.49	0.72	0.63	2.60	1.23	1.20
$\bar{e}_p(1)$	2.63	1.49	1.87	1.99	1.34	1.35	2.63	1.44	0.80	2.33	1.22	0.62	-0.32	2.28	-0.36	1.59
$P_0(e_p)$	0.99	0.99	0.99	0.99	0.94	0.99	0.99	0.99	0.99	0.99	0.99	0.88	0.74	0.99	0.99	0.99
$e_p(1)$	2.03	2.89	2.35	1.86	0.37	2.22	0.87	1.88	0.99	3.36	0.49	0.72	0.63	2.60	-1.23	1.20
$n = 2$, $k(2) = 0.25$, $\sigma^2_{\exp} = 0.39$																
$e_p(1)$	2.03	2.89	2.35	1.86	0.37	2.22	0.87	1.88	0.99	3.36	0.49	0.72	0.63	2.60	-1.23	1.20
$\bar{e}_p(2)$	2.37	2.16	1.52	2.14	0.96	1.73	1.34	1.51	1.45	3.03	0.72	0.30	-0.87	2.06	-0.36	1.36
$P_0(e_p)$	0.99	0.99	0.99	0.99	0.97	0.99	0.99	0.99	0.99	0.99	0.94	0.73	0.94	0.99	0.99	0.99
$e_p(2)$	2.03	2.89	2.35	1.86	0.37	2.22	0.87	1.88	0.99	3.36	0.49	0.72	-0.63	2.60	-1.23	1.20
$n = 3$, $k(3) = 0.37$, $\sigma^2_{\exp} = 0.21$																
$e_p(2)$	2.03	2.89	2.35	1.86	0.37	2.22	0.87	1.88	0.99	3.36	0.49	0.72	-0.63	2.60	-1.23	1.20
$\bar{e}_p(3)$	2.20	2.33	1.63	1.96	0.74	1.51	1.28	1.51	1.49	3.13	0.78	0.07	-0.87	2.37	-0.51	1.42
$P_0(e_p)$	0.99	0.99	0.99	0.99	0.94	0.99	0.99	0.99	0.99	0.99	0.95	0.56	0.94	0.99	0.99	0.99
$e_p(3)$	2.03	2.89	2.35	1.86	0.37	2.22	0.87	1.88	0.99	3.36	0.49	0.72	-0.63	2.60	-1.23	1.20

[a] $e_p(1)$, $e_p(2)$, $e_p(3)$: values of e_p used in the computation of the mean value of e_p. $e_p(n) = \bar{e}_p(n)$ if $P_0(e_p) \geq 0.94$; $e_p(0) = e_p$; $e_p(n) = e_p(n-1)$ if $P_0(e_p) < 0.94$. $k(n) = \Lambda_{17}/D_{16}$; Λ_{17} is calculated with the values of $\bar{e}_p(n-1)$.

[b] Values with negative signs are shown in italics.

TABLE 9.9. Allowed Sets of Signs [a]

p =	1	2	3	4	5	6	7	8	9	10	11	12	13	14	15
$k(-t_1, -t_p)$	−3.14	−1.99	−3.72	−3.92	−3.08	−3.26	−4.66	−2.99	−3.66	−6.74	−3.09	−3.50	−4.61	−5.23	−3.73
	−1.99	−4.93	−9.82	−5.92	−5.15	−6.51	−3.65	−6.20	−4.81	−6.49	−5.17	−4.90	−6.35	−2.63	−4.28
.	−3.72	−9.82	−2.05	−1.48	−1.93	−4.34	−1.68	−2.93	−2.61	−2.53	−2.69	−2.07	−1.70	−2.79	−1.85
.	−3.92	−5.92	−1.48	−0.66	−0.90	−1.69	−1.47	−0.85	−1.14	−3.41	−0.53	−0.90	−2.62	−1.17	−0.94
.	−3.08	−5.15	−1.93	−0.90	*0.28*	−0.76	−0.52	−0.13	*0.05*	−2.83	*0.14*	*0.11*	−1.21	−0.40	−0.01
	−3.26	−6.51	−4.34	−1.69	−0.76	−0.65	−0.81	−0.86	−1.16	−3.26	−0.65	−0.93	−0.09	−0.87	−1.03
	−4.66	−3.65	−1.68	−1.47	−0.52	−0.81	−0.29	−0.09	−0.53	−3.07	−0.37	−0.45	−2.01	−1.84	−0.68
	−2.99	−6.20	−2.93	−0.85	−0.13	−0.86	−0.09	−1.51	*0.01*	−2.00	−0.16	−0.30	−1.34	−0.60	−0.17
	−3.66	−4.81	−2.61	−1.14	*0.05*	−1.16	−0.53	*0.01*	−1.16	−2.80	−0.11	−0.12	−1.44	−0.46	−0.15
	−6.74	−6.49	−2.53	−3.41	−2.83	−3.26	−3.07	−2.00	−2.80	−2.67	−2.61	−2.43	−4.82	−2.71	−3.33
	−3.09	−5.17	−2.69	−0.53	*0.14*	−0.65	−0.37	−0.16	−0.11	−2.61	*0.29*	*0.15*	−1.16	−0.40	*0.12*
	−3.50	−4.90	−2.07	−0.90	*0.11*	−0.93	−0.45	−0.30	−0.12	−2.43	*0.15*	*0.28*	−1.00	−0.29	*0.05*
	−4.61	−6.35	−1.70	−2.62	−1.21	−0.09	−2.01	−1.34	−1.44	−4.82	−1.16	−1.00	−1.07	−1.83	−1.29
	−5.23	−2.63	−2.79	−1.17	−0.40	−0.87	−1.84	−0.60	−0.46	−2.71	−0.40	−0.29	−1.83	−1.84	−0.65
$k(-t_{15}, -t_p)$	−3.73	−4.28	−1.85	−0.94	−0.01	−1.03	−0.68	−0.17	−0.15	−3.33	*0.12*	*0.05*	−1.29	−0.65	*0.16*

[a] Positive values (allowed sets) are shown in italics.

next steps. The final set of 15 signs will be the approximate set. After deletion of $p = 12$, the value of k changes from $k = 0.37$ to $k = 0.41$.

(3) Next, we proceed to the selection of allowed sets by inequalities. The values k^{s-} of $k = \Delta_{16}/D_{15}$ are calculated when we switch in the approximate set one sign t_p, $s = 1$, then two signs (t_p, t_q), $s = 2$, then three signs (t_p, t_q, t_r), $s = 3$, then $\cdots s$ signs. Table 9.9 indicates the values of $k^{2-}(-t_p, -t_q)$; the values $k^{1-}(-t_p)$ are the diagonal elements. Positive values $k^{3-}(-t_p, -t_q, -t_r)$ have not been found: the number of allowed sets is very small.

(4) Among the allowed sets, the most probable is that one that leads to the maximum value of the Δ_{m+1} determinant. Table 9.10 shows that one of the 11 allowed sets is associated with a very high probability; it is the correct set including two negative invariants.

TABLE 9.10. Probabilities of Allowed Sets of Signs

p	e_p	Correct		Incorrect				
1	2.03							
2	2.89	+	+	+	+	+	+	+
3	2.35	+	+	+	+	+	+	+
4	1.86	+	+	+	+	+	+	+
5	0.37	+	+	+	−	+	+	−
6	2.22	+	+	+	+	+	+	+
7	0.87	+	+	+	+	+	+	+
8	1.88	+	+	+	+	+	+	+
9	0.99	+	+	+	+	+	+	+
10	3.36	+	+	+	+	+	+	+
11	0.49	+	−	+	+	+	−	−
12	0.63	−	−	+	−	−	+	−
13	2.60	+	+	+	+	+	+	+
14	1.23	−	−	−	−	−	−	−
15	1.20	+	+	+	+	−	+	+
Δ_{m+1}/D_m		0.411	0.285	0.281	0.277	0.156	0.154	0.144
$p(\Delta_{m+1})$		0.9940	0.0024	0.002	0.0016	$\sim$0	$\sim$0	$\sim$0

Appendix 9A.5. Hilbert Space and Gram Determinants

We use the Karle–Hauptman determinants to introduce the Hilbert space concept. Let us associate a unit vector e_j with each atom in a Hilbert space of N dimensions. We define a vector V_p in this space, associated with a reciprocal vector H in the following way:

$$V_p = \sum_{j=1}^{N} [(n_j)^{1/2} \exp(2\pi i H_p \cdot r_j)] e_j$$

The scalar expression in the bracket is the jth coordinate of V_p. We assume also that the N unit vectors e_j are orthonormal. Then the scalar product $V_p \cdot V_q$ is the unitary structure factor $U_{H_p - H_q}$:

$$V_{pq} = (V_p \cdot V_q) = \sum_i \sum_j [(n_i)^{1/2} \exp(2\pi i H_p \cdot r_i)]$$
$$\times [(n_j)^{1/2} \exp(2\pi i H_q \cdot r_j)]^* e_i \cdot e_j$$
$$= \sum_j n_j \exp 2\pi i (H_p - H_q) \cdot r_j = U_{H_p - H_q}$$

since

$$e_i \cdot e_j = \delta_{ij}$$

These results can be written in a more general form:

$$V_p = \sum_{j=1}^{n} X_{jp} e_j, \qquad p = 1, \ldots, m$$
$$V_{pq} = V_p \cdot V_q = \sum_{j=1}^{n} X_{jp} X_{jq}^*$$

The determinant formed by the $m \times m$ matrix V_{pq} is called a Gram determinant. By a convenient choice of the coordinates X_p, we can produce Gram determinants useful in specific crystallographic problems. We now give two examples.

9A.5.1. Neutron Diffraction Determinants

$$X_{jp} = (n_j)^{1/2} \exp(2\pi i H_p \cdot r_j), \qquad p = 1, \ldots, m$$
$$X_{j,m+1} = \frac{n_j'}{(n_j)^{1/2}} \exp(2\pi i H_p \cdot r_j), \qquad p = m + 1$$

where n_j is the X-ray scattering factor, which is nonnegative, and n_j' is the neutron scattering factor, which may be negative. We obtain thus the following nonnegative determinant (Tsoucaris, 1970b):

$$\Delta_m = \begin{vmatrix} U_{11} & U_{12} & \cdots & \vdots & U'_{1,m+1} \\ U_{21} & U_{22} & \cdots & \vdots & U'_{2,m+1} \\ \vdots & \vdots & & \vdots & \vdots \\ \hline U'_{m+1,1} & U'_{m+1,2} & \cdots & & A \end{vmatrix} \geq 0$$

with

$$|\mathbf{V}_{m+1}|^2 = A = \sum_j \frac{|n_j'|^2}{n_j}$$

$$U_{\mathbf{H}} = \sum_{j=1}^{N} n_j \exp(2\pi i \mathbf{H} \cdot \mathbf{r}_j)$$

$$U_{\mathbf{H}}' = \sum_{j=1}^{N} n_j' \exp(2\pi i \mathbf{H} \cdot \mathbf{r}_j)$$

with

$$\sum_j n_j = \sum_j n_j' = 1$$

9A.5.2. The "Moduli-Model" Determinant

For $p = 1, \ldots, m$, we define

$$\mathbf{V}_p = \sum_{i,j=1}^{N} \{(n_i n_j)^{1/2} \exp[2\pi i \mathbf{H}_p \cdot (\mathbf{r}_i - \mathbf{r}_j)]\}\mathbf{e}_{ij}$$

For $\mathbf{V}_{m+1}$, the above double summation is extended to N_f, the number of atoms in the fragment, instead of N.

Then

$$\mathbf{V}_p \cdot \mathbf{V}_q = |U_{\mathbf{H}_p - \mathbf{H}_q}|^2, \qquad p, q = 1, \ldots, m$$

and

$$\mathbf{V}_p \cdot \mathbf{V}_{m+1} = |C_{\mathbf{H}_{m+1} - \mathbf{H}_p}|^2$$
$$= \text{contribution of the sole fragment}$$

References

Bertaut, E. F. (1956). *Acta Crystallogr.* **9**, 455–460.

Braun, P. B., Hornstra, J. J., and Leenhouts, J. I. (1969). *Philips Res. Rep.* **24**, 85.

Brunie, S., and Tsoucaris, G. (1974). *Cryst. Struct. Commun.* **3**, 481–484.

Castellano, E., Podjarny, A., and Navaza, J. (1973). *Acta Crystallogr. Sect. A* **29**, 609–615.

Cochran, W. (1955). *Acta Crystallogr.* **8**, 473–478.

Cooper, A., Norton, D. A., and Hauptman, H. (1969). *Acta Crystallogr. Sect. B* **25**, 814–828.

Diamond, R. (1963). *Acta Crystallogr.* **16**, 627–639.

Eller, G. von (1962). *Acta Crystallogr.* **15**, 590–595.

Gassman, J. (1976). *Acta Crystallogr. Sect.* **32**, 274–280.

Gifkins, M. (1972). Thesis, University of York, York, England.

Goedkoop, J. A. (1950). *Acta Crystallogr.* **3**, 374–378.

Harker, D., and Kasper, J. S. (1948). *Acta Crystallogr.* **1**, 70–75.

Hauptman, H. (1971). *Z. Kristallogr.* **134**, 28.

Heinerman, J. J. L., and Kroon, J. (1976). *Acta Crystallogr. Sect. A* **32**, 115–119.

Hirshfeld, F. L. (1968). *Acta Crystallogr. Sect. A* **24**, 301–311.

Hughes, E. W. (1953). *Acta Crystallogr.* **6**, 871.

Karle, J. (1966). *Acta Crystallogr.* **21**, 273–276.

Karle, J. (1972). *Acta Crystallogr. Sect. B* **28**, 3362–3369.

Karle, J., and Hauptman, H. (1950). *Acta Crystallogr.* **3**, 181–187.

Karle, J., and Karle, I. L. (1966). *Acta Crystallogr.* **21**, 849–859.

Kitaigorodski, A. I. (1961). *Theory of Crystal Structure Determination*, Consultants Bureau, New York.

Klug, A. (1958). *Acta Crystallogr.* **11**, 515–543.

Knossow, M. (1975). Thèse de 3ème cycle, Université de Paris, Paris.

Knossow, M., Rango, C. de, Mauguen, Y., Sarrazin, M. and Tsoucaris, G. (1977). *Acta Crystallogr. Sect. A* **33**, 119–125.

Lajzerowicz, J., and Lajzerowicz, J. (1966). *Acta Crystallogr.* **21**, 8–12.

Main, P. (1974). Personal communication.

Mauguen, Y. (1972). Unpublished.

Mauguen, Y., Rango, C. de, and Tsoucaris, G. (1973). *Acta Crystallogr. Sect. A* **29**, 574–578.

Messager, J. C., and Tsoucaris, G. (1972). *Acta Crystallogr. Sect. A* **28**, 482–484.

Podjarny, A. D., and Yonath, A. (1977). *Acta Crystallogr. Sect. A* **33**, 655.

Podjarny, A. D., Yonath, A., and Traub, W. (1976). *Acta Crystallogr. Sect. A* **32**, 281–292.

Rango, C. de (1969). Thèse, Université de Paris, Paris.

Rango, C. de, Tsoucaris, G., and Zelwer, C. (1974). *Acta Crystallogr. Sect. A* **30**, 342–353.

Rango, C. de, Mauguen, Y., and Tsoucaris, G. (1975a). *Acta Crystallogr. Sect. A* **31**, 227–233.

Rango, C. de, Mauguen, Y., and Tsoucaris, G. (1975b). Xth International Congress of Crystallography, Amsterdam.

Sarrazin, M. (1970). CECAM Workshop reports, Orsay, France.

Sarrazin, M. (1971). Personal communication.

Sarrazin, M. (1973). Personal communication.

Tollin, P., and Cochran, W. (1964). *Acta Crystallogr.* **17**, 1322–1324.

Taylor, D. J., and Woolfson, M. M. (1975). Xth International Congress of Crystallography, Amsterdam, S. 16.

Tsoucaris, G. (1969). *C. R. Acad. Sci. Ser. B* **268**, 875.

Tsoucaris, G. (1970a). *Acta Crystallogr. Sect. A* **26**, 492–499.

Tsoucaris, G. (1970b). *Acta Crystallogr. Sect. A* **26**, 499–501.

Wilson, A. J. C. (1949). *Acta Crystallogr.* **2**, 318–321.

Woolfson, M. M. (1954). *Acta Crystallogr.* **7**, 61–64.

Molecular Replacement Method

PATRICK ARGOS
and MICHAEL G. ROSSMANN

10.1. Introduction

The term molecular replacement encompasses a variety of techniques designed to utilize the occurrence of a molecule, molecular subunit, or molecular fragment in more than one crystallographic environment for the purpose of phase refinement. The limited size of nature's building blocks and the necessity of their aggregation into molecules with closed point symmetries (Crick and Watson, 1956; Monod *et al.*, 1965) provide the basis of the subject and its application to macromolecules. For example, a particular compound may be induced to crystallize in different forms, or a given molecule might be composed of several identical subunits.

Physicists have often employed the concept of data combination resulting from multiple observations of a given object to produce an image superior to that obtained from any single observation. Typical examples include diffraction gratings, such as a crystal, where an object is repeated by translation, and electron microscope image reconstruction processes, where different but otherwise identical objects are observed many times. In molecular replacement an object is repeated by noncrystallographic symmetry, which is local in character and thus cannot pertain to the infinite crystal. Such concepts were first explored by Rossmann and Blow in 1962. Their ideas were developed in three papers: the first dealt with the rotational

PATRICK ARGOS and MICHAEL G. ROSSMANN • Department of Biological Sciences, Purdue University, West Lafayette, Indiana 47907.

relationship between molecules (Rossmann and Blow, 1962), the second with the corresponding determination of translational vectors (Rossmann et al., 1964), and the third with the phase problem (Rossmann and Blow, 1963). A review of the existing literature to 1972 was compiled by Rossmann (1972a) in a book that also republished all relevant papers. Other reviews, each covering a specific portion of the field, were given by Blow (1976), Colman et al. (1976), and Tollin (1976) at the 1975 Prague conference of the International Union of Crystallography.

Only in the last few years have studies of larger biological aggregates (Matthews, 1976) become possible, resulting in an increasing library of biological folds (Levitt and Chothia, 1976). Furthermore, it has become apparent that many structures with functional conservation (Rossmann and Liljas, 1974; Kretsinger, 1972, 1976) are similar. Since the molecular replacement method utilizes the redundant information in noncrystallographically related molecules and allows the direct solution of unknown structures by comparison to those already available, the application of the technique is now becoming widespread, making the current review timely.

Noncrystallographic symmetry between equivalent points of two copies can be expressed as

$$\mathbf{x_2} = [\mathbf{C}]\mathbf{x_1} + \mathbf{d} \tag{10.1}$$

where $[\mathbf{C}]$ is a rotation matrix and $\mathbf{d}$ is a translation vector. The nine elements of the rotation matrix are a function of three angular parameters, and the translation vector has meaning only if the origin of the position vectors $\mathbf{x_1}$ and $\mathbf{x_2}$ is defined. The relationship (10.1) becomes a crystallographic operation in the special case that it is valid throughout the crystal lattice, as well as within the confining envelopes of molecules 1 and 2. The rotation matrix $[\mathbf{C}]$ can be either proper or improper.[†] A proper rotation represents a closed point group as in the tetrameric glyceraldehyde-3-phosphate dehydrogenase molecule with 222 symmetry. Improper symmetry is exemplified in some crystals of hexokinase where the monomers are related by a 156° rotation and a 13.8-Å translation (Steitz et al., 1976). Relationships between different crystal forms of the same molecule are obviously improper.

The major tool in solving structures of biological macromolecules has been the isomorphous replacement technique introduced for proteins

---

[†] The terms proper and improper as used here were introduced by Rossmann (1972a); however, they are sometimes used in group theory, where an improper rotation is a symmetry operation that involves a change of hand.

by Perutz in 1954 (Green *et al.*, 1954; Bragg and Perutz, 1954). Since molecular replacement is gaining prominence as a useful adjunct to isomorphous replacement, it is instructive to contrast the relative merits of these methods.

The quantity of data that needs to be collected for each method is roughly equivalent. This is not surprising as the information required to solve a particular problem should be independent of the method used in its solution. For the isomorphous replacement method, it is necessary to collect data of the native copy as well as two or more heavy-atom derivative sets. In the molecular replacement method, the amount of data is proportional to the number of noncrystallographically related copies since each unit proportionately increases the volume of the real cell.

The isomorphous replacement technique requires the production of a variety of isomorphous heavy-atom derivatives, a task which is often difficult. Once accomplished, the computational devices to solve the phase problems are quite straightforward; for reviews see Blundell and Johnson, 1976; Phillips, 1966; Jansonius, 1976. In contrast, the physical requirements of molecular replacement are relatively easy to accomplish, since biological molecules frequently crystallize in different forms and since larger aggregates are mostly built of identical subunits. However, the problem of deconvoluting the data to solve the phase problem is difficult.

The degree of similarity among covalent structures in different environments is a significant consideration for the molecular replacement technique, just as is the nature of isomorphism for the isomorphous replacement technique. Lack of equivalences between noncrystallographic copies has generally been found only at resolutions better than 3 Å in regions close to noncrystallographic symmetry operators (Tulinsky *et al.*, 1973; Hodgkin, 1974). Where heavy atoms at chemically identical sites have been refined independently of noncrystallographic symmetry, they obey the molecular symmetry to within experimental error (Buehner *et al.*, 1974; Bricogne, 1976a; Champness *et al.*, 1976). This is not to demean the effect of the environment or to indicate that oligomers must have perfect symmetry, but only to emphasize that such effects are not readily detectable and generally are of little concern in the initial structure solution.

Perutz not only introduced isomorphous replacement to the study of macromolecules but also took the first tentative steps toward the use of molecular replacement. A special case of molecular replacement occurs in the study of the shrinkage stages of hemoglobin (Perutz, 1954). As the unit cell size decreases with the removal of water, the reciprocal lattice expands and alters the position at which the molecular transform is sampled.

By collecting data for several shrinkage stages, Perutz constructed a continuous transform along lines of reciprocal space with centric phases. By suitable interpolation, he was able to ascertain the sign of the molecular transform over many of the centric $h0l$ reflections. The various shrinkage stages for the same hemoglobin molecule in different salt and water environments are equivalent to different crystal forms. In the more general molecular replacement case, the molecular transform is similarly over-sampled at intervals less than discrete increments in the reciprocal lattice, but not necessarily on centric lines as in Perutz's application. This filling-in of the transform provides an alternative view of molecular replacement. Indeed, Shannon (1949) has shown that a continuous transform is derivable from intensities corresponding to points on a reciprocal lattice of a real cell with a doubled volume.

Another critically important use of the availability of multiple images occurs in the reconstruction of electron micrographs (Crowther and Klug, 1975). Its use in the analysis of biomolecular structure has made considerable impact. However, this review is limited explicitly to the analysis of X-ray diffraction data in the presence of noncrystallographic symmetry.

10.2. Preliminary Theoretical Considerations

Consider a molecule placed in any crystal cell (**h**), within which any point shall be designated by **x**. Let the corresponding structure factors be $F_\mathbf{h}$. It is then possible to compute the structure factors $F_\mathbf{p}$ for another cell (**p**) into which the same molecule has been placed N times related by the crystallographic symmetry operators $[\mathbf{C}_1], \mathbf{d}_1; [\mathbf{C}_2], \mathbf{d}_2; \ldots; [\mathbf{C}_N], \mathbf{d}_N$. Let the electron density at a point $\mathbf{y}_1$ in the first crystallographic asymmetric unit be spatially related to the point $\mathbf{y}_n$ in the nth asymmetric unit of the **p** crystal such that

$$\varrho(\mathbf{y}_n) = \varrho(\mathbf{y}_1) \tag{10.2}$$

where

$$\mathbf{y}_n = [\mathbf{C}_n]\mathbf{y}_1 + \mathbf{d}_n \tag{10.3}$$

From the definition of a structure factor

$$F_\mathbf{p} = \sum_{n=1}^{N} \int_U \varrho(\mathbf{y}_n) \exp(2\pi i \mathbf{p} \cdot \mathbf{y}_n) \, d\mathbf{y}_n \tag{10.4}$$

where the integral is taken over the volume U of one molecule. But since each molecule is identical as expressed in equation (10.2) and since (10.3) can be substituted in equation (10.4), we have

$$F_p = \sum_{n=1}^{N} \int_U \varrho(\mathbf{y}_1) \exp\{2\pi i \mathbf{p} \cdot ([C_n]\mathbf{y}_1 + \mathbf{d}_n)\} \, d\mathbf{y}_1 \qquad (10.5)$$

Now let the molecule in the **h** crystal be related to the molecule in the first asymmetric unit of the **p** crystal by the noncrystallographic symmetry operation

$$\mathbf{x} = [C]\mathbf{y} + \mathbf{d} \qquad (10.6)$$

which implies

$$\varrho(\mathbf{x}) = \varrho(\mathbf{y}_1) = \varrho(\mathbf{y}_2) = \cdots \qquad (10.7)$$

Furthermore, in the **h** cell

$$\varrho(\mathbf{x}) = \frac{1}{V_h} \sum_h F_h \exp(-2\pi i \mathbf{h} \cdot \mathbf{x}) \qquad (10.8)$$

and thus, by combining with (10.6) and (10.7)

$$\varrho(\mathbf{y}_1) = \frac{1}{V_h} \sum_h F_h \exp\{-2\pi i(\mathbf{h}[C] \cdot \mathbf{y}_1 + \mathbf{h} \cdot \mathbf{d})\} \qquad (10.9)$$

Now using (10.5), (10.8), and (10.9) it can be shown that

$$F_p = \frac{U}{V_h} \sum_h F_h \sum_{n=1}^{N} G_{hpn} \exp[2\pi i(\mathbf{p} \cdot \mathbf{S}_n - \mathbf{h} \cdot \mathbf{S})] \qquad (10.10)$$

where

$$U G_{hpn} = \int_U \exp\{2\pi i(\mathbf{p}[C_n] - \mathbf{h}[C]) \cdot \mathbf{u}\} \, d\mathbf{u} \qquad (10.11)$$

S is a chosen molecular origin in the **h** crystal, and $\mathbf{S}_n$ is the corresponding molecular position in the nth asymmetric unit of the **p** crystal. Thus

$$\mathbf{S}_n = [C_n]\mathbf{S}_1 + \mathbf{d}_n$$

and

$$\mathbf{S} = [C]\mathbf{S}_1 + \mathbf{d}$$

The definition of the integral G uses the running variable **u** such that $\mathbf{y}_1 = \mathbf{S}_1 + \mathbf{u}$. If this integral is bounded by a sphere of radius R, its value can be expressed as

$$G_{hpn} = \frac{3[\sin(2\pi \mathbf{H} \cdot \mathbf{R}) - 2\pi \mathbf{H} \cdot \mathbf{R} \cos(2\pi \mathbf{H} \cdot \mathbf{R})]}{(2\pi \mathbf{H} \cdot \mathbf{R})^3} \qquad (10.12)$$

where

$$\mathbf{H} = \mathbf{p}[C_n] - \mathbf{h}[C]$$

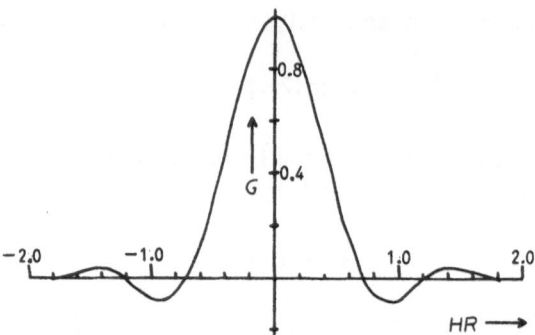

FIGURE 10.1. Shape of the interference function G for a spherical envelope of radius R at a distance H from the reciprocal-space origin.

The relationship between $\mathbf{H} \cdot \mathbf{R}$ and G is shown in Fig. 10.1. The form of the G function is of critical importance for the evaluation of the basic expression (10.10), which will be used in subsequent sections of the review.

10.3. Rotation Function

10.3.1. Fundamentals

The problem of finding the relative orientations of two molecular copies referred to known crystal axes will be pursued in this section. The basic principles were stated by Rossmann and Blow (1962):

Consider a structure of two identical units which are in different orientations. The Patterson function of such a structure consists of three parts. There will be the self-Patterson vectors of one unit, being the set of interatomic vectors which can be formed within that unit, with appropriate weights. The set of self-Patterson vectors of the other unit will be identical, but they will be rotated from the first due to the different orientation. Finally, there will be the cross-Patterson vectors, or set of interatomic vectors which can be formed from one unit to another. The self-Patterson vectors of the two units will all lie in a volume extending from the origin by the overall dimensions of the units. Some or all of the cross-Patterson vectors will lie outside this volume.

Suppose the Patterson function is now superposed on a rotated version of itself. There will be no particular agreement except when one set of self-Patterson vectors of one unit has the same orientation as the self-Patterson vectors from the other unit. In this position, we would expect a maximum of agreement or "overlap" between the two.

Similarly the superposition of the molecular self-Patterson derived from different crystal forms can provide the relative orientation of the two crystals when the molecules are aligned.

The overlap was defined by Rossmann and Blow (1962) as

$$R(\theta_1, \theta_2, \theta_3) = \int_U P_1(\mathbf{x}) \cdot P_2(\mathbf{y}) \, d\mathbf{x} \qquad (10.13)$$

where $P_1(\mathbf{x})$ designates the value of the Patterson at $\mathbf{x}$ for crystal $\mathbf{h}$, and $P_2(\mathbf{y})$ designates the value of the Patterson at $\mathbf{y}$ for crystal $\mathbf{p}$, $\theta_1, \theta_2, \theta_3$ are three angles which rotate the Patterson P_2 from its initial arbitrary position, and the integral is taken over a volume U, usually a sphere centered on the Patterson origin. It is clear that R, which is termed the rotation function (Rossmann and Blow, 1962), will take on large values when peaks in P_1 are superimposed on peaks in P_2; this can occur when the self-Patterson vectors match. When searching for the relationship between identical molecular structures within one crystal form, P_2 is simply a rotated version of P_1.

The usual procedure is to orthogonalize the axes in both Pattersons and to define the initial situation when corresponding axes are superimposed. If the $\mathbf{p}$ crystal self-Patterson is then rotated and again superimposed on the $\mathbf{h}$ crystal Patterson, the point $\mathbf{x}$ will now lie atop the point $\mathbf{y}$, where

$$\mathbf{y} = [C]\mathbf{x} \qquad (10.14)$$

and $[C]$ is a rotation matrix whose elements are a function of the angles $(\theta_1, \theta_2, \theta_3)$.

While it would be possible to evaluate R by interpolation in P_2 and forming the point-by-point product with P_1 within the volume U for every combination of θ_1, θ_2, and θ_3, such a process is tedious and requires large computer storage for the Pattersons. Instead, the process is usually performed in reciprocal space where the number of independent structure amplitudes which form the Pattersons is about one-thirtieth of the number of Patterson grid points. Thus, the computation of a rotation function is carried out directly on the structure amplitudes, while the overlap definition (10.13) simply serves as a physical basis for the technique.

10.3.2. Reciprocal-Space Expression

The derivation of the reciprocal-space expression depends on the expansion of each Patterson either as a Fourier summation, the conventional approach of Rossmann and Blow (1962), or as a sum of spherical harmonics, Crowther's analysis (1972). The conventional and mathematically easier treatment is discussed presently, but the reader is referred also to Section 10.3.8 for Crowther's elegant approach. The latter leads to a rapid technique for performing the computations which is about one hundred times faster than conventional methods (Blow, 1976).

Let

$$P_1(\mathbf{x}) = \sum_{\mathbf{h}} |F_{\mathbf{h}}|^2 \exp(2\pi i \mathbf{h} \cdot \mathbf{x})$$

and

$$P_2(\mathbf{y}) = \sum_{\mathbf{p}} |F_p|^2 \exp(2\pi i \mathbf{p} \cdot \mathbf{y})$$

(10.15)

From (10.14) and (10.15) it follows that

$$P_2(\mathbf{y}) = \sum_{\mathbf{p}} |F_p|^2 \exp(2\pi i \mathbf{p}[\mathbf{C}] \cdot \mathbf{x})$$

and hence by substitution in (10.13)

$$R(\theta_1, \theta_2, \theta_3) = \int_U \left[\sum_{\mathbf{h}} |F_{\mathbf{h}}|^2 \exp(2\pi i \mathbf{h} \cdot \mathbf{x})\right]\left\{\sum_{\mathbf{p}} |F_p|^2 \exp(2\pi i \mathbf{p}[\mathbf{C}] \cdot \mathbf{x})\right\} d\mathbf{x}$$

$$= \sum_{\mathbf{h}} |F_{\mathbf{h}}|^2 \left(\sum_{\mathbf{p}} |F_p|^2 G_{\mathbf{h}\mathbf{p}}\right)$$

(10.16)

where $G_{\mathbf{h}\mathbf{p}}$ is generally the spherical interference function (10.12) in Section 10.2 with its argument $\mathbf{H} = \mathbf{h} + \mathbf{p}[\mathbf{C}]$. Statement (10.16) represents the rotation function in reciprocal space. If $\mathbf{h}' = [\mathbf{C}^T]\mathbf{p}$ in the argument of $G_{\mathbf{h}\mathbf{p}}$, then $\mathbf{h}'$ can be seen as the point in reciprocal space to which $\mathbf{p}$ is rotated by $[\mathbf{C}]$. Only for the integral reciprocal-lattice points close to $\mathbf{h}'$ will $G_{\mathbf{h}\mathbf{p}}$ be of an appreciable size. Thus, the number of significant terms is greatly reduced in the summation over $\mathbf{p}$ for every value of $\mathbf{h}$, making the computation of the rotation function manageable.

The radius of integration R within which the self-Patterson is considered should be approximately equal to the molecular diameter. Since R is roughly equal to the length of a lattice translation, the separation of reciprocal-lattice points is about $1/R$. Hence, when $\mathbf{H}$ is equal to one reciprocal-lattice separation, $HR \simeq 1$, and G is thus quite small. Indeed, all terms with $HR > 1$ might well be neglected. Thus, in general, the only terms that need be considered are those where $-\mathbf{h}'$ is within one lattice point of $\mathbf{h}$. However, in dealing with a small molecular fragment for which R is small compared to the unit cell dimensions, more reciprocal-lattice points must be included for the summation over $\mathbf{p}$ in the rotation function expression (10.16).

In practice, the equation

$$\mathbf{h} + \mathbf{h}' = 0$$

that is

$$[C^T]\mathbf{p} = -\mathbf{h}$$

or

$$\mathbf{p} = [C^T]^{-1} \cdot -\mathbf{h} \tag{10.17}$$

determines $\mathbf{p}$, given a set of Miller indices $\mathbf{h}$. This will give a nonintegral set of Miller indices. The terms included in the inner summation of (10.16) will be integral values of $\mathbf{p}$ around the nonintegral lattice point found by solving (10.17).

Details of the conventional program were given by Tollin and Rossmann (1966) and follow the principles outlined above. Various strategies are also discussed as to which crystal form should be the first ($\mathbf{h}$) and second ($\mathbf{p}$) Patterson. Rossmann and Blow (1962) noted that the factor $\sum_\mathbf{p} |F_\mathbf{p}|^2 G_{\mathbf{hp}}$ in expression (10.16) represents an interpolation of the square transform of the self-Patterson of the second ($\mathbf{p}$) crystal. Thus the rotation function is a sum of the products of the two molecular transforms taken over all the $\mathbf{h}$ reciprocal-lattice points. Lattman and Love (1970) therefore computed the molecular transform explicitly and stored it in the computer, sampling it as required by the rotation operation.

10.3.3. Matrix Algebra

The initial step in the rotation function procedure involves the orthogonalization of both crystal systems. Thus if fractional coordinates in the first crystal system are represented by $\mathbf{x}$, these can be orthogonalized by a matrix $[\beta]$ to give the coordinates $\mathbf{X}$ in units of length (Fig. 10.2); that is,

$$\mathbf{X} = [\beta]\mathbf{x}$$

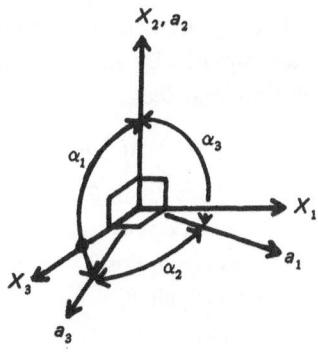

FIGURE 10.2. Relationship of the orthogonal axes X_1, X_2, X_3 to the crystallographic axes a_1, a_2, a_3.

If the point X is rotated to the point Y, then

$$Y = [\rho]X \tag{10.18}$$

where $[\rho]$ represents the rotation matrix relating the two vectors in the orthogonal system. Finally Y is converted back to fractional coordinates measured along the oblique cell dimension in the second crystal by

$$y = [\alpha]Y$$

Thus, by substitution

$$\begin{aligned} y &= [\alpha][\rho]X \\ &= [\alpha][\rho][\beta]x \end{aligned} \tag{10.19}$$

and by comparison with (10.14) it follows that

$$[C] = [\alpha][\rho][\beta]$$

Figure 10.2 shows the mode of orthogonalization used by Rossmann and Blow (1962), which is now generally preferred. With their definition it can be shown that

$$[\alpha] = \begin{pmatrix} 1/(a_1 \sin \alpha_3 \sin \omega) & 0 & 0 \\ 1/(a_2 \tan \alpha_1 \tan \omega) - 1/(a_2 \tan \alpha_3 \sin \omega) & 1/a_2 & -1/(a_2 \tan \alpha_1) \\ -1/(a_3 \sin \alpha_1 \tan \omega) & 0 & 1/(a_3 \sin \alpha_1) \end{pmatrix}$$

and

$$[\beta] = \begin{pmatrix} a_1 \sin \alpha_3 \sin \omega & 0 & 0 \\ a_1 \cos \alpha_3 & a_2 & a_3 \cos \alpha_1 \\ a_1 \sin \alpha_3 \cos \omega & 0 & a_3 \sin \alpha_1 \end{pmatrix}$$

where $\cos \omega = (\cos \alpha_2 - \cos \alpha_1 \cos \alpha_3)/(\sin \alpha_1 \sin \alpha_3)$ with $0 \leq \omega < \pi$. For a Patterson compared with itself $[\alpha] = [\beta]^{-1}$.

Both spherical $(\varkappa, \psi, \phi)$ and Eulerian $(\theta_1, \theta_2, \theta_3)$ angles are used in evaluating the rotation function. The usual definitions employed are given diagrammatically in Figs. 10.3 and 10.4. They give rise to the following rotation matrices.

(a) *Matrix* $[\rho]$ *in terms of Eulerian angles* $\theta_1, \theta_2, \theta_3$:

$$\begin{pmatrix} -\sin \theta_1 \cos \theta_2 \sin \theta_3 + \cos \theta_1 \cos \theta_3 & \cos \theta_1 \cos \theta_2 \sin \theta_3 + \sin \theta_1 \cos \theta_3 & \sin \theta_2 \sin \theta_3 \\ -\sin \theta_1 \cos \theta_2 \cos \theta_3 - \cos \theta_1 \sin \theta_3 & \cos \theta_1 \cos \theta_2 \cos \theta_3 - \sin \theta_1 \sin \theta_3 & \sin \theta_2 \cos \theta_3 \\ \sin \theta_1 \sin \theta_2 & -\cos \theta_1 \sin \theta_2 & \cos \theta_2 \end{pmatrix}$$

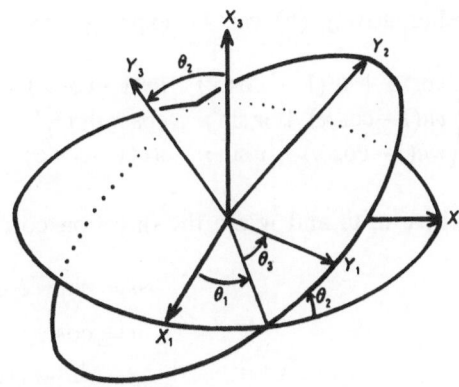

FIGURE 10.3. Eulerian angles $\theta_1, \theta_2, \theta_3$ relating the rotated axes Y_1, Y_2, Y_3 to the original unrotated orthogonal axes X_1, X_2, X_3.

and

(b) Matrix $[\rho]$ in terms of rotation angle $\varkappa$ and the spherical polar coordinates ψ, ϕ:

$$\begin{pmatrix} \cos \varkappa + \sin^2 \psi \cos^2 \phi(1 - \cos \varkappa) & \sin \psi \cos \psi \cos \phi(1 - \cos \varkappa) & -\sin^2 \psi \cos \phi \sin \phi (1 - \cos \varkappa) \\ & + \sin \psi \sin \phi \sin \varkappa & + \cos \psi \sin \varkappa \\ \sin \psi \cos \psi \cos \phi(1 - \cos \varkappa) & \cos \varkappa + \cos^2 \psi(1 - \cos \varkappa) & -\sin \psi \cos \psi \sin \phi(1 - \cos \varkappa) \\ - \sin \psi \sin \phi \sin \varkappa & & - \sin \psi \cos \phi \sin \varkappa \\ -\sin^2 \psi \sin \phi \cos \phi(1 - \cos \varkappa) & -\sin \psi \cos \psi \sin \phi(1 - \cos \varkappa) & \cos \varkappa + \sin^2 \psi \sin^2 \phi(1 - \cos \varkappa) \\ - \cos \psi \sin \varkappa & + \sin \psi \cos \phi \sin \varkappa & \end{pmatrix}$$

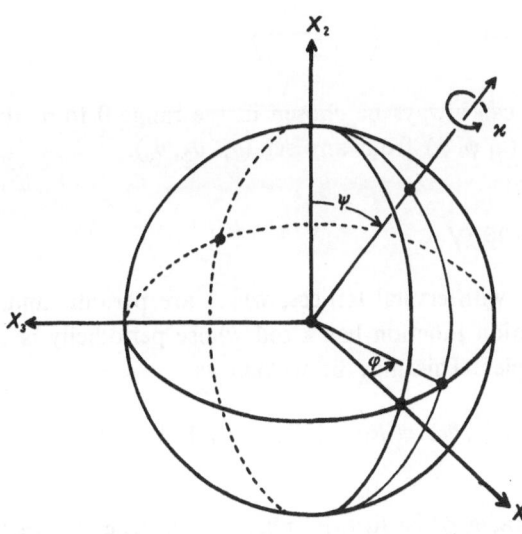

FIGURE 10.4. Variables ψ and φ are polar coordinates which specify a direction about which the axes may be rotated through an angle $\varkappa$.

Alternatively (b) can be expressed as

$$\begin{pmatrix} \cos\varkappa + u^2(1-\cos\varkappa) & uv(1-\cos\varkappa)-w\sin\varkappa & uw(1-\cos\varkappa)+v\sin\varkappa \\ vu(1-\cos\varkappa)+w\sin\varkappa & \cos\varkappa + v^2(1-\cos\varkappa) & vw(1-\cos\varkappa)-u\sin\varkappa \\ wu(1-\cos\varkappa)-v\sin\varkappa & wv(1-\cos\varkappa)+u\sin\varkappa & \cos\varkappa + w^2(1-\cos\varkappa) \end{pmatrix}$$

where u, v, and w are the direction cosines of the rotation axis given by

$$u = \sin\psi \cos\phi$$

$$v = \cos\psi$$

$$w = -\sin\psi \sin\phi$$

This latter form also demonstrates that the trace of a rotation matrix is $2\cos\varkappa + 1$.

The relationship between the two sets of variables established by comparison of the elements of the two matrices yields

$$\cos(\varkappa/2) = \cos(\theta_2/2)\cos\left(\frac{\theta_1 + \theta_3}{2}\right)$$

$$\tan\phi = -\cot(\theta_2/2)\sin\left(\frac{\theta_1 + \theta_3}{2}\right)\sec\left(\frac{\theta_1 - \theta_3}{2}\right)$$

$$\cos\phi\tan\psi = \cot\left(\frac{\theta_1 - \theta_3}{2}\right)$$

Since ϕ and ψ can always be chosen in the range 0 to π, these equations suffice to find $(\varkappa, \psi, \phi)$ from any set $(\theta_1, \theta_2, \theta_3)$.

10.3.4. Symmetry

In analogy with crystal lattices, which are periodic and contain symmetry, the rotation function has a cell whose periodicity is 2π in any one of its three angles. This may be written as

$$R(\theta_1, \theta_2, \theta_3) \equiv R(\theta_1 + 2\pi n_1, \theta_2 + 2\pi n_2, \theta_3 + 2\pi n_3)$$

or

$$R(\varkappa, \psi, \phi) \equiv R(\varkappa + 2\pi n_1, \psi + 2\pi n_2, \phi + 2\pi n_3)$$

where n_1, n_2 and n_3 are integers. A redundancy in the definition of either set of angles leads to the equivalence of the following points:

$$R(\theta_1, \theta_2, \theta_3) \equiv R(\theta_1 + \pi, -\theta_2, \theta_3 + \pi) \quad \text{in Eulerian space}$$

or

$$R(\varkappa, \psi, \phi) \equiv R(\varkappa, 2\pi - \psi, \phi + \pi) \quad \text{in polar space}$$

These relationships imply an n glide plane perpendicular to θ_2 for Eulerian space or a φ glide plane perpendicular to ψ in polar space.

In addition, the Laue symmetry of the two Pattersons themselves must be considered. This problem was first discussed by Rossmann and Blow (1962) and later systematized by Tollin *et al.* (1966) and Burdina (1970, 1971, 1973). The rotation function will have the same value whether the Patterson density at $\mathbf{X}$ or $[\mathbf{T}_i]\mathbf{X}$ in the first crystal is multiplied by the Patterson density at $\mathbf{Y}$ or $[\mathbf{T}_j]\mathbf{Y}$ in the second crystal. $[\mathbf{T}_i]$ and $[\mathbf{T}_j]$ refer to the ith and jth crystallographic rotations in the orthogonalized coordinate systems of the first and second crystal, respectively. Hence, from (10.18)

$$([\mathbf{T}_j]\mathbf{Y}) = [\rho]([\mathbf{T}_i]\mathbf{X})$$

or

$$\mathbf{Y} = [\mathbf{T}_j^T][\rho][\mathbf{T}_i]\mathbf{X}$$

Thus, it is necessary to find angular relationships which satisfy the relation

$$[\rho] = [\mathbf{T}_j^T][\rho][\mathbf{T}_i]$$

for given Patterson symmetries. Tollin *et al.* (1966) show that the Eulerian angular equivalences can be expressed in terms of the Laue symmetries of each Patterson (Table 10.1).

The example given by Tollin *et al.* (1966) is instructive in the use of Table 10.1. They consider the determination of the Eulerian space group when P_1 has symmetry *Pmmm* and P_2 has symmetry *P2/m*. These Pattersons contain the proper rotation groups 222 and 2 (parallel to b), respectively. Inspection of Table 10.1 shows that these symmetries produce the following Eulerian relationships.

(a) In the first crystal (*Pmmm*):

$$\theta_1\theta_2\theta_3 \rightarrow \pi + \theta_1, -\theta_2, \pi + \theta_3 \qquad \text{(onefold axis)}$$
$$\theta_1\theta_2\theta_3 \rightarrow \pi - \theta_1, \pi + \theta_2, \theta_3 \qquad \text{(twofold axis parallel to } b)$$
$$\theta_1\theta_2\theta_3 \rightarrow \pi + \theta_1, \theta_2, \theta_3 \qquad \text{(twofold axis parallel to } c)$$

TABLE 10.1. Eulerian Symmetry Elements for All Possible Types of Space-Group
Rotations

Axis	Direction	First crystal	Second crystal
1		$(\pi + \theta_1, -\theta_2, \pi + \theta_3)$	$(\pi + \theta_1, -\theta_2, \pi + \theta_3)$
2	[010]	$(\pi - \theta_1, \pi + \theta_2, \theta_3)$	$(\theta_1, \pi + \theta_2, \pi - \theta_3)$
2	[001]	$(\pi + \theta_1, \theta_2, \theta_3)$	$(\theta_1, \theta_2, \pi + \theta_3)$
4	[001]	$(-\pi/2 + \theta_1, \theta_2, \theta_3)$	$(\theta_1, \theta_2, \pi/2 + \theta_3)$
3	[001]	$(-2\pi/3 + \theta_1, \theta_2, \theta_3)$	$(\theta_1, \theta_2, 2\pi/3 + \theta_3)$
6	[001]	$(-\pi/3 + \theta_1, \theta_2, \theta_3)$	$(\theta_1, \theta_2, \pi/3 + \theta_3)$
2 [a]	[110]	$(3\pi/2 - \theta_1, \pi - \theta_2, \pi + \theta_3)$	$(\pi + \theta_1, \pi - \theta_2, -3\pi/2 - \theta_3)$

[a] This axis is not unique (that is, it can always be generated by two other unique axes), but
is included for completeness.

(b) In the second crystal ($P2/m$):

$$\theta_1\theta_2\theta_3 \to \pi + \theta_1, -\theta_2, \pi + \theta_3 \qquad \text{(onefold axis)}$$

$$\theta_1\theta_2\theta_3 \to \theta_1, \pi + \theta_2, \pi - \theta_3) \qquad \text{(twofold axis parallel to } b)$$

Upon combination of these symmetry operators two cells result, each of
space group $Pbcb$ (Fig. 10.5). The asymmetric unit within which the rotation
function need be evaluated is found from a knowledge of the Eulerian space
group. In the above example, the limits of the asymmetric unit are $0 \le \theta_1$
$\le \pi/2$; $0 \le \theta_2 \le \pi$; and $0 \le \theta_3 \le \pi/2$.

Nonlinear transformations occur when using Eulerian symmetries for
threefold axes along [111] (as in the cubic system) or when using polar
coordinates. Hence, Eulerian angles are far more suitable for a derivation
of the limits of the rotation function asymmetric unit. However, when
searching for given molecular axes, where some plane of $\varkappa$ need be explored,
polar angles are more useful.

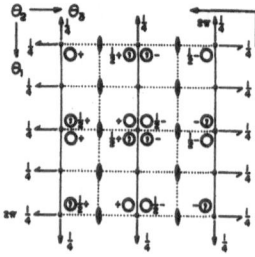

FIGURE 10.5. Rotation-space group diagram for rotation function of a $Pmmm$ Patterson function (P_1) against
a $P2/m$ Patterson function (P_2). The Eulerian angles
$\theta_1, \theta_2, \theta_3$ repeat themselves after an interval of 2π.
Heights above the plane are given in fractions of a
revolution.

10.3.5. Sampling and Background

If the origins are retained in the Pattersons, their product will form a high but constant plateau on which the rotation function peaks are superimposed; this leads to a small apparent peak-to-noise ratio. The effect is eliminated by removal of the origins through a modification of the Patterson coefficients. Irrespective of origin removal, a significant peak is one which is more than three rms deviations from the mean background.

As in all continuous functions sampled at discrete points, a convenient grid size must be chosen. Small intervals result in an excessive computing burden, while large intervals might miss peaks. Furthermore, equal increments of angles do not represent equal changes in rotation, which can result in distorted peaks (Lattman, 1972). In general, a crude idea of a useful sampling interval can be obtained by considering the angle necessary to move one reciprocal-lattice point onto its neighbor at the extremity of the resolution limit. This interval is given by

$$\Delta\theta = d_{min}/2R$$

10.3.6. Locked Rotation Function and Klug Peaks

In general the interpretation of the rotation function is straightforward. Early examples are those of insulin (Dodson *et al.*, 1966) (Fig. 10.6) and chymotrypsin (Blow *et al.*, 1964). In many structures, noncrystallographic symmetry axes are in special directions relative to the crystallographic axes (Fig. 10.6). This makes interpretation of the rotation diagram particularly easy as the peaks are doubled on mirror planes of the rotation function. The physical reason for this special packing in the crystal is not well understood, but presumably results from the same phenomenon that leads to the utilization of molecular axes in crystal symmetry.

Without these special circumstances, the interpretation of the rotation function can be more difficult. If a molecule possesses more than one noncrystallographic symmetry axis, then searching for all the molecular axes simultaneously by use of the locked rotation function (Rossmann, 1972b) can make the interpretation task easier as exemplified in the structure determination of glyceraldehyde-3-phosphate dehydrogenase (Fig. 10.7).

The case of α-chymotrypsin, which has two monomers per asymmetric unit in the monoclinic cell, is particularly instructive in detecting multiple symmetry elements in the rotation function (Fig. 10.8). The molecules are related by a dimer axis, perpendicular to the crystallographic 2_1 axis, such that their product produces a twofold translation axis perpendicular to both of them. Thus, in the rotation function there are three $\varkappa = 180°$ peaks; that is, the distribution representing the sum of the four self-Pattersons in the unit cell has 222 symmetry, of which one axis is crystallographic, the second corresponds to the dimer axis, and the third is generated by the combination of the former two axes. This latter peak has been referred to (Johnson *et al.*, 1975) as a Klug peak in recognition of Aaron Klug's contribution to the interpretation of the satellite tobacco necrosis virus rotation function (Åkervall *et al.*, 1971). Litvin (1975) has attempted to predict Klug peaks in a general way, although he only analyzes them for a special case (Eventoff and Gurskaya, 1975).

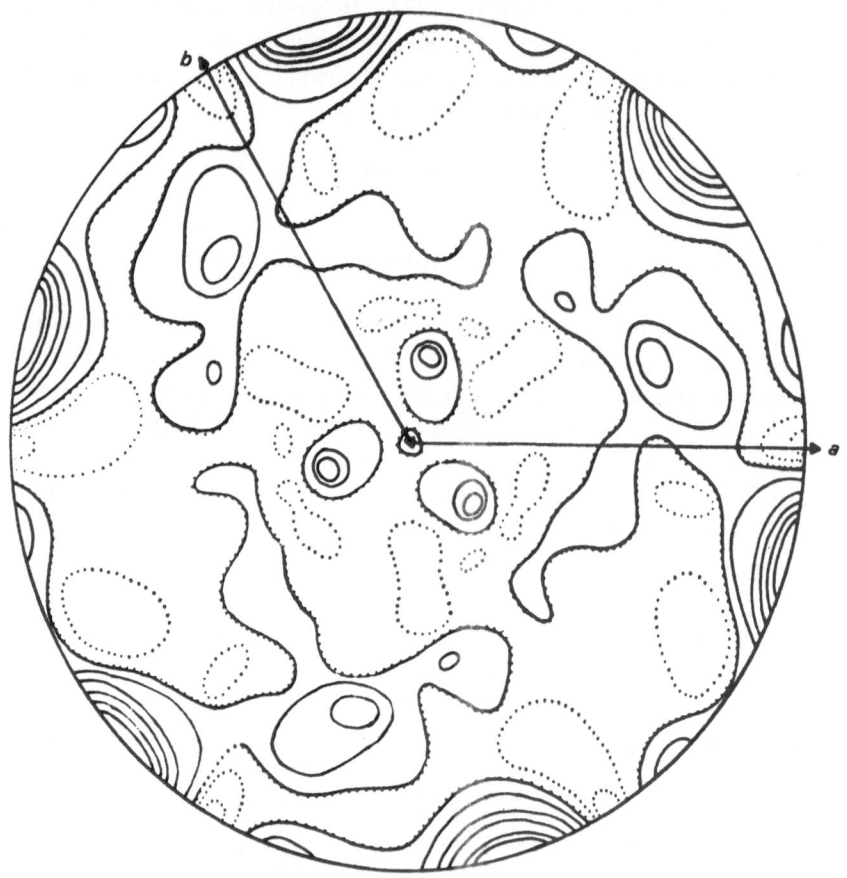

FIGURE 10.6. Stereogram of $R(180°, \psi, \varphi)$ for 2 Zn insulin showing noncrystallographic twofold axes perpendicular to the rhombohedral threefold axis.

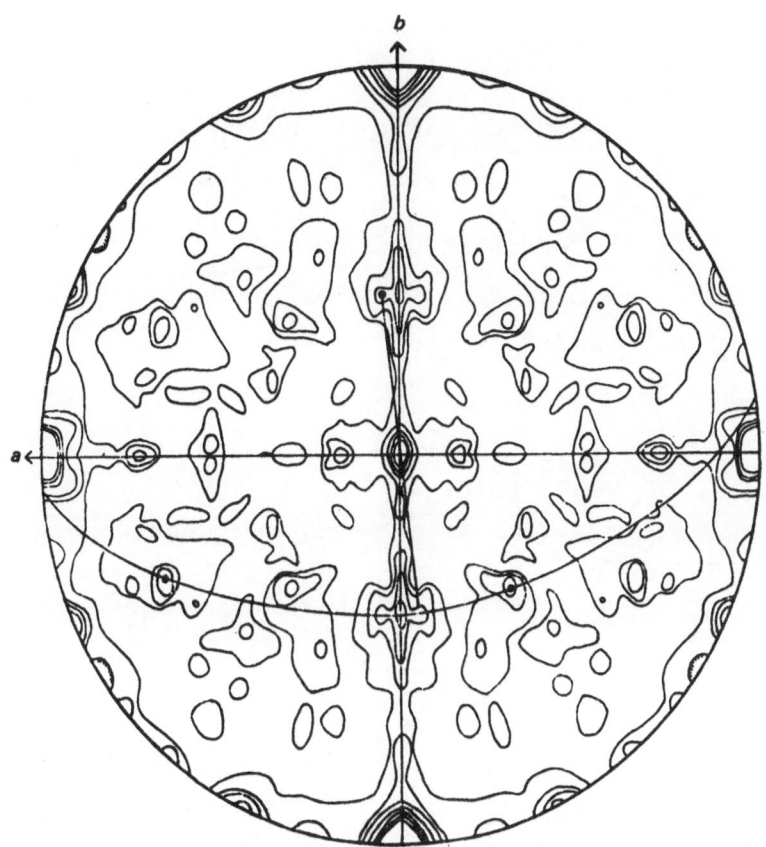

FIGURE 10.7. Rotation function of tetrameric glyceraldehyde-3-phosphate dehydrogenase, at 6-Å resolution, showing the presence of only one set of three mutually perpendicular peaks.

10.3.7. Recognizing Known Fragments

The rotation function can be used not only to match two unknown self-Pattersons within the same crystal (self-rotation function) or in different crystals (cross-rotation function), but can also compare an unknown structure with a known structure (Tollin and Rossmann, 1966). For this purpose, the known structure is introduced into an artificially large unit cell for the calculation of structure factors. The large unit cell avoids overlap of the self-Pattersons at neighboring origins, increases the signal strength in the Patterson, and retains some phase information of the known molecule. The cross-rotation function then determines the orientation of the unknown

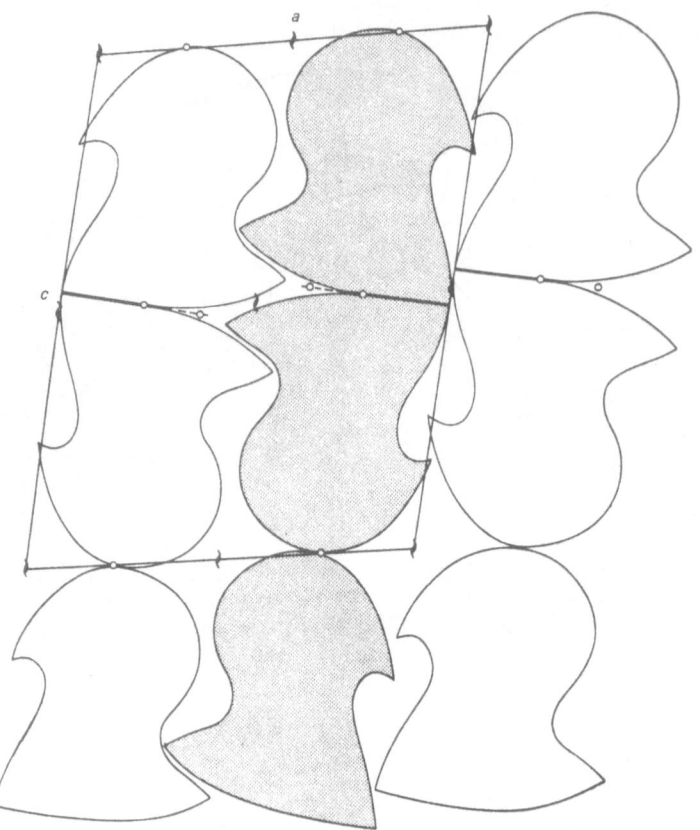

FIGURE 10.8. Diagram to show the symmetry of the molecular arrangement in α-chymotrypsin. The shaded molecules are at approximately $y = \frac{1}{4}$; the unshaded molecules are at approximately $y = 0$.

molecule with respect to the known orientation in the artificial cell. Such procedures are followed by a determination of translation vectors and calculation of structure factors (Section 10.2) which provide phases to be applied to the amplitudes of the unknown structure for an electron density map. In this mode, the cross-rotation function has been used frequently to solve structures of a given molecule whose fold had previously been discovered in other species or isozymes. A similar procédure has been suggested by Hoppe (1957) and has been used by Huber (1965, 1970), Hornsta (1970), and Katsube et al. (1966).

The rotation function has also been used in the solution of small molecules. Nordman (1966, 1972) and Nordman and Schilling (1970) searched for rigid fragments by comparing a point Patterson distribution of the known fragment with the real Patterson, in contrast to the overlap function calculations given by equation (10.13). Tollin and co-workers (Tollin, 1970, 1976; Tollin and Cochran, 1964; Watson et al., 1965) use a special technique to determine the orientation of the plane of flat molecules by searching the Patterson for a corresponding plane of dense vectors. Their $I(\theta, \varphi)$ function (Tollin and Cochran, 1964; Watson et al., 1965) corresponds to equation (10.13) except that the zero-order spherical harmonic G_{hp} is replaced by a Bessel function. As this article deals primarily with the application to biological macromolecules, little attempt is made here to enlarge on the procedures used in small-molecule crystallography except to note similarity of technique.

10.3.8. Fast Rotation Function

Unfortunately, the rotation function computations can be extremely time-consuming by conventional methods. Sasada (1964) developed a technique for finding rapidly the maximum of a given peak by looking at the slope of the rotation function. A major breakthrough came when Crowther (1972) recast the rotation function in a manner suitable for rapid computation. Only a brief outline of Crowther's fast rotation function is given here. Details are found in the original text (Crowther, 1972) and his computer program description.

Since the rotation function correlates spherical volumes of a given Patterson density with rotated versions of either itself or another Patterson density, it is likely that a more natural form for the rotation function will involve spherical harmonics rather than the Fourier components $|F_{\mathbf{h}}|^2$ of the crystal. Thus, if the two Patterson densities $P_1(r, \psi, \phi)$ and $P_2(r, \psi, \phi)$ are expanded within the spherical volume of radius less than a limiting value of a, then

$$P_1(r, \psi, \phi) = \sum_{lmn} a^*_{lmn} j_l(k_{ln}r)\hat{Y}_l^{m*}(\psi, \phi)$$

and

$$P_2(r, \psi, \phi) = \sum_{l'm'n'} b_{l'm'n'} j_{l'}(k_{l'n'}r)\hat{Y}_{l'}^{m'}(\psi, \phi)$$

the rotation function would then be defined as

$$R = \int_{\text{sphere}} P_1(r, \psi, \phi)\mathscr{R}P_2(r, \psi, \phi)r^2 \sin \psi \, dr \, d\psi \, d\phi$$

Here $\hat{Y}_l{}^m(\psi, \varphi)$ is the normalized spherical harmonic of order l; $\hat{j}_l(k_{ln}r)$ is the normalized spherical Bessel function of order l; a_{lmn}, b_{lmn} are complex coefficients; and $\mathscr{R}P_2(r, \psi, \phi)$ represents the rotated second Patterson. The rotated spherical harmonic can then be expressed in terms of the Eulerian angles $\theta_1, \theta_2, \theta_3$ as

$$\mathscr{R}(\theta_1, \theta_2, \theta_3)\hat{Y}_l{}^m(\psi, \phi) = \sum_{q=-l}^{l} D_{qm}^l(\theta_1, \theta_2, \theta_3)\hat{Y}_l{}^q(\psi, \phi)$$

where

$$D_{qm}^l(\theta_1, \theta_2, \theta_3) = \exp(iq\theta_3)\, d_{qm}^l(\theta_2)\exp(im\theta_1)$$

and $d_{qm}^l(\theta_2)$ are the matrix elements of the three-dimensional rotation group. It follows that

$$R(\theta_1, \theta_2, \theta_3)$$
$$= \int_{\text{sphere}} \sum_{lmn} a_{lmn}^* \hat{j}_l(k_{ln}r)\hat{Y}_l{}^{m*}(\psi, \phi) \sum_{l'm'n'} b_{l'm'n'}\hat{j}_{l'}(k_{l'n'}r)\mathscr{R}\hat{Y}_{l'}{}^{m'}(\psi, \phi)\, dV$$

and, substituting for the rotated harmonics,

$$R(\theta_1, \theta_2, \theta_3) = \int_{\text{sphere}} \sum_{lmn} a_{lmn}^* \hat{j}_l(k_{ln}r)\hat{Y}_l{}^{m*}(\psi, \phi) \sum_{l'm'n'} b_{l'm'n'}\hat{j}_{l'}(k_{l'n'}r)$$
$$\times \sum_q D_{qm'}^{l'}\hat{Y}_{l'}{}^q(\psi, \phi)\, dV$$

which, after introduction of suitable orthonormality relationships, reduces to

$$R(\theta_1, \theta_2, \theta_3) = \sum_{lmm'n} a_{lmn}^* b_{lm'n} D_{m'm}^l(\theta_1, \theta_2, \theta_3)$$

Since the radial summation over n is independent of the rotation,

$$c_{lmm'} = \sum_n a_{lmn}^* b_{lmn}$$

and hence

$$R(\theta_1, \theta_2, \theta_3) = \sum_{lmm'} c_{lmm'} D_{m'm}^l(\theta_1, \theta_2, \theta_3)$$

or

$$R(\theta_1, \theta_2, \theta_3) = \sum_{mm'} \left[\sum_l c_{lmm'} d_{m'm}^l(\theta_2) \right] \exp[i(m'\theta_3 + m\theta_1)]$$

The coefficients $c_{lmm'}$ refer to a particular pair of Patterson densities and are independent of the rotation. The coefficients $D_{m'm}^l$, containing the whole rotational part, refer to rotations of spherical harmonics and are independent of the particular Patterson densities. Since the summations over m and m' represent a Fourier synthesis, rapid calculation is possible.

As polar coordinates rather than Eulerian angles provide a more graphic interpretation of the rotation function, Tanaka (1977) has recast the initial definition as

$$R(\theta_1, \theta_2, \theta_3)$$

$$= \int_{\text{sphere}} [\mathscr{R}(\theta_1, \theta_2, \theta_3 = 0)P_1(r, \psi, \phi)][\mathscr{R}(\theta_1, \theta_2, \theta_3)P_2(r, \psi, \phi)] \, dV$$

$$= \int_{\text{sphere}} [P_1(r, \psi, \phi)][\mathscr{R}^{-1}(\theta_1, \theta_2, \theta_3 = 0)\mathscr{R}(\theta_1, \theta_2, \theta_3)P_2(r, \psi, \phi)] \, dV$$

He shows that the polar coordinates are now equivalent to $\varkappa = \theta_3$, $\psi = \theta_2$, and $\phi = \theta_1 - \pi/2$. The rotation function can then be expressed as

$$R(\varkappa, \psi, \phi) = \sum_{lmm'} \left(\sum_n a^*_{lmn} b_{lm'n} \right) \sum_q d^l_{qm}(\psi) \, d^l_{qm'}(\psi)(-1)^{(m'-m)}$$
$$\times \exp[i(\varkappa q)] \exp[i(m' - m)\phi]$$

permitting fast calculation of the rotation function in polar coordinates.

10.4. Translation Problem

10.4.1. Introduction

Considerable thought has been lavished on the determination of the rotation parameters of noncrystallographic symmetry and on the solution of the phase problem given both rotation and translation parameters. In contrast, the translation problem has received little attention, probably owing to its inherent lack of precise meaning. Colman *et al.* (1976) have written a review of available techniques whose interrelationships had been examined by Karle (1972).

Rossmann *et al.* (1964) differentiated between the precise and imprecise components of translation (Fig. 10.9). The precise component represents translation parallel to a rotation axis. The imprecise component is perpendicular to the rotation axis and is dependent on the definition of an origin. Only when the approximate molecular structure and its orientation are available can the position of a molecular origin be determined (Crowther and Blow, 1967; Tollin, 1966, 1969a).

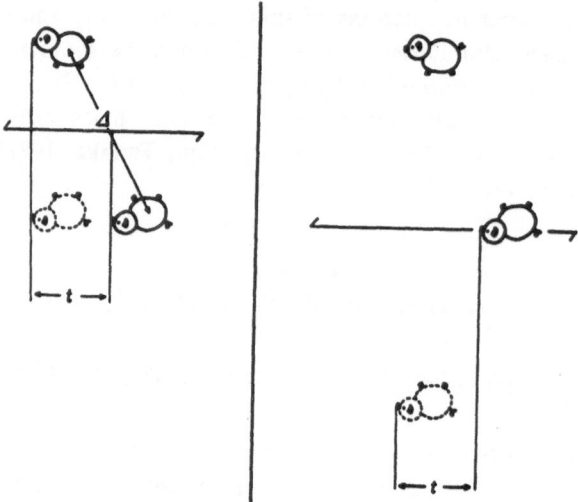

FIGURE 10.9. Position of the twofold rotation axis that relates the two piglets is completely arbitrary. The diagram on the left shows the situation when the translation is parallel to the rotation axis. The diagram on the right has an additional component of translation perpendicular to the rotation axis, but the component parallel to the axis remains unchanged.

The general translation function can be defined as

$$T(\mathbf{S}_x, \mathbf{S}_y) = \int_U \varrho_1(\mathbf{x}) \cdot \varrho_2(\mathbf{y}) \, d\mathbf{x}$$

where T is a six-variable function given by each of the three components that define $\mathbf{S}_x$ and $\mathbf{S}_y$. Following the same procedure used for the rotation function derivation, Fourier summations are substituted for $\varrho_1(\mathbf{x})$ and $\varrho_2(\mathbf{y})$. It can then be shown that

$$T(\mathbf{S}_x, \mathbf{S}_y) = \int_U \left\{ \frac{1}{V_h} \sum_{\mathbf{h}} |F_{\mathbf{h}}| \exp[i(\alpha_{\mathbf{h}} - 2\pi\mathbf{h} \cdot \mathbf{x})] \right\}$$
$$\times \left\{ \frac{1}{V_p} \sum_{\mathbf{p}} |F_{\mathbf{p}}| \exp[i(\alpha_{\mathbf{p}} - 2\pi\mathbf{p} \cdot \mathbf{y})] \right\} d\mathbf{x}$$

Using the substitution $\mathbf{y} = [\mathbf{C}]\mathbf{x} + \mathbf{d}$ and simplifying

$$T(\mathbf{S}_x, \mathbf{S}_y) = \frac{1}{V_h V_p} \sum_{\mathbf{h}} \sum_{\mathbf{p}} |F_{\mathbf{h}}| |F_{\mathbf{p}}| \exp[i(\alpha_{\mathbf{h}} + \alpha_{\mathbf{p}} - 2\pi\mathbf{p} \cdot \mathbf{d})]$$
$$\times \int_U \exp\{i[-2\pi(\mathbf{h} + \mathbf{p}[\mathbf{C}]) \cdot \mathbf{x}]\} \, d\mathbf{x}$$

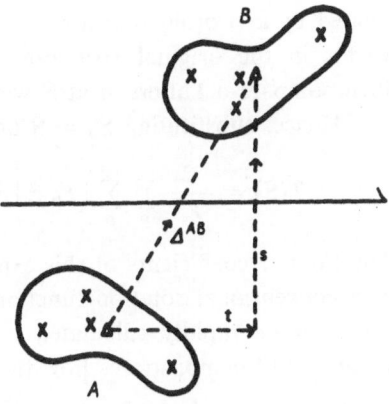

FIGURE 10.10. Crosses represent atoms in a two-dimensional model structure. The triangles are the points chosen as approximate centers of molecules A and B. Δ^{AB} has components $\mathbf{t}$ and $\mathbf{s}$ parallel and perpendicular to the screw rotation axis, respectively.

The integral is the diffraction function $G_{\mathbf{hp}}$ defined in Section 10.2. If the integration is taken over the volume U centered at $\mathbf{S}_x$ and $\mathbf{S}_y$, it follows that

$$T(\mathbf{S}_x, \mathbf{S}_y) = \frac{2}{V_{\mathbf{h}} V_{\mathbf{p}}} \sum_{\mathbf{h}} \sum_{\mathbf{p}} |F_{\mathbf{h}}| |F_{\mathbf{p}}| G_{\mathbf{hp}} \cos[\alpha_{\mathbf{h}} + \alpha_{\mathbf{p}} - 2\pi(\mathbf{h} \cdot \mathbf{S}_x + \mathbf{p} \cdot \mathbf{S}_y)] \tag{10.20}$$

10.4.2. Neither Structure is Known

Consider the determination of the precise translation vector parallel to a rotation axis between two identical molecules of unknown structure. For simplicity, let the noncrystallographic axis be a diad (Fig. 10.10). Figure 10.11 shows the corresponding Patterson of the hypothetical point atom structure. Opposite sets of cross-Patterson vectors in Fig. 10.11 are

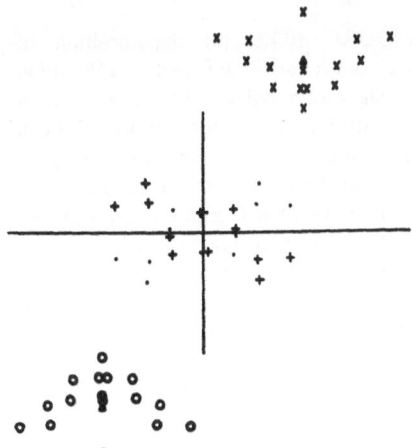

FIGURE 10.11. Vectors arising from the structure in Fig. 10.10. The self-vectors of molecules A and B are represented by $+$ and $\cdot$; the cross-vectors from molecules A to B and B to A by $\times$ and $\bigcirc$. Triangles mark the position of $+\Delta^{AB}$ and $-\Delta^{AB}$.

related by a twofold rotation and a translation equal to twice the precise vector in the original structure. A suitable translation function would then compare a Patterson at S with the rotated Patterson at −S.

Hence, substituting $S_x = S$ and $S_y = -S$ in (10.20),

$$T(\mathbf{S}) = \frac{2}{V^2} \sum_{\mathbf{h}} \sum_{\mathbf{p}} |F_\mathbf{h}|^2 |F_\mathbf{p}|^2 G_{\mathbf{hp}} \cos[2\pi(\mathbf{h} - \mathbf{p}) \cdot \mathbf{S}] \qquad (10.21)$$

The Fourier coefficients of this expression can be evaluated by adaptation of a conventional rotation function program and the translation function can then be rapidly calculated by typical Fourier synthesis techniques. Figure 10.12 demonstrates how the precise and imprecise translation com-

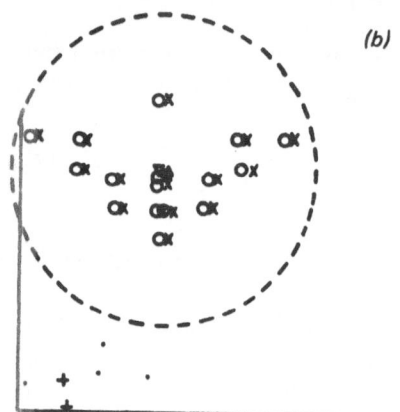

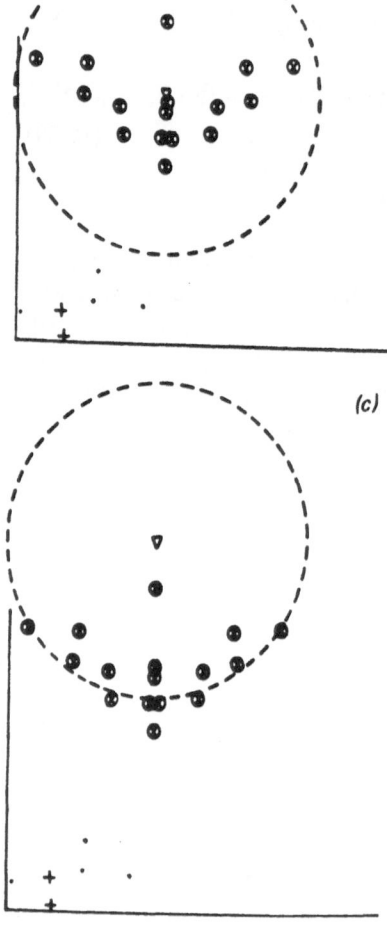

FIGURE 10.12. (a) Superposition of vectors around $+\Delta^{AB}$ and $-\Delta^{AB}$ within a sphere of suitable radius, after rotating the latter about the rotation axis through the Patterson origin and translating by 2t parallel to the axis. (b) Similar superposition with a wrong choice of the precise parameter t producing no significant vector coincidences. (c) Similar superposition but with the imprecise parameter s chosen poorly, producing good vector coincidences although some fall outside the sphere of integration.

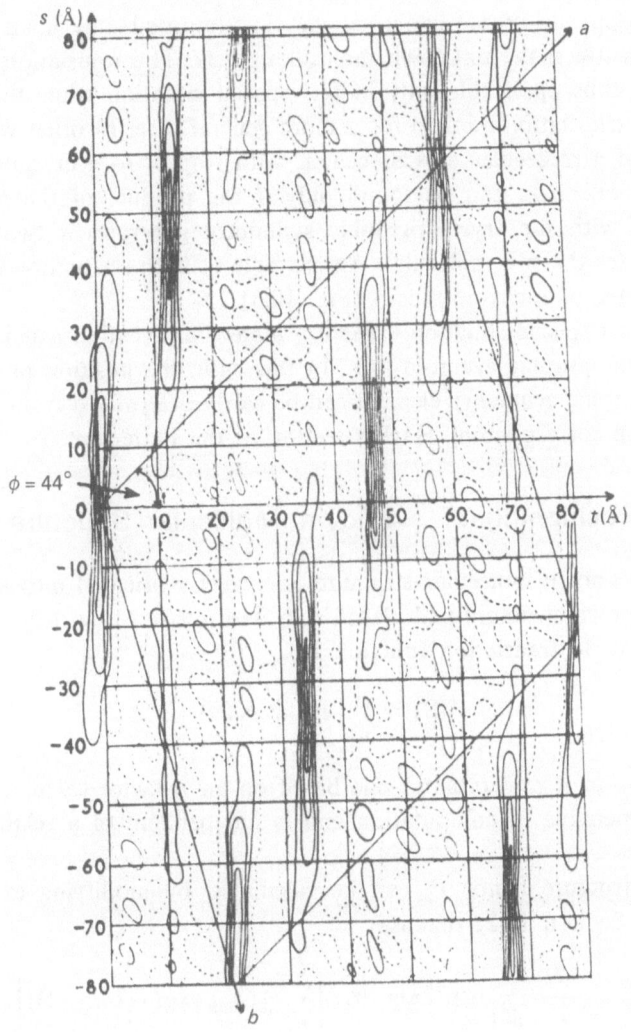

FIGURE 10.13. Translation function of 2 Zn insulin in the hexagonal plane $z = 0$. The direction of the noncrystallographic twofold axis is drawn horizontally. It lies in the *ab* plane of the hexagonal cell at $44°$ with respect to **a**. Thus precise distances (**t**) are measured along this axis and imprecise distances (**s**) are perpendicular to the axis (see Figs. 10.9 and 10.10). The translation function $T(\mathbf{S})$ is large near each of the three lattice points in the cell showing zero translation along the molecular dimer axis.

ponents effect the nature of the translation function, an actual example of which is given in Figure 10.13.

The opposite cross-vectors can be superimposed only if an evenfold rotation between the unknown molecules exists. The translation function (10.21) is thus applicable only in this special situation while there is no published translation method for a more general case. In other words, the position of a molecular axis or center, with respect to a crystallographic symmetry operator, can be found only if the product of the molecular symmetry with the crystallographic symmetry generates a twofold axis which relates the two molecules. An example of such a situation occurs in the structure of α-chymotrypsin (Fig. 10.8).

Another special situation exists if a molecular evenfold axis is parallel to a crystallographic evenfold axis. In this case, the position of the non-crystallographic symmetry element can be easily determined from the large peak in the corresponding Harker section of the Patterson.

10.4.3. Positioning of a Known Molecular Structure

If the known molecular structure is correctly oriented into a cell (**p**) of an unknown structure and placed at **S** with respect to a defined origin, then a suitable translation function is

$$T(\mathbf{S}) = \sum_{\mathbf{p}} | F_{\mathbf{p},o} |^2 F_{\mathbf{p}}(\mathbf{S}) \tag{10.22}$$

This definition is preferable to one based on an R-factor calculation as it is more amenable to computation and is independent of a relative scale factor.

The structure factor $F_{\mathbf{p}}$ can be calculated by modifying expression (10.10) in Section 10.2; that is,

$$F_{\mathbf{p}} = \frac{U}{V_{\mathbf{h}}} \sum_{n=1}^{N} \exp(2\pi i \mathbf{p} \cdot \mathbf{S}_n) \left[\sum_{\mathbf{h}} F_{\mathbf{h}} G_{\mathrm{hpn}} \exp(-2\pi i \mathbf{h} \cdot \mathbf{S}) \right]$$

Let

$$A_{\mathbf{p},n} e^{i\gamma n} = \sum_{\mathbf{h}} F_{\mathbf{h}} G_{\mathrm{hpn}} \exp(-2\pi i \mathbf{h} \cdot \mathbf{S})$$

which are the coefficients of the molecular transform for the known molecule placed into the nth asymmetric unit of the **p** cell. Thus

$$F_{\mathbf{p}} = \frac{U}{V_{\mathbf{h}}} \sum_{n=1}^{N} A_{\mathbf{p},n} \exp[i(\gamma_n + 2\pi \mathbf{p} \cdot \mathbf{S}_n)]$$

or

$$F_{\mathbf{p}} = \frac{U}{V_{\mathrm{h}}} \sum_{n=1}^{N} A_{\mathbf{p},n} \exp[i(\gamma_n + 2\pi \mathbf{p}_n \cdot \mathbf{S})] \tag{10.23}$$

where $\mathbf{p}_n = [C_n^T]\mathbf{p}$ and $\mathbf{S} = \mathbf{S}_1$. Hence

$$| F_{\mathbf{p}} |^2 = \frac{U}{V_{\mathrm{h}}^2} \sum_n \sum_m A_{\mathbf{p},n} A_{\mathbf{p},m} \exp\{i[2\pi(\mathbf{p}_n - \mathbf{p}_m) \cdot \mathbf{S} + (\gamma_n - \gamma_m)]\}$$

and then from (10.22)

$$T(\mathbf{S}) = \frac{U}{V_{\mathrm{h}}^2} \sum_{\mathbf{p}} \sum_n \sum_m | F_{\mathbf{p},o} |^2 A_{\mathbf{p},n} A_{\mathbf{p},m} \exp\{i[2\pi(\mathbf{p}_n - \mathbf{p}_m) \cdot \mathbf{S} + (\gamma_n - \gamma_m)]\} \tag{10.24}$$

which is a Fourier summation with known coefficients $[| F_{\mathbf{p},o} |^2 A_{\mathbf{p},n} A_{\mathbf{p},m} \times e^{i(\gamma_n - \gamma_m)}]$ such that $T(\mathbf{S})$ will be a maximum at the correct molecular position.

Terms with $n = m$ in expression (10.24) can be omitted as they are independent of $\mathbf{S}$, and thus provide only a constant toward the value of $T(\mathbf{S})$. For terms with $n \neq m$, the indices take on special values. For instance, if the $\mathbf{p}$ cell is monoclinic with its unique axis parallel to $\mathbf{b}$ such that $\mathbf{p}_1 = (p, q, r)$ and $\mathbf{p}_2 = (\bar{p}, q, \bar{r})$, then $\mathbf{p}_1 - \mathbf{p}_2$ would be $(2p, 0, 2r)$. Hence $T(\mathbf{S})$ would be a two-dimensional function consistent with the physical requirement that the translation component, parallel to the twofold monoclinic axis, is arbitrary.

Crowther and Blow (1967) show that if F_M are the structure factors of a known molecule correctly oriented within the cell of the unknown structure at an arbitrary molecular origin, then (altering the notation very slightly from above)

$$T(\mathbf{S}) = \sum_{\mathbf{p}} | F_o |^2 (\mathbf{p}) F_M(\mathbf{p}) F_M^*(\mathbf{p}[C]) \exp(-2\pi i \mathbf{p} \cdot \mathbf{S}) \tag{10.25}$$

where $[C]$ is a crystallographic symmetry operator relative to which the molecular origin is to be determined. This is of the same form as (10.24) but concerns the special case where the $\mathbf{h}$ cell, into which the known molecule was placed, has the same dimensions as the $\mathbf{p}$ cell. Expression (10.25) can also be derived from (10.20) by setting $F_{\mathbf{h}} = | F_o |^2$ and observing that $G_{\mathbf{hp}} = 1$ when $\mathbf{h} = \mathbf{p}$ and $G_{\mathbf{hp}} = 0$ when $\mathbf{h} \neq \mathbf{p}$.

Tollin's Q function (1966) and Crowther and Blow's translation function (1967) are both of the form given by (10.25). Indeed, Tollin (1969a) showed that both techniques are essentially identical. Their derivations, unlike the arguments given here, depend on deconvoluting the Patterson function.

In practice, spurious peaks can easily arise in translation functions, particularly if the known starting model is not defined with sufficient accuracy (Burnett and Rossmann, 1971; Colman et al., 1976). These peaks may result from the fourth power nature of the function, which tends to deemphasize all but a few terms.

R-factor calculations are sometimes used, especially when the position of a molecule need only be determined in one dimension, such as along a crystallographic symmetry axis (Eventoff et al., 1975; Bott and Sarma, 1976).

10.4.4. Both Structures are Known

If an initial set of poor phases, for example from a single isomorphous replacement derivative, is available and the rotation function has given the orientation of a noncrystallographic rotation axis, it is possible to search the electron density map systematically to determine the translation axis position. The translation function must, therefore, measure the quality of superposition of the poor electron density map on itself. Hence $S_x = S_y = S$ and the function (10.20) now becomes

$$T(\mathbf{S}) = \frac{2}{V_h^2} \sum_{\mathbf{h}} \sum_{\mathbf{p}} |F_{\mathbf{h}}| |F_{\mathbf{p}}| G_{\mathbf{hp}} \cos[\alpha_{\mathbf{h}} + \alpha_{\mathbf{p}} - 2\pi(\mathbf{h} + \mathbf{p}) \cdot \mathbf{S}]$$

This real-space translation function has been used successfully to determine the intermolecular diad axis for α-chymotrypsin (Blow et al., 1964) and to verify the position of immunoglobulin domains (Colman and Fehlhammer, 1976).

10.4.5. Use of Heavy Atoms to Determine a Molecular Center

If a heavy-atom derivative is available for a molecular aggregate whose symmetry is not incorporated into the crystal symmetry, then the position of the molecular center may be found in a straightforward manner. Initially, the heavy-atom positions are determined from the self-vectors (see Section 10.6). Then the vector set is generated and compared with the Patterson for every possible position of the molecular center in the crystallographic asymmetric unit (Argos and Rossmann, 1974; Argos et al., 1976). Buehner et al. (1974) used this principle to determine, by visual inspection of the Patterson, the molecular center of tetrameric glyceraldehyde-3-phosphate dehydrogenase.

10.4.6. Use of Packing Considerations

By considering the packing of the molecule in a cell, many possible positions of the molecular center can be eliminated. For instance, it is clearly impossible to have molecular centers too close to a rotation axis. When the molecular structure is unknown, the assumption of a spherical envelope does not allow a particularly precise determination. If, however, a molecular structure or envelope is known together with the molecular orientation, more precise computations can be made by scoring for the overlapped molecular volumes with a suitable algorithm. Hendrickson and Ward (1976) successfully used this technique in the solution of the hemerythrin structure. The peak position was subsequently refined by *R*-factor calculation, a procedure feasible for small volume explorations.

If the solvent contains low concentrations of salt, it is sometimes possible to consider the molecule as a uniform sphere at very low resolution. In this case, simple structure-factor plots of very-low-order reflections can help in narrowing the possible molecular center positions (Eventoff *et al.*, 1976).

10.5. Phase Determination

10.5.1. Introduction

Even though Muirhead *et al.* (1967) and Matthews *et al.* (1967) had used real-space averaging to improve the clarity of electron density maps, the molecular replacement technique had been considered an essentially *ab initio* phase-determining method until approximately 1973. The arbitrary techniques of Rossmann and Blow (1963, 1964) and the chancy methods of Main and Rossmann (1966) were beset with problems such as defining an envelope, determining a hand, and truncation errors. Furthermore, there existed no immediate applications to real problems. Crowther (1967, 1969) elegantly applied the techniques of linear algebra which provided insight into the properties of the molecular replacement equations and endeavored to answer such controversies as the uniqueness of their solutions. Unfortunately, the large matrices made any extensive applications prohibitive, although Crowther's techniques were used in preliminary studies of tobacco mosaic virus protein (Jack, 1973) and tomato bushy stunt virus (Harrison and Jack, 1975; Winkler *et al.*, 1977).

A radical departure occurred when Buehner *et al.* (1974) used real-space averaging to interpret the single isomorphous replacement map of lobster glyceraldehyde-3-phosphate dehydrogenase. Furthermore, Colman (1974), in a special case, and Bricogne (1974), quite generally, showed rigorously the equivalence of the real- and reciprocal-space methods. This was followed by the successful application of Bricogne's real-space computing techniques (Bricogne, 1976b) to *B. stearothermophilus* glyceraldehyde-3-phosphate dehydrogenase (Wonacott and Bricogne, 1976), tobacco mosaic virus protein (Champness *et al.*, 1976) and tomato bushy stunt virus (Winkler *et al.*, 1977). Similar methods were also used in Uppsala on satellite tobacco necrosis virus (Strandberg and Lentz, 1976), at Purdue University on lactate dehydrogenase isozymes (Eventoff *et al.*, 1975; Musick *et al.*, 1976) and southern bean mosaic virus (Johnson *et al.*, 1976), and at Yale University on hexokinase (Fletterick and Steitz, 1976). Since real-space averaging gave an immediate visual benefit, molecular replacement was now strongly emphasized as a support to isomorphous replacement. Nevertheless, the time required to sort the unit cell grid positions, to associate noncrystallographically related densities, and to back-transform the averaged electron density is still exceedingly formidable. Reciprocal space methods are presently being developed (Caspar, 1976; unpublished results from the authors' laboratory) which may be faster than the real-space averaging, thus swinging back the molecular replacement pendulum.

10.5.2. Reciprocal-Space Equations

The reciprocal-space equations as derived in Section 10.2 are

$$F_{\mathbf{p}} = \frac{U}{V_{\mathbf{h}}} \sum_{\mathbf{h}} F_{\mathbf{h}} \sum_{n=1}^{N} G_{\mathbf{h}\mathbf{p}n} \exp[2\pi i(\mathbf{p} \cdot \mathbf{S}_n - \mathbf{h} \cdot \mathbf{S})]$$

for all **p**. By setting

$$B_{\mathbf{h}\mathbf{p}} = \frac{U}{V_{\mathbf{h}}} \sum_{n=1}^{N} G_{\mathbf{h}\mathbf{p}n} \exp[-2\pi i(\mathbf{h} \cdot \mathbf{S} + \mathbf{p} \cdot \mathbf{S}_n)] \qquad (10.26)$$

the molecular replacement equations can be rewritten as

$$F_{\mathbf{p}}{}^{*} = \sum_{\mathbf{h}} B_{\mathbf{h}\mathbf{p}} F_{\mathbf{h}} \qquad (10.27)$$

or in matrix notation

$$\mathbf{F}^{*} = [\mathbf{B}]\mathbf{F}$$

which is the form of the equations used by Crowther (1967). Main and Rossmann (1966) showed that one set of independent equations is generated for each noncrystallographic asymmetric unit, corresponding to a different molecular choice of S.

Main (1967) used these equations to solve a simple hypothetical structure of four crystallographically independent, but chemically identical, molecules in space group $P1$ with 10 carbon atoms per molecule. The procedure used (Main, 1967; Rossmann and Blow, 1963, 1964) is instructive in the physical principles of molecular replacement. Initially, the only known phase is that of $F(000)$. This provides large coefficients in the equations for very-low-order terms where the G function has strong interactions with near neighbors in reciprocal space (Section 10.3). Thus, to initiate the phasing process, it is necessary to have a reasonable estimate of $F(000)$, which implies that the amplitudes need to be placed on an absolute scale. A negative sign for $F(000)$ would alter the phase determination by π for every reflection, in accordance with Babinet's principle (Wood, 1934). The approximate innermost phases can now be substituted into the reciprocal-space equations which then provide improved values for α_p. The resultant new phases can be assigned to the left-hand sides and the process repeated.

A more systematic approach to the solution of (10.27) was derived by Crowther (1967). There are $(2N + 1)$ equations of the form (10.27) if N is the number of unique non-Friedel-related reflections. If, then, $\mathbf{F}$ is considered a complex column vector of Friedel-related structure factors such that

$$\mathbf{F} = \begin{pmatrix} F_{-N} \\ \vdots \\ 0 \\ \vdots \\ F_N \end{pmatrix}$$

and since $[\mathbf{B}]$ is a $(2N + 1) \times (2N + 1)$ complex matrix,

$$[\mathbf{T}]\mathbf{F} = \mathbf{F}^* \tag{10.28}$$

where

$$[\mathbf{T}] = \begin{pmatrix} 0 & \cdots & & & 1 \\ & & & \cdot & \\ \vdots & & 1 & & \vdots \\ & \cdot & & & \\ 0 & \cdots & & & 1 \end{pmatrix}$$

If $[H]$ is defined as $[T][B]$, then from (10.27) and (10.28),

$$[B]F = [T]F$$

and

$$[H]F = F \qquad (10.29)$$

$[B]$ is Hermitian along its second diagonal and thus $[H]$ is of Hermitian type along its principal diagonal as they are related by the T matrix, which merely interchanges diagonals. The expression (10.29) is a concise form of the reciprocal-space equations and provides a format to attempt their solution directly.

Since the expression (10.29) is a set of $(2N + 1)$ homogeneous linear equations, a general solution will be a linear combination of the several independent solutions. If (10.29) is considered a general eigenvalue problem such that $[H]F = \lambda F$, there will be $(2N + 1)$ eigenvalues λ_j. Crowther (1967) showed that the eigenvalues must lie in the range 0 to 1; however only those with $\lambda_j = 1$ will yield linearly independent solutions. If there are m unit eigenvalues for $[H]$, then F can be expressed as a linear combination of the corresponding m orthonormal eigenvectors u_j. Then

$$F = \sum_{j=1}^{m} \mu_j u_j$$

where all μ_j must be real since the electron density $\varrho(x)$ is itself a linear combination of the transforms of the eigenvectors u_j determinable from $[H]$. As only the intensities of the reflections are available, the μ_j values have to be obtained from the set of N nonlinear equations

$$|F_p|^2 = \sum_{j=1}^{m} \sum_{k=1}^{m} \mu_j \mu_k u_{jp}^* u_{kp}$$

where $m < N$. The coefficients $u_{jp}^* u_{kp}$ are difficult to compute, especially if N is large. Crowther (1967) thus suggests an iterative procedure that attempts to calculate structure-factor phases rather than the μ_j parameters. This is accomplished by a cyclic process similar to that described by Main and Rossmann (1966).

Since the phases at any one cycle are only approximate, the structure factors F will contain components of allowed and nonallowed eigenvectors; that is, their transform will not express the noncrystallographic symmetry exactly. F can then be written as

$$F = \sum_{j=1}^{m} \mu_j u_j + \sum_{j=m+1}^{2N+1} \mu_j u_j$$

where there are m allowed elements.

Multiplication of F by $[H]$ will leave the allowed components ($\lambda \simeq 1$) unchanged and reduce the effect of unallowed elements. However, the resulting moduli of $[H]F$ will have to be individually rescaled to equal the observed amplitudes. The structure-factor column vector for the next cycle of refinement and extension will then be given by

$$F_{new} = [S_{new}][H]F_{old}$$

where $[S]$ is a diagonal rescaling matrix.

Applications of reciprocal-space molecular replacement have been primarily on model structures. However, Jack (1973) used Crowther's analysis of the H matrix directly to determine the eigenvalues and their vectors analytically. This process gave signs for 840 centric reflections for the projection of the 17-fold tobacco mosaic virus protein disk, representing 7.0-Å resolution data. These signs compared well with those determined by single isomorphous replacement and were also checked by computing a heavy-atom difference Fourier. Harrison and Jack (1975) applied the same process to low-resolution tomato bushy stunt virus data for the refinement of a crude phase set based on electron microscope image reconstruction and on single isomorphous replacement data. In the tomato bushy stunt virus, unlike the tobacco mosaic virus protein, the low-resolution reflections were only pseudocentric.

10.5.3. Real-Space Molecular Replacement

Molecular replacement in real space consists of the following steps: (a) calculation of electron density based on a starting phase set and observed amplitudes; (b) averaging of this density among the noncrystallographic asymmetric units or molecular copies in several crystal forms, a process which defines a molecular envelope as the averaging is only valid within the range of the noncrystallographic symmetry; (c) reconstructing the unit cell based on averaged density in every noncrystallographic asymmetric unit; (d) calculating structure factors from the reconstructed cell; (e) combining the new phases with others to obtain a weighted best-phase set; and (f) returning to step (a) at the previous or an extended resolution.

A method dependent on the computation of skew planes through the electron density distribution has been used when the molecular point group can be incorporated into the electron density grid on which the averaged molecule is represented (Buehner et al., 1974; Argos et al., 1975). Skew planes are computed perpendicular to one of the molecular axes and on a grid suitable for the given point group. In averaging, grid points and sections are appropriately associated. For instance, for a molecular trimer calculated on a trigonal grid, points at XY; $-Y, X-Y$; $Y-X, -X$ must be averaged on each section of Z. When reconstructing the averaged electron density at cell grid points for subsequent transformation into reciprocal space, skew planes must again be taken through the averaged molecule. Structure factors computed from the reconstructed cell suffer from the need to perform two interpolations to produce averaged electron density with respect to the molecular and crystallographic axes.

This technique was used originally by Cullis *et al.* (1961) and Muirhead *et al.* (1967) in studies of hemoglobin and by Matthews *et al.* (1967) on α-chymotrypsin. In both cases, there were only two noncrystallographic copies. Buehner *et al.* (1974) used this technique on the 222 symmetric glyceraldehyde-3-phosphate dehydrogenase molecule.

Bricogne (1976a,b) and Johnson (1978) devised a faster methodology based on sorting grid points to associate those densities to be averaged. As this procedure was performed for every grid point in the original asymmetric unit of the cell, only one interpolation was required.

Bricogne's double-sorting technique involves generating real-space nonintegral points ($\mathbf{D}_i$) which are related to integral grid points ($\mathbf{I}_i$) in the cell asymmetric unit through the noncrystallographic symmetry operators. The elements of the set $\mathbf{D}_i$ are then brought back to their equivalent points in the cell asymmetric unit ($\mathbf{D}_i'$) and sorted by their proximity to two real-space sections. The set $\mathbf{I}_i'$, calculated on a finer grid than $\mathbf{I}_i$ and stored in the computer memory two sections at a time, is then used for linear interpolation to determine the density values at $\mathbf{D}_i'$ which are successively stored and summed in the related array $\mathbf{I}_i$. A count is kept of the number of densities received at each $\mathbf{I}_i$, resulting in a final averaged aggregate when all real-space sections have been utilized. The density to be assigned outside the molecular envelope (defined with respect to the set $\mathbf{I}_i$) is determined by averaging the density of all unused points in $\mathbf{I}_i$. The grid interval for the set $\mathbf{I}_i'$ should be about one-sixth of the resolution to avoid serious errors from interpolation (Bricogne, 1976b). The grid point separation in the set $\mathbf{I}_i$ need only be sufficient for standard interpolation of electron density, that is about one-third of the resolution.

Structure factors calculated from the averaged electron density in the reconstructed cell (obtained by either the skew plane or the double-sorting technique) can then be weighted (Sim, 1959, 1960) and combined with the joint phase probability curve obtained from isomorphous replacement data (Rossmann and Blow, 1961; Hendrickson and Lattman, 1970). In calculating the phase probability curve for the molecular replacement results, it will be necessary to compare the calculated structure factors with the observed amplitudes. This comparison will involve a scale and temperature factor to compensate for the increasing error of interpolation with resolution, as well as the concomitant decreasing isomorphous figures of merit.

Phase improvement by real-space averaging between noncrystallographically related units has been used with considerable success in a variety of enzyme and virus studies (see Section 10.7). Most of these structures involve the presence of more than two copies within one crystallographic asymmetric unit. Fletterick and Steitz (1976; Steitz et al., 1976) have, for the first time, demonstrated the method in the structure determination of hexokinase where more than one crystal form is available.

Phase extension has been applied to the known protein structure of lobster glyceraldehyde-3-phosphate dehydrogenase (Argos et al., 1975) where ten cycles of real-space averaging extended phases from a 6-Å single isomorphous replacement set to 4.9-Å resolution. The extension after each cycle was approximately one reciprocal-lattice point.

Ab initio phasing of glyceraldehyde-3-phosphate dehydrogenase (Argos et al., 1975) was successfully attempted by initially filling the known envelope with uniform density and then gradually extending phases to 6.3-Å resolution. The phases not only gave an improved electron density map, but were of sufficient quality to locate heavy atoms. Assumption of the envelope centered at a general position in the $P2_12_12_1$ unit cell provided sufficient acentric bias to select the correct enantiomorph.

Johnson et al. (1976) assumed a uniform sphere for the southern bean mosaic virus, calculated phases to 35-Å resolution and extended them to 22.5 Å through icosahedral averaging. Although the phases remained centric, it did show the quasi $T = 3$ symmetry which had not been assumed. In the case of satellite tobacco necrosis virus, where the direction of the noncrystallographic axes are general within the $C2$ unit cell, selection of an enantiomorph is feasible even if only a simple spherical envelope is assumed.

10.5.4. Equivalence of Real- and Reciprocal-Space Molecular Replacement

Main (1967) first showed the equivalence of the real- and reciprocal-space methods. Colman (1974) applied Shannon's theorem to prove the same concept in the presence of proper rotations. Bricogne (1974), however, has given a rigorous proof of the equivalence of real-space averaging and the molecular replacement equations (Crowther, 1967; Main and Rossmann, 1966). He first derives the equations in real space via a linear operator formalism representing the averaging process and rebuilding of the independent molecule in the various crystal forms. A Fourier transformation of the real-space equations then yields Crowther's reciprocal-space molecular replacement equations. Although Bricogne's treatment is mathematically sophisticated and rigorous, a less complex yet complete analysis is presented here in the same notation as used throughout this article to further its pedagogic goals.

Let $\varrho(\mathbf{x}_n)$ $(n = 1, 2, \ldots, N)$ be the electron densities at equivalent points in the N different crystallographic copies. Hence the mean electron density as determined in practice will be

$$\langle \varrho(\mathbf{x}) \rangle = \frac{1}{N} \sum_{n=1}^{N} \varrho(\mathbf{x}_n) \tag{10.30}$$

where $\mathbf{x}$ can be in any of the N subunits related by a proper rotation.

When the original cell is rebuilt with N copies of this averaged density, then the structure factor will be given by

$$F_p = \sum_{n=1}^{N} \int_U \langle \varrho(\mathbf{x}_n) \rangle \exp(2\pi i \mathbf{p} \cdot \mathbf{x}_n) \, d\mathbf{x} \tag{10.31}$$

where U is the volume within the cell for which the noncrystallographic symmetry is valid. Outside the volume U, the summation (10.30) is no longer strictly valid, a property frequently expressed in real-space averaging as a molecular envelope. As

$$\varrho(\mathbf{x}_n) = \frac{1}{V} \sum_{\mathbf{h}} F_{\mathbf{h}} \exp(-2\pi i \mathbf{h} \cdot \mathbf{x}_n) \tag{10.32}$$

for $n = 1$ to N, and since the noncrystallographic symmetry can be expressed as

$$\mathbf{x}_n = [\mathbf{C}_n]\mathbf{x}_1 + \mathbf{d}_n \tag{10.33}$$

it follows that

$$F_p = \frac{1}{VN} \sum_{\mathbf{h}} F_{\mathbf{h}} \exp(-2\pi i \mathbf{h} \cdot \mathbf{d}_n) \sum_{n=1}^{N} \int_U \exp\{2\pi i(-\mathbf{h}[\mathbf{C}_n] + \mathbf{p}) \cdot \mathbf{x}\} \, d\mathbf{x}$$

In a manner similar to that used in Section 10.2, this can be rewritten as

$$F_p = \frac{U}{V} \sum_{\mathbf{h}} F_{\mathbf{h}} \sum_{n=1}^{N} G_{\mathbf{h}pn} \exp[2\pi i(\mathbf{p} - \mathbf{h}) \cdot \mathbf{S}] \tag{10.34}$$

where $\mathbf{S}$ is the position vector of the molecular center. Expression (10.34) is identical to the reciprocal-space equation as expressed in (10.10) of Section 10.2 for the single-crystal form assumed here.

The derivation above can be generalized to include improper crystallographic operators and more than one crystal form. Bricogne (1974) has handled this most general situation.

10.6. Noncrystallographic Symmetry and Heavy-Atom Searches

It has been shown that the isomorphous and molecular replacement methods can assist each other in phase determination. Unfortunately, the number of heavy-atom sites also increases proportionally to the number of noncrystallographic asymmetric units. The determination of heavy-atom positions is usually difficult unless at least one derivative is available with only one or two well-substituted sites. This condition cannot be met as the complexity of the molecule increases. For instance, the number of heavy-atom sites should be multiples of 60 for satellite tobacco necrosis virus where one icosahedral virus particle occurs per crystallographic asymmetric unit. The isomorphous replacement technique thus appears inoperative for a larger oligomer. Fortunately, the noncrystallographic symmetry can be used to find the heavy atoms, since the noncrystallographic asymmetric unit itself is likely to contain only one or two well-substituted heavy-atom sites.

There are two approaches that might be taken. The first utilizes approximate phases from some other source; the second requires no phase information. In the first case, a poor difference map can be improved by averaging among the noncrystallographic asymmetric units (Argos et al., 1975). In the second case, a systematic interpretation of the Patterson is required, as described below.

Heavy-atom sites can be generated from arbitrary trial positions in the noncrystallographic asymmetric unit (see Fig. 10.14) containing one or more molecular subunits, given the orientation of the noncrystallographic symmetry axes. Self-vectors within the molecule are calculated and compared systematically with the actual Patterson distribution. The largest peak in the search map will establish the heavy-atom sites with respect to the molecular center (Argos and Rossmann, 1976). Similarly the coordinates of the molecular center with respect to the crystallographic axes can be established by calculating both self- and cross-vectors within and between the molecules and compared with the difference Patterson synthesis. This method has been demonstrated in the determination of the major heavy-atom sites for the 222 symmetric lobster glyceraldehyde-3-phosphate dehydrogenase (Argos et al., 1975), and the icosahedral satellite

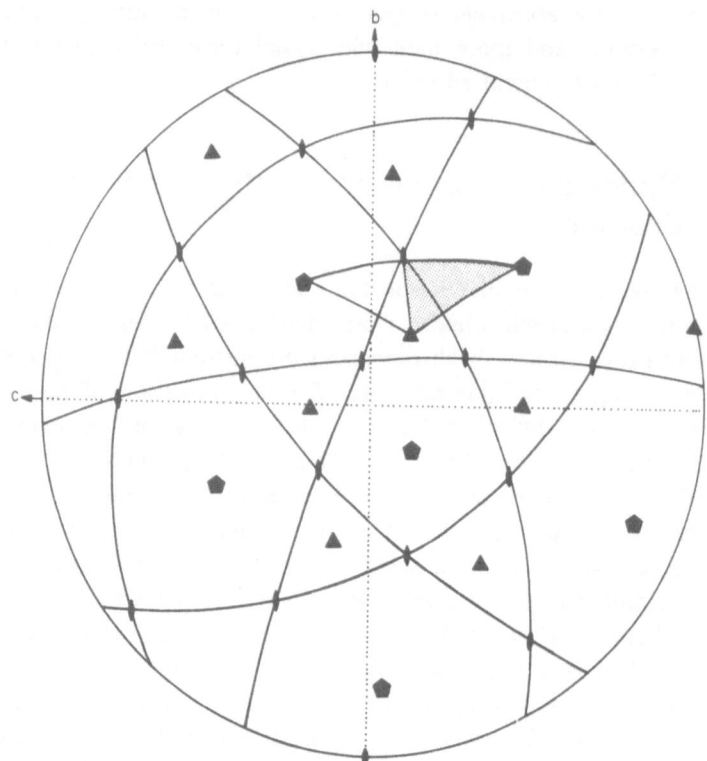

FIGURE 10.14. Distribution of symmetry elements for one satellite tobacco necrosis virus particle (532 symmetry) in its monoclinic cell. Trial heavy-atom positions are taken within the shaded noncrystallographic asymmetric unit of search. The Patterson asymmetric unit is half the size of the icosahedral asymmetric unit due to the inability of the Patterson to distinguish enantiomorphs.

tobacco necrosis (Lentz *et al.*, 1976; Argos and Rossmann, 1976) and southern bean mosaic (unpublished results) viruses.

The magnitude of the computing operation increases rapidly with the number of noncrystallographic asymmetric units. However, if the particle center is assumed known so that only self-vectors need be calculated, and if cross-vectors between molecules can be considered as a roughly uniform background (Argos and Rossmann, 1976), the following criterion is useful:

$$C = \sum_{i=1}^{N} P_i - NK$$

Here the value of the Patterson P_i is taken only once at each of the N grid points where one or more self-vectors appear and K is the average value of the Patterson. The right-most term provides a simple multiplicity correction.

Gilbert and Klug (1974) noted that there will be a piling up of vectors in the Patterson at a radius equal to the diameter of the circle in real space determined by the locus of heavy-atom sites. Their observation was useful in obtaining heavy-atom sites for the tobacco mosaic virus protein disk.

After initial heavy-atom positions have been obtained it becomes necessary to refine their parameters for use in isomorphous replacement phase determination. A variety of standard least-squares procedures, used in the structural determination of proteins, have been devised for this purpose. It is possible to incorporate both crystallographic and noncrystallographic symmetry into such refinement procedures as was noted by Harrison and Jack (1975) and discussed by Rossmann (1976). Maslen (1968a,b) used a similar procedure in the refinement of small molecules in the presence of noncrystallographic symmetry.

The structure-factor expression

$$f_{\mathbf{h}} = \sum_{j=1}^{J} Z_j \exp\left[-B_j\left(\frac{\sin\theta}{\lambda}\right)^2\right] \sum_{m=1}^{M} \exp\{2\pi\mathbf{h}([\mathbf{T}_m]\mathbf{S} + \mathbf{d}_m)\}$$
$$\times \sum_{n=1}^{N} \exp(2\pi i\mathbf{h}[\mathbf{R}_{mn}]\mathbf{x}_j)$$

is for J independent heavy atoms in each noncrystallographic asymmetric unit, where Z_j and B_j are the occupancy and shape factors of the jth atoms at $\mathbf{x}_j$ with respect to the molecular center at $\mathbf{S}$. The latter is defined with respect to the crystal origin. $[\mathbf{T}_m]$ and $\mathbf{d}_m$ represent the mth crystallographic symmetry operator and

$$[\mathbf{R}_{mn}] = [\mathbf{T}_m][\alpha][\rho_n][\beta]$$

Here $[\rho_n]$ is the nth noncrystallographic symmetry operator and $[\alpha]$ and $[\beta]$ are the orthogonalizing matrices as defined in Section 10.3. Given the known orientation and position of the noncrystallographic symmetry operators, $f_{\mathbf{h}}$ can now be readily differentiated with respect to the components of the noncrystallographically independent heavy-atom parameters. The resulting expressions are then used in the conventional manner within the least-squares isomorphous replacement procedure. Since averaging requires not only the accurate knowledge of the molecular orientation but also the position of the molecular center $\mathbf{S}$, refinement of the latter can also be incorporated into the least-squares procedure.

A different view was taken in the refinement of the heavy atoms of tobacco mosaic virus disk protein (Champness et al., 1976) where the

TABLE 10.2. Applications of Noncrystallographic

Structure	Authors and date	Rotation function		Conditions		Translation function	
		Self [a]	Cross [a]	Resolution, Å	Program [b]	Nature of comparison [c]	Method [d]
Horse oxyhemoglobin	Rossmann and Blow, 1962	UU		10–6	S		
Ninhydrin	Tollin and Cochran, 1964		UE		T	UK	T
Deoxyadenosine	Tollin and Cochran, 1964		UE		T	UK	T
Human oxyhemoglobin	Prothero and Rossmann, 1964		UU	10–6	S		
α-Chymotrypsin	Blow et al., 1964	UU, K		6.0	S	UU KK	R R
Villastonine	Nordman and Kumra, 1965		UP		N	UK	N
Insulin: 2 Zn and 4 Zn	Dodson et al., 1966	UU	UU	6.0	S	UU	R
Human deoxy-hemoglobin	Muirhead et al., 1967						
α-Chymotrypsin	Matthews et al., 1967						
4-Acetyl-2'-fluorobiphenyl	Young et al., 1968		UE		T	UK	T
Thymidine	Young et al., 1969		UE		T	UK	T
Seal myoglobin	Tollin, 1969b		UE	5.0	S	UK	T
A variety of small molecules	Huber, 1970		UP		H	UK	S

[a] K: Analysis of Klug peaks; L: locked rotation function analysis; UA: rotation function of unknown with respect to known atomic model; UE: rotation function of unknown with respect to known electron density; UP: rotation function of unknown versus point Patterson distribution; UU: rotation function of unknown versus unknown.

[b] C: Crowther's fast rotation function (Crowther, 1972); H: Huber's rotation function approach (Huber, 1965); L: Lattman's modification of Rossmann and Blow approach (Lattman and Love, 1970); N: Nordman's point Patterson comparison (Nordman, 1966); P: molecular packing analysis; S: standard program based on Rossmann and Blow approach (Rossmann and Blow, 1962); T: Tollin and Cochran's $I(\theta, \varphi)$ function (Tollin and Cochran, 1964).

[c] KK: Known versus known structures; UK: unknown versus known structures; UU: unknown versus unknown structures.

[d] A: Molecular origin can be placed in an arbitrary position; C: Patterson comparison technique of

Symmetry in the Determination of Molecular Structure

Phase determination		Heavy-atom determination	Non-crystallographic symmetry	Crystallographic symmetry	Number of copies in crystallographic asymmetric unit [f]	Comments
Type [e]	Residual (%)					
			pseudo 222	$C2$	½ M	
KM				$P2_1$	1 M	
KM				$P2_1$	1 M	
				$P4_12_12$		Comparison with horse oxyhemoglobin using coincidence of crystallographic twofolds
			2	$P2_1$	2 M	
KM				$P2_1$	1 M	Illustrative of the use of a known molecular fragment
			32	$R3$	2 S	The two crystal forms contain similar packing arrangements
PI			2	$P2_1$	1 M	Real-space averaging and recalculation of structure factors from averaged density
PI			2	$P2_1$	2 M	Real-space averaging
KM				$P2_1/c$		
KM				$P2_12_12_1$		
KM				$A2$	1 M	Sperm whale myoglobin used as model
KM						Use of known molecular fragments

continued overleaf

Crowther and Blow (1967) or as generalized by Rossmann (1972b); H: occupance of large Harker section peaks when noncrystallographic and crystallographic evenfold axes are parallel; N: Nordman's point Patterson comparison (Nordman, 1966); P: molecular packing analysis; R: Rossmann's convolution of two functions (Rossmann, 1972a, pp. 22–25) expressed for special situations (cf. Rossmann *et al.*, 1964 or Blow *et al.*, 1964); S: *R*-factor searches; T: Tollin's *Q* function (Tollin, 1966) [Note that Tollin has shown the equivalence of C and T (Tollin, 1969a)].

[e] AB: *Ab initio* phase determination; KM: use of known model to provide an initial phase set. When available the residual given compares the calculated and observed structure amplitudes for this case; PI: phase improvement; PX: phase extension.

[f] M: Represents molecules; S: represents subunits.

[g] Residual obtained after refinement.

TABLE 10.2 (*continued*)

Structure	Authors and date	Rotation function				Translation function	
		Self [a]	Cross [a]	Conditions		Nature of comparison [c]	Method [d]
				Resolution, Å	Program [b]		
Rabbit aldolase	Eagles *et al.*, 1969	UU		8.5	S	UU	H
Lamprey hemoglobin	Lattman and Love, 1970		UA	12–7	L		
6β-Acetyl-i-cholesterol	Harrison and Joynson, 1970		UA		S	UK	T
Triclinic hen egg-white lysozyme	Joynson *et al.*, 1970		UA	6.0	S		A
Satellite tobacco necrosis virus	Åkervall *et al.*, 1971; Klug, 1971	UU, L, K		12.3	S		
Yellow fin tuna myoglobin	Lattman *et al.*, 1971		UA	6.0	L	UK	C
Demetallized concanavalin A	Jack *et al.*, 1971	UU		6.0	S	UU	R
Hexon unit of adenovirus	Franklin *et al.*, 1971	UU		35–10	S		
E. coli L-asparaginase	Epp *et al.*, 1971	UU			H	UU	H
Lactone derivative, $C_{10}H_{16}O_3$	Burnett and Rossmann, 1971		UA		S	UK	T
Lobster glyceraldehyde-3-phosphate dehydrogenase	Rossmann *et al.*, 1972	UU, L	UE	6.0	S	UK	R
Pig insulin	Rossmann and Hodgkin, 1972	UU, L	UU	6.0	S		
Sperm whale myoglobin	Nordman, 1972		UP	1.5	N	UK	N
Satellite tobacco necrosis virus	Rossmann *et al.*, 1973			12.3		UU	R
Tobacco mosaic virus protein	Jack, 1973			7.0			
Soluble malate dehydrogenase	Hill *et al.*, 1973	UU		7.9	S		

Phase determination		Heavy-atom determina-tion	Non-crystallo-graphic symmetry	Crystallo-graphic symmetry	Number of copies in crystallo-graphic asymmetric unit [f]	Comments
Type [e]	Residual (%)					
			222	$P2_1$	1 M	
				$P2_12_12_1$	1 M	Comparison with sperm whale myoglobin
KM				$P2_1$	1 M	Use of known molecular fragment
KM	43			$P1$	1 M	Tetragonal lysozyme struc-ture provided model
			532	$C2$	60 S	Octahedrally distributed Klug peaks dominate the rotation function of ico-sahedral particles. First discovery of this effect.
KM	49			$P2_12_12_1$	1 M	Comparison with sperm whale myoglobin
			2	$P2_122_1$	$\frac{1}{2}$ M	
			6 or 32	$P2_13$	2 S	
			222	$C2$	1 M	
KM				$P2_12_12_1$	1 M	Use of known molecular fragment
			222	$P2_12_12_1$	1 M	Illustrates the locked rota-tion function. Attempted comparison with lactate dehydrogenase
			32	$P2_1$	6 S	Comparison with rhombo-hedral pig insulin
				$P2_1$	1 M	Use of rigid models to determine helical and heme orientations and positions
			532	$C2$	60 S	
AB			17	$P22_12_1$	34 S	Crowther's linear analysis of molecular replacement equations for centric pro-jection
			2	$P2_12_12$	1 M	

continued overleaf

TABLE 10.2 (*continued*)

Structure	Authors and date	Rotation function		Conditions		Translation function	
		Self [a]	Cross [a]	Resolution, Å	Program [b]	Nature of comparison [c]	Method [d]
Pig M_4 lactate dehydrogenase	Hackert et al., 1973	UU	UE	6.0	S	UK	R, S
Wheat germ agglutinin	Wright, 1974						
Bovine carboxypeptidase B	Schmid et al., 1974; Schmid and Herriott, 1976		UA	12–5.5	L	UK	C
Phe tRNA	Quigley et al., 1974			4.0	P	UK	P
Satellite tobacco necrosis virus	Lentz and Strandberg, 1974	UU, L		10.0	S		
Cytochrome b_{562}	Czerwinski and Mathews, 1974	UU		4.3	S		A
Human F_c antibody fragment	Colman et al., 1974; Colman et al., 1976	UU	UA	10–5	L		
Lobster glyceraldehyde-3-phosphate dehydrogenase	Buehner et al., 1974						
Lobster glyceraldehyde-3-phosphate dehydrogenase	Argos and Rossmann, 1974						
Hagfish insulin	Cutfield et al., 1974		UA	∞–10, 10–6	S	UK	S
Sickle cell deoxyhemoglobin in PEG	Wishner et al., 1975		UA	5.0	L	UK	C
Deoxyhemoglobin A in PEG	Ward et al., 1975		UA	10–6	L	UK	C

Phase determination Type[e]	Phase determination Residual (%)	Heavy-atom determination	Non-crystallographic symmetry	Crystallographic symmetry	Number of copies in crystallographic asymmetric unit[f]	Comments
PI, KM	38		222	$P22_12_1$	$\frac{1}{2}$ M	Comparison with dogfish M_4 lactate dehydrogenase and phase improvement by real-space averaging
		γ	2_{screw}	$C2$	2 S	Heavy atom positions showed intermolecular twofold screw axis with translational component of 6.4 Å
KM	39			$P3_1$ or $P3_2$	1 M	Comparison with bovine carboxypeptidase A
KM	42			$P2_1$	1 M	Comparison with ortho-rhombic crystal form
			532	$C2$	60 S	Refinement of orientation of viral particle by best fit to rotation function peaks
			2	$P1$	2 M	Iron atoms determined with anomalous dispersion verifying twofold orientation
			2	$P2_12_12_1$	2 S	
PI		γ	222	$P2_12_12_1$	1 M	Heavy-atom determination aided by knowledge of noncrystallographic symmetry. SIR phases improved by real-space averaging
		γ	222	$P2_12_12_1$	1 M	Heavy-atom and molecular center determined by systematic Patterson search
KM	49	γ	2	$P4_12_12$ or $P4_32_12$	1 S	Comparison with known pig insulin structure resulting in phases to find heavy atoms
KM	46		pseudo 222	$P2_1$	1 M	Use of known horse oxy-hemoglobin structure
KM	41		pseudo 222	$P2_12_12$	1 M	Comparison with horse oxy-hemoglobin

continued overleaf

TABLE 10.2 (*continued*)

Structure	Authors and date	Rotation function				Translation function	
		Self [a]	Cross [a]	Conditions		Nature of comparison [c]	Method [d]
				Resolution, Å	Program [b]		
Proteus vulgaris L-asparaginase	Lee *et al.*, 1975	UU		6.3	S	UU	H
Southern bean mosaic virus	Johnson *et al.*, 1975	UU, L, K		22.0	S		
Tomato bushy stunt virus	Harrison and Jack, 1975			16.0			
Bence–Jones Au protein	Fehlhammer *et al.*, 1975; Colman *et al.*, 1976		UA	10–5	C	UK	S
Bovine β-trypsin	Fehlhammer and Bode, 1975; Colman *et al.*, 1976		UP	8–2.5		UK	C, S
Bovine liver catalase	Eventoff and Gurskaya, 1975	UU, L, K		10.0	S		
Lobster glyceraldehyde-3-phosphate dehydrogenase	Argos *et al.*, 1975						
Hemoglobin Kansas	Anderson, 1975	UU	UA	25–4	C	UK	C
Phascolopsis gouldii hemerythrin	Ward, Hendrickson and Klippenstein, 1975		UE	5.5	C	UK	S, P
Pig H₄ lactate dehydrogenase	Eventoff *et al.*, 1975	UU	UE	6.0	S	UK	R, S
Human lysozyme	Nixon and North, 1976		UA	10–6	S	UK	S
B. stearothermophilus glyceraldehyde-3-phosphate dehydrogenase	Wonacott and Bricogne, 1976; Bricogne, 1976a,b	UU		5.0	S		
Satellite tobacco necrosis virus	Lentz *et al.*, 1976						

Phase determination		Heavy-atom determina-tion	Non-crystallo-graphic symmetry	Crystallo-graphic symmetry	Number of copies in crystallo-graphic asymmetric unit [f]	Comments
Type [e]	Residual (%)					
			222	$P2_1$	1 M	
			532	$R32$	30 S	
PI			532	$I23$	5 S	Phase improvement by real-space averaging and analysis of Crowther's H matrix
KM	31			$P6_122$	1 M	Use of known Bence–Jones Rei structure
KM	28 [g]			$P2_12_12_1$	1 M	Use of known trypsin structure
			222	$P3_121$ or $P3_221$	1 M	
AB, PX		√	222	$P2_12_12_1$	1 M	Real-space averaging of electron density and heavy-atom difference electron density
KM	37		Pseudo 222	$P2_1$	1 M	Comparison with known he-moglobin structure
KM	43		422	$P422$	2 S	Use of known monomeric myohemerythrin electron density
PI, KM	38		222	$C2$	1 M	Use of known dogfish M_4 lactate dehydrogenase structure. Real-space averaging
KM	49			$P2_12_12_1$	1 M	Comparison with hen egg-white lysozyme
PI			222	$P2_1$	1 M	Real-space averaging
PI		√	532	$C2$	60 S	Heavy atoms found in sys-tematic Patterson search. Real-space averaging (private communication)

continued overleaf

TABLE 10.2 (*continued*)

Structure	Authors and date	Rotation function		Conditions		Translation function	
		Self [a]	Cross [a]	Resolu-tion, Å	Pro-gram [b]	Nature of com-parison [c]	Method [d]
Mouse testes lactate dehydrogenase	Musick et al., 1976	UU	UE	7.0	S		A
Southern bean mosaic virus	Johnson et al., 1976						
Yeast hexokinase	Fletterick and Steitz, 1976; Steitz et al., 1976		UE	3.5			
Liver alcohol dehydrogenase ternary complex	Eklund and Brändén, 1976	UU	UE		S		A
Myeloma protein Kol	Colman and Fehlhammer, 1976		UA	5.0	C	UK	C
Tobacco mosaic virus disk protein	Champness et al., 1976			6.0			
Rabbit phos-phorylase a	Bartels and Colman, 1976	UU, L, K		3.0	C	UU	R
Bovine trypsinogen	Bode et al., 1976		UA	2.5	H, C	UK	T, S
Turkey egg-white lysozyme	Bott and Sarma, 1976		UA	5.0	S	UK	S, P
Desulfovibrio vulgaris rubredoxin	Adman et al., 1976		UA	10–6	L		
Human fetal deoxy-hemoglobin	Frier and Perutz, 1977		UA	20–4	C		
Wheat germ agglutinin	Wright, 1977						
Tomato bushy stunt virus	Winkler et al., 1977						

Phase determination		Heavy-atom determination	Non-crystallographic symmetry	Crystallographic symmetry	Number of copies in crystallographic asymmetric unit [f]	Comments
Type [e]	Residual (%)					
PI, KM	37		222	$P1$	1 M	Use of known dogfish M_4 lactate dehydrogenase structure. Real-space averaging
AB, PX			532	$R32$	10 S	Real-space averaging using a spherical envelope
PI		γ	2_{screw}	(a) $P2_12_12_1$ (b) $P2_12_12_1$	2 S 1 S	Use of heavy atoms to determine rotational and translational relationships, and refined by electron density superposition. Real-space averaging between different crystal forms
KM			2	$P1$	1 M	Comparison with apo liver alcohol dehydrogenase structure
				$P3_121$ or $P3_221$	1 M	Excision of a domain which was recognized by cross-rotation function against the Bence–Jones protein (Rei V_k) and mouse Mc PC603 Fab fragment
PI			17	$P22_12_1$	34 S	Real-space averaging
			222	$P2_1$	1 M	
KM	40			$P3_121$	1 M	Comparison with known bovine trypsin structure
KM	47			$P6_122$	1 M	Comparison with known hen egg-white lysozyme
KM	45			$P2_1$	1 M	Translation obtained from native anomalous dispersion difference Patterson
KM	34	γ	2	$P2_12_12_1$	1 M	Molecular center found from heavy-atom positions. Use of human deoxyhemoglobin structure
PI			2_{screw}	$C2$	2 S	Real-space averaging
PI			532	$I23$	5 S	Real-space averaging

parameters in each of the 17 identical subunit pairs were purposely kept independent. Similar considerations prevailed in the heavy-atom refinements of lobster glyceraldehyde-3-phosphate dehydrogenase (Buehner *et al.*, 1974). Thus, the noncrystallographic subunits of the protein which act as a matrix for heavy-atom attachment were shown to be locally equivalent in both cases.

10.7. Applications of Molecular Replacement

Table 10.2 summarizes structural results that depend, at least in part, on the use of noncrystallographic symmetry. While it is hoped that the table is exhaustive in listing all applications to proteins, nucleic acids, and viruses, some published examples might have unfortunately and accidentally been omitted. The authors are aware of recent examples that have not yet been sufficiently documented for inclusion in Table 10.2. References describing small molecule structures have been included where they represent particularly illustrative examples of the molecular replacement method. A summary of the major types of applications is given in Table 10.3 and plotted in Figure 10.15 as a function of each year since 1962, the year in which the basic principles were formulated (Rossmann and Blow, 1962).

The rapid increase in the use of noncrystallographic symmetry since 1970 can be attributed to at least four factors.

(1) Previous to about 1970 considerable controversy existed as to the theoretical foundations of the method. However, molecular replacement has now become an accepted technique due to many successes of the rotation function, the translation function in special cases, and real-space averaging for phase improvement.

(2) The number of protein determinations has increased rapidly (Matthews, 1976) since 1965 with the solution of hen egg-white lysozyme.

(3) The number of crystalline oligomeric proteins, and even viruses, being studied has increased rapidly since about 1973, providing greater familiarity with the use of the combined techniques of isomorphous and molecular replacement.

(4) Better technology in the areas of computers, oscillation photography (Arndt *et al.*, 1973; Arndt and Wonacott, 1976), X-ray generators and beam focusing devices (Harrison, 1968) have provided the hardware for the large data collection task for macromolecules with high noncrystallographic symmetry.

TABLE 10.3. Number of Applications of Noncrystallographic Symmetry Made in the Determination of Biological Macromolecules Every Year

Year	Self-rotation function	Use of known model to help solve similar structures	Phase determination or improvement with non-crystallographic symmetry	Heavy-atom determination in presence of non-crystallographic symmetry	Total
1962	1	—	—	—	1
1963	—	—	—	—	0
1964	1	1	—	—	2
1965	—	—	—	—	0
1966	1	—	—	—	1
1967	—	—	2	—	2
1968	—	—	—	—	0
1969	1	1	—	—	2
1970	—	2	—	—	2
1971	4	1	—	—	5
1972	2	1	—	—	3
1973	2	1	2	1	6
1974	3	3	1	2	9
1975	4	6	2	—	12
1976	4	7	8	3	22

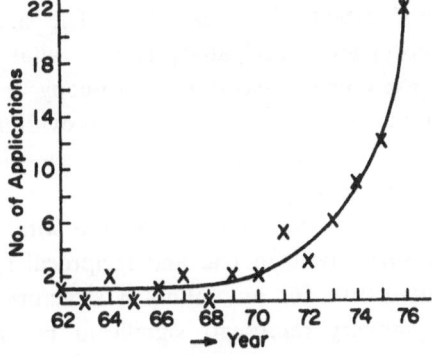

FIGURE 10.15. Plot of the number of applications of the molecular replacement method in the determination of macromolecular structures as a function of each year since 1962.

10.8. Conclusions

The molecular replacement technique has begun to make significant contributions to the solution of biomolecular structures. Yet, there is no case where a structure has been determined to even 6-Å resolution without the aid of isomorphous replacement. Nevertheless, there are now many examples where the isomorphous replacement technique was supplemented with molecular replacement before an interpretable electron density map was obtained. Yet the most widely used application of molecular replacement has been in the solution of unknown structures from similar but known structures.

The first requirement of molecular replacement, the determination of the rotational parameters, is normally achieved with relative ease. The second requirement, the determination of translational parameters, is frequently found to be difficult and still presents formidable unsolved problems when no extra structural information is available. It is, however, in the phase determination stage where the challenges present the greatest glamor for future work.

Molecular replacement phase determination was originally derived in reciprocal space. More recently, real-space averaging has been used primarily for phase improvement; in this form molecular replacement has had substantial success, although severe computational problems and ill-understood constraints on the solution still exist.

Data from at least one single isomorphous derivative is still a necessary prerequisite in solving a completely unknown structure exhibiting noncrystallographic symmetry. The heavy atoms of an isomorphous derivative are often helpful in determining the position of the molecular symmetry axes when no useful translation function exists. The averaging of a single isomorphous replacement electron density map not only makes the map interpretable, but can provide the molecular envelope for phase refinement and extension of resolution. The starting phase set, determined from a heavy-atom derivative, is also vital in establishing a consistent hand. Occasionally, electron microscopy can act as an aid or alternative to isomorphous replacement in providing some crude starting phases (Harrison and Jack, 1975). In contrast, *ab initio* phase determination is beset by these very problems.

With the present state of affairs, *ab initio* phasing is still beyond the horizon both in real and reciprocal space. Yet the necessity of only one derivative for an unknown structure with sufficient noncrystallographic symmetry represents significant progress, without which the larger aggregates now being studied would not be soluble. As more oligomeric biomolecular complexes and viruses are crystallized, the impetus will exist to develop further the present molecular replacement techniques into direct methods.

Acknowledgments

We are deeply grateful to the many colleagues who have assisted our task by contributing reprints, preprints, and ideas to this chapter. We are enormously indebted to Sharon Wilder, who turned our crude manuscript into a presentable and organized chapter. We would also like to thank Drs. John Johnson and Donald Musick for interest and stimulating discussion during the preparation of this review. This writing was supported by the National Science Foundation (grant No. BMS74-23537) and the National Institutes of Health (grants Nos. GM 10704 and AI 11219).

References

Adman, E. T., Sieker, L. C., and Jensen, L. H., (1976). *J. Biol. Chem.* **251**, 3801–3806.

Åkervall, K. B., Strandberg, B., Rossmann, M. G., Bengtsson, U., Fridborg, K., Johannisen, H., Kannan, K. K., Petef, G., Öberg, B., Eaker, D., Hjertén, S., Rydén, L., and Moring, I. (1971). *Cold Spring Harbor Symp. Quant. Biol.* **36**, 469–488.

Anderson, L. (1975). *J. Mol. Biol.* **94**, 33–49.

Argos, P., and Rossmann, M. G. (1974). *Acta Crystallogr. Sect. A* **30**, 672–677.

Argos, P., and Rossmann, M. G. (1976). *Acta Crystallogr. Sect. B* **32**, 2975–2979.

Argos, P., Ford, G. C., and Rossmann, M. G. (1975). *Acta Crystallogr. Sect. A* **31**, 499–506.

Argos, P., Rossmann, M. G., and Ford, G. C. (1976). In *Crystallographic Computing Techniques*, Eds. F. R. Ahmed, K. Huml, and B. Sedlacek, Munksgaard, Copenhagen, pp. 222–228.

Arndt, U. W., and Wonacott, A. J. (1976). *The Rotation Method in Crystallography*, North-Holland, Amsterdam.

Arndt, U. W., Champness, J. N., Phizackerley, R. P., and Wonacott, A. J. (1973). *J. Appl. Crystallogr.* **6**, 457–463.

Bartels, K., and Colman, P. M. (1976), *Biophys. Struct. Mech.* **2**, 43–59.

Blow, D. M. (1976). In *Crystallographic Computing Techniques*, Eds. F. R. Ahmed, K. Huml, and B. Sedlacek, Munksgaard, Copenhagen, pp. 229–238.

Blow, D. M., Rossmann, M. G., and Jeffery, B. A. (1964). *J. Mol. Biol.* **8**, 65–78.

Blundell, T. L., and Johnson, L. C. (1976). *Protein Crystallography*, Academic Press, New York.

Bode, W., Fehlhammer, H., and Huber, R. (1976). *J. Mol. Biol.* **106**, 325–335.

Bott, R., and Sarma, R. (1976). *J. Mol. Biol.* **106**, 1037–1046.

Bragg, L., and Perutz, M. F. (1954). *Proc. R. Soc. London* **A225**, 315–329.

Bricogne, G. (1974). *Acta Crystallogr. Sect. A* **30**, 395–405.

Bricogne, G. (1976a). In *Crystallographic Computing Techniques*, Eds. F. R. Ahmed, K. Huml, and B. Sedlacek, Munksgaard, Copenhagen, pp. 239–247.

Bricogne, G. (1976b). *Acta Crystallogr. Sect. A* **32**, 832–847.

Buehner, M., Ford, G. C., Moras, D., Olsen, K. W., and Rossmann, M. G. (1974). *J. Mol. Biol.* **82**, 563–585.

Burdina, V. I. (1970). *Kristallografiya* **15**, 623–630.

Burdina, V. I. (1971). *Soviet Phys. Crystallogr.* **15**, 545–550.

Burdina, V. I. (1973). *Kristallografiya* **18**, 694–700.

Burnett, R. M., and Rossmann, M. G. (1971). *Acta Crystallogr. Sect. B* **27**, 1378–1387.

Caspar, D. L. D. (1976). Private communication.

Champness, J. N., Bloomer, A. C., Bricogne, G., Butler, P. J. G., and Klug, A. (1976). *Nature (London)* **259**, 20–24.

Colman, P. M. (1974). *Z. Kristallogr.* **140**, 344–349.

Colman, P. M., and Fehlhammer, H. (1976). *J. Mol. Biol.* **100**, 278–282.

Colman, P. M., Epp, O., Fehlhammer, H., Bode, W., Schiffer, M., Lattman, E. E., and Jones, T. A. (1974). *FEBS Lett.* **44**, 194–199.

Colman, P. M., Fehlhammer, H., and Bartels, K. (1976). In *Crystallographic Computing Techniques*, Eds. F. R. Ahmed, K. Huml, and B. Sedlacek, Munksgaard, Copenhagen, pp. 248–258.

Crick, F. H. C., and Watson, J. D. (1956). *Nature (London)* **177**, 473–475.

Crowther, R. A. (1967). *Acta Crystallogr.* **22**, 758–764.

Crowther, R. A. (1969). *Acta Crystallogr. Sect. B* **25**, 2571–2580.

Crowther, R. A. (1972). In *The Molecular Replacement Method*, Ed. M. G. Rossmann, Gordon & Breach, New York, pp. 173–178.

Crowther, R. A., and Blow, D. M. (1967). *Acta Crystallogr.* **23**, 544–548.

Crowther, R. A., and Klug, A. (1975). *Ann. Rev. Biochem.* **44**, 161–182.

Cullis, A. F., Muirhead, H., Perutz, M. F., Rossmann, M. G., and North, A. C. T. (1961). *Proc. R. Soc. London* **A265**, 15–38.

Cutfield, J. F., Cutfield, S. M., Dodson, E. J., Dodson, G. G., and Sabesan, M. N. (1974). *J. Mol. Biol.* **87**, 23–30.

Czerwinski, E. W., and Mathews, F. S. (1974). *J. Mol. Biol.* **86**, 49–57.

Dodson, E. J., Harding, M. M., Hodgkin, D. C., and Rossmann, M. G. (1966). *J. Mol. Biol.* **16**, 227–241.

Eagles, P. A. M., Johnson, L. N., Joynson, M. A., McMurray, C. H., and Gutfreund, H. (1969). *J. Mol. Biol.* **45**, 533–544.

Eklund, H., and Brändén, C. I. (1976). Private communication.

Epp, O., Steigemann, W., Formanek, H., and Huber, R. (1971). *Eur. J. Biochem.* **20**, 432–437.

Eventoff, W., and Gurskaya, G. V. (1975). *J. Mol. Biol.* **93**, 55–62.

Eventoff, W., Hackert, M. L., and Rossmann, M. G. (1975). *J. Mol. Biol.* **98**, 249–258.

Eventoff, W., Tanaka, N., and Rossmann, M. G. (1976). *J. Mol. Biol.* **103**, 799–801.

Fehlhammer, H., and Bode, W. (1975). *J. Mol. Biol.* **98**, 683–692.

Fehlhammer, H., Schiffer, M., Epp, O., Colman, P. M., Lattman, E. E., Schwager, P., and Steigemann, W. (1975). *Biophys. Struct. Mech.* **1**, 139–146.

Fletterick, R. J., and Steitz, T. A. (1976). *Acta Crystallogr. Sect. A* **32**, 125–132.

Franklin, R. M., Harrison, S. C., Pettersson, U., Philipson, L., Brändén, C. I., and Werner, P. E. (1971). *Cold Spring Harbor Symp. Quant. Biol.* **36**, 503–510.

Frier, J. A., and Perutz, M. F. (1977). *J. Mol. Biol.* **112**, 97–112.

Gilbert, P. F. C., and Klug, A. (1974). *J. Mol. Biol.* **86**, 193–207.

Green, D. W., Ingram, V. M., and Perutz, M. F. (1954). *Proc. R. Soc. London* **A225**, 287–307.

Hackert, M. L., Ford, G. C., and Rossmann, M. G. (1973). *J. Mol. Biol.* **78**, 665–673.

Harrison, H. R., and Joynson, M. A. (1970). *Acta Crystallogr. Sect. A* **26**, 692–694.

Harrison, S. C. (1968). *J. Appl. Crystallogr.* **1**, 84–90.

Harrison, S. C., and Jack, A. (1975). *J. Mol. Biol.* **97**, 173–191.

Hendrickson, W. A., and Lattman, E. E. (1970). *Acta Crystallogr. Sect. B* **26**, 136–143.

Hendrickson, W. A., and Ward, K. B. (1976). *Acta Crystallogr. Sect. A* **32**, 778–780.

Hill, E. J., Tsernoglou, D., and Banaszak, L. J. (1973). *Acta Crystallogr. Sect. B* **29**, 921–922.

Hodgkin, D. C. (1974). *Proc. R. Soc. London* **B186**, 191–215.

Hoppe, W. (1957). *Elektrochem.* **61**, 1076–1079.

Hornsta, J. (1970). In *Crystallographic Computing*, Ed. F. R. Ahmed, Munksgaard, Copenhagen, pp. 103–109.

Huber, R. (1965). *Acta Crystallogr.* **19**, 353–356.

Huber, R. (1970). In *Crystallographic Computing*, Ed. F. R. Ahmed, Munksgaard, Copenhagen, pp. 96–102.

Jack, A. (1973). *Acta Crystallogr. Sect. A* **29**, 545–554.

Jack, A., Weinzierl, J., and Kalb, A. J. (1971). *J. Mol. Biol.* **58**, 395–398.

Jansonius, J. N. (1976). In *Crystallography of Molecular Biology*, Ed. B. Strandberg, Conference held at the *Ettore Majorana* Center for Scientific Culture, Erice, Sicily.

Johnson, J. E. (1978). *Acta Crystallogr. Sect. B* **34**, 575–577.

Johnson, J. E., Argos, P., and Rossmann, M. G. (1975). *Acta Crystallogr. Sect. B* **31**, 2577–2583.

Johnson, J. E., Akimoto, T., Suck, D., Rayment, I., and Rossmann, M. G. (1976). *Virology* **75**, 394–400.

Joynson, M. A., North, A. C. T., Sarma, V. R., Dickerson, R. E., and Steinrauf, L. K. (1970). *J. Mol. Biol.* **50**, 137–142.

Karle, J. (1972). *Acta Crystallogr. Sect. B* **28**, 820–824.

Katsubi, Y., Sasada, Y., and Kakudo, M. (1966). *Bull. Chem. Soc. Japan* **39**, 2572–2576.

Klug, A. (1971). *Cold Spring Harbor Symp. Quant. Biol.* **36**, 483–487.

Kretsinger, R. H. (1972). *Nature New Biol.* **240**, 85–86.

Kretsinger, R. H. (1976). *Ann. Rev. Biochem.* **45**, 239–266.

Lattman, E. E. (1972). *Acta Crystallogr. Sect. B* **28**, 1065–1068.

Lattman, E. E., and Love, W. E. (1970). *Acta Crystallogr. Sect. B* **26**, 1854–1857.

Lattman, E. E., Nockolds, C. E., Kretsinger, R. H., and Love, W. E. (1971). *J. Mol. Biol.* **60**, 271–277.

Lee, B., Yang, H. J., Henry, G. M., Seymour, J. P., and Chibata, I. (1975). *J. Biol. Chem.* **250**, 6228–6231.

Lentz, P. J., Jr., and Strandberg, B. (1974). *Acta Crystallogr. Sect. A* **30**, 552–559.

Lentz, P. J., Jr., Strandberg, B., Unge, T., Vaara, I., Borell, A., Fridborg, K., and Petef, G. (1976). *Acta Crystallogr. Sect. B* **32**, 2979–2983.

Levitt, M., and Chothia, C. (1976). *Nature (London)* **261**, 552–558.

Litvin, D. B. (1975). *Acta Crystallogr. Sect. A* **31**, 407–416.

Main, P. (1967). *Acta Crystallogr.* **23**, 50–54.

Main, P., and Rossmann, M. G. (1966). *Acta Crystallogr.* **21**, 67–72.

Maslen, E. N. (1968a). *Acta Crystallogr. Sect. B* **24**, 1165–1170.

Maslen, E. N. (1968b). *Acta Crystallogr. Sect. B* **24**, 1170–1172.

Matthews, B. W. (1976). *Annu. Rev. Phys. Chem.* **27**, 493–523.

Matthews, B. W., Sigler, P. B., Henderson, R., and Blow, D. M. (1967). *Nature (London)* **214**, 652–656.

Monod, J., Changeux, J. P., and Wyman, J. (1965). *J. Mol. Biol.* **12**, 88–118.

Muirhead, H., Cox, J. M., Mazzarella, L., and Perutz, M. F. (1967). *J. Mol. Biol.* **28**, 117–156.

Musick, W. D. L., Adams, A. D., Rossmann, M. G., Wheat, T. E., and Goldberg, E. (1976). *J. Mol. Biol.* **104**, 659–668.

Nixon, P. E., and North, A. C. T. (1976). *Acta Crystallogr. Sect. A* **32**, 320–325.

Nordman, C. E. (1966). *Trans. Amer. Crystallogr. Assoc.* **2**, 29–38.

Nordman, C. E. (1972). *Acta Crystallogr. Sect. A* **28**, 134–143.

Nordman, C. E., and Kumra, S. K. (1965). *J. Amer. Chem. Soc.* **87**, 2059–2060.

Nordman, C. E., and Schilling, J. W. (1970). In *Crystallographic Computing*, Ed. F. R. Ahmed, Munksgaard, Copenhagen, pp. 110–114.

Perutz, M. F. (1954). *Proc. Soc. London* **A225**, 264–286.

Phillips, D. C. (1966). In *Advances in Structure Research by Diffraction Methods*, Eds. R. Brill, and R. Mason, Interscience Publishers, New York, pp. 73–140.

Prothero, J. W., and Rossmann, M. G. (1964). *Acta Crystallogr.* **17**, 768–769.

Quigley, G. J., Suddath, F. L., McPherson, A., Jr., Kim, J. J., Sneden, D., and Rich, A. (1974). *Proc. Natl. Acad. Sci. U.S.A.* **71**, 2146–2150.

Rossmann, M. G. (1972a). *The Molecular Replacement Method*, Gordon and Breach, New York.

Rossmann, M. G. (1972b). *J. Mol. Biol.* **64**, 246–249.

Rossmann, M. G. (1976). *Acta Crystallogr. Sect. A* **32**, 774–777.

Rossmann, M. G., and Blow, D. M. (1961). *Acta Crystallogr.* **14**, 641–647.

Rossmann, M. G., and Blow, D. M. (1962). *Acta Crystallogr.* **15**, 24–31.

Rossmann, M. G., and Blow, D. M. (1963). *Acta Crystallogr.* **16**, 39–45.

Rossmann, M. G., and Blow, D. M. (1964). *Acta Crystallogr.* **17**, 1474–1475.

Rossmann, M. G., and Hodgkin, D. C. (1972). In *The Molecular Replacement Method*, Ed. M. G. Rossmann, Gordon and Breach, New York, pp. 36–38.

Rossmann, M. G., and Liljas, A. (1974). *J. Mol. Biol.* **85**, 177–181.

Rossmann, M. G., Blow, D. M., Harding, M. M., and Coller, E. (1964). *Acta Crystallogr.* **17**, 338–342.

Rossmann, M. G., Ford, G. C., Watson, H. C., and Banaszak, L. J. (1972). *J. Mol. Biol.* **64**, 237–249.

Rossmann, M. G., Åkervall, K., Lentz, P. J., Jr., and Strandberg, B. (1973). *J. Mol. Biol.* **79**, 197–204.

Sasada, Y. (1964). *Acta Crystallogr.* **17**, 611–612.

Schmid, M. F., and Herriott, J. R. (1976). *J. Mol. Biol.* **103**, 175–190.

Schmid, M. F., Lattman, E. E., and Herriott, J. R. (1974). *J. Mol. Biol.* **84**, 97–101.

Shannon, C. E. (1949). *Proc. Inst. Radio Eng. N. Y.* **37**, 10–14.

Sim, G. A. (1959). *Acta Crystallogr.* **12**, 813–815.

Sim, G. A. (1960). *Acta Crystallogr.* **13**, 511–512.

Steitz, T. A., Fletterick, R. J., Anderson, W. F., and Anderson, C. M. (1976). *J. Mol. Biol.* **104**, 197–222.

Strandberg, B., and Lentz, P. J., Jr., (1976). Private communication.

Tanaka, N. (1977). *Acta Crystallogr. Sect. A* **33**, 191–193.

Tollin, P. (1966). *Acta Crystallogr.* **21**, 613–614.

Tollin, P. (1969a). *Acta Crystallogr. Sect. A* **25**, 376–377.

Tollin, P. (1969b). *J. Mol. Biol.* **45**, 481–490.

Tollin, P. (1970). In *Crystallographic Computing*, Ed. F. R. Ahmed, Munksgaard, Copenhagen, pp. 90–95.

Tollin, P. (1976). In *Crystallographic Computing Techniques*, Eds. F. R. Ahmed, K. Huml, and B. Sedlacek, Munksgaard, Copenhagen, pp. 212–221.

Tollin, P., and Cochran, W. (1964). *Acta Crystallogr.* **17**, 1322–1324.

Tollin, P., and Rossmann, M. G. (1966). *Acta Crystallogr.* **21**, 872–876.

Tollin, P., Main, P., and Rossmann, M. G. (1966). *Acta Crystallogr.* **20**, 404–407.

Tulinsky, A., Vandlen, R. L., Morimoto, C. N., Mani, N. V., and Wright, L. H. (1973). *Biochemistry* **12**, 4185–4192.

Ward, K. B., Hendrickson, W. A., and Klippenstein, G. L. (1975). *Nature (London)* **257**, 818–821.

Ward, K. B., Wishner, B. C., Lattman, E. E., and Love, W. E. (1975). *J. Mol. Biol.* **98**, 161–177.

Watson, D. G., Sutor, D. J., and Tollin, P. (1965). *Acta Crystallogr.* **19**, 111–124.

Winkler, F. K., Schutt, C. E., Harrison, S. C., and Bricogne, G. (1977). *Nature (London)* **265**, 509–513.

Wishner, B. C., Ward, K. B., Lattman, E. E., and Love, W. E. (1975). *J. Mol. Biol.* **98**, 179–194.

Wright, C. S. (1974). *J. Mol. Biol.* **87**, 835–841.

Wright, C. S. (1977). *J. Mol. Biol.* **111**, 439–458.

Wonacott, A. J., and Bricogne, G. (1976). Private communication.

Wood, R. W. (1934). *Physical Optics*, Macmillan, New York, p. 268.

Young, D. W., Tollin, P., and Sutherland, H. (1968). *Acta Crystallogr. Sect. B* **24**, 161–167.

Young, D. W., Tollin, P., and Wilson, H. R. (1969). *Acta Crystallogr. Sect. B* **25**, 1423–1432.

Index